MERVIN L. KEEDY
Purdue University

MARVIN L. BITTINGER
Indiana University – Purdue University at Indianapolis

INTRODUCTORY ALGEBRA
A MODERN APPROACH

SECOND EDITION

ADDISON-WESLEY PUBLISHING COMPANY

Reading, Massachusetts • Menlo Park, California
Amsterdam • London • Don Mills, Ontario • Sydney

Third printing, May 1976

ISBN 0-201-03727-0
CDEFGHIJ-MU-7987

PREFACE TO THE SECOND EDITION

The users of the first edition have been so enthusiastic in their response and so helpful in offering constructive criticism, the authors first wish to say "thank you."

WHAT'S NEW IN THE SECOND EDITION?

Suggestions from Users. In various places errors have been corrected, changes in wording have been made, boldface type has been used, and exercises have been added, deleted, or changed. Chapter 4 in the first edition, which was rather long, has been divided into two chapters, and Chapters 5 and 6 of the first edition have been combined. There is also a new, more detailed index, and the headings of chapters, sections, exercise sets, and answers have been redesigned to make them easier to use.

The Metric System. To aid transition to the metric system, we have used metric units in approximately 50% of the applied problems. Students do not need to know what the metric units are in order to solve the problems, but if they wish they can take a "short course" on the metric system using the newly-added appendix on "The Metric System" and the familiarization material on the inside front cover.

SUGGESTIONS FOR USING THIS BOOK

In the preface to the first edition, it was pointed out that there are many ways in which this book can be used, and that flexibility in that regard is one of its important features. The book has been very effective in lecture, math lab, and independent study situations.

Worth special mention is a teaching method developed by some users that works well in classes of all sizes, notably large ones. The instructor does not lecture, but makes assignments which students do on their own, including working exercise sets. The following class period the instructor spends answering questions. Students have an additional day or two to polish their homework before handing it in. In the meantime, they are working on the next assignment. This method has the advantage of providing individualization while at the same time keeping the class together and working as a group. It also minimizes the number of instructor man-hours required.

SUPPLEMENTS THAT ACCOMPANY THE SECOND EDITION

In order to further enhance flexibility in using this book, the following (optional) supplements are now available:

- *Audio-Tape Cassettes* are available for use in audio-tutorial or math lab situations.
- *An Instructor's Manual* contains commentary and a sample course syllabus.
- *A Test Booklet* contains five alternate forms of each chapter test and the final examination, with answers spaced for easy grading.
- *An Answer Booklet* contains the answers to all of the exercises.
- *A Student's Solutions Booklet* contains worked out solutions to all of the margin exercises.

ACKNOWLEDGEMENTS

The authors especially wish to thank Donald Evans, Betty Crawshaw, Dave Buckley and Don Knorr of Polk Community College for their detailed and helpful suggestions for revision. Thanks also to Richard Spangler, Tacoma Community College, for his valued assistance, and to Mary Kay Smith, University of Georgia; C.W. Davis, Columbus College; Thomas McGannon, Olive Harvey College; Martha Breitweiser, Kellogg Community College; and Josephine Hamer, Drew University, for their help in developing scripts for the audio-tape cassettes.

October, 1974

M.L.K.
M.L.B.

PREFACE TO THE FIRST EDITION

This book covers beginning algebra. It is intended for use by students who have not been exposed to the subject. There is strong emphasis on the development of algebraic skills, but the "why's" of algebra also receive significant attention.

It is the emphasis on the "why's" that prompts the authors to call this book "modern". We feel that emphasis on understanding is essentially what makes a book such as this modern, rather than the use of such things as excessive set notation and sophisticated theory.

In this book, as well as its companion volumes, the authors bring to bear not only their own experience in teaching remedial courses but also the suggestions and recommendations of instructors in scores of two-year institutions across the country.

One of the principal and distinguishing features of the book is its design and format. Each page has an outer margin, which is used for materials of several types. For each lesson the objectives are stated in behavioral terms at the top of the page. These can easily be seen by the student and they tell him clearly what is expected of him in terms of performance. We hope that this will help answer the question all-too-often heard, "What material are we responsible for?" The most important items in the margins, however, are the sample or developmental exercises. These are placed with the text development so that the student can become involved actively in the development of the topic and gain some practice on exercises of the type he will be expected to do as homework for the lesson. The text refers to these exercises in the margins at the appropriate places.

There are many ways in which this book can be used. Flexibility in this regard is indeed one of its important features. The instructor who wishes to use it as he would an ordinary textbook can do so very easily. All he need do is ignore, and have students ignore, the exercises in the margins.

If an instructor wishes to use the lecture method primarily, but would like to bring some student-centered activity into the class, he can very easily do so. He would merely stop lecturing and have the students do the exercises in the margins at the appropriate times. On the other hand, the book is well suited for use in a learning laboratory situation. Because of its design, it can be used by a student with minimal instructor guidance. Yet it retains the flavor of the ordinary textbook, without the often deadly quality of the programed textbook.

This book contains some other features not usually found in a college textbook. There are tests at the ends of chapters, in addition to a final examination. Besides these, there is a pretest which can be used diagnostically. The exercise sheets which the student removes from the book are designed for quick and easy grading or scoring. The answers in the *Instructor's Manual* are arranged so that they match the spacing on the exercise sheets. The *Manual* also contains alternative forms of the chapter tests and the final examination. The book contains a great number of exercises (about 3500), and the authors have attempted to use language sparingly, so that the student has a maximal chance to learn the mathematics by reading it.

The material herein is suitable for a semester or quarter course for students who have not been exposed to algebra. This is the second in a series of books written in the same style. The preceding and following books are *Arithmetic: A Modern Approach*, and *Intermediate Algebra: A Modern Approach*.

The authors wish to thank the numerous people who helped make this book what it is. These include the many instructors in two-year colleges with whom we visited and who made many constructive suggestions. Professors Jerry Ball of Chabot College, Ralph Mansfield of Chicago City College, Loop Campus, and others made many helpful suggestions for improving the manuscript, and the staff of the Addison-Wesley Publishing Company has prodded and encouraged us most appropriately. Last, but not least, we thank our wives for their patience and helpful encouragement.

M. L. K.
M. L. B.

January, 1971

CONTENTS

viii   Contents

PRETEST

ANSWERS

The purpose of this test is to determine your background in algebra and help you or your instructor decide where to begin your study. When you have completed the test, read the test analysis on the answer page at the end of the book. The answers are given along with page numbers that refer you to the material for that question. Remove the test from the book and begin.

1.

2.

Chapter 1

1. What does x^2 represent if x represents 3?

2. What does $x^2 + 2$ represent if x represents 3?

3.

3. What is the reciprocal of $\dfrac{x}{2}$?

4. Which symbol, $>$ or $<$, should be inserted to make a true statement?

3   12

4.

Chapter 2

5. Which symbol, $>$ or $<$, should be inserted to make a true statement?

-8 -2

6. Simplify.

$|-12|$

5.

6.

7. Add.

$-7 + 3$

8. Subtract.

$12 - (-20)$

7.

9. Add.

$-3 + 12 + (-8) + 2$

10. Simplify.

$-3 \cdot 2 + 7 \cdot 3 - (-2 \cdot 5)$

8.

9.

10.

ANSWERS

11. _____

12. _____

13. _____

14. _____

15. _____

16. _____

17. _____

18. _____

19. _____

20. _____

11. Divide and simplify.

$$-\frac{3}{5} \div \frac{5}{7}$$

12. Multiply.

$$(-3)(-2)\left(\frac{1}{2}\right)(-1)$$

13. Rename, using a negative exponent.

$$\frac{1}{4^3}$$

14. Simplify.

$$x^{-4}x^2$$

15. Simplify.

$$\frac{y^4}{y^{-3}}$$

Chapter 3

16. Solve.

$$4x + 5 = 3x + 8$$

17. Solve.

$$4(x - 1) - 3(x + 3) = -11$$

18. The sum of two consecutive odd integers is 64. What are the integers?

Chapter 4

19. Collect like terms.

$$3x^2 + 2x - 5x^2 + 3x - 2 + 4$$

20. Arrange in descending order.

$$3x - 2 + 5x^4 - 4x^2$$

NAME

CLASS

ANSWERS

21. Add.

$4x^3 - 3x^2 + 2x - 5$

$5x^2 - 4x + 8$

$7x^2 - 12x - 2$

22. Subtract.

$(8x^2 - 5x + 2) - (-3x^2 + 4)$

21.

22.

23. Add.

$(3y^4 + 2y^2 - 5) + (7y^2 - 3y + 7)$

24. Multiply.

$3x(2x^2 + 5)$

23.

25. Multiply.

$(x - 2)(x^2 + 2x - 3)$

26. Multiply.

$(y + 3)(y - 3)$

24.

25.

Chapter 5

27. Factor.

$6x - 12$

28. Factor.

$9y^2 - 1$

26.

27.

29. Factor.

$x^2 + 3x - 10$

30. Factor.

$2y^2 - y - 6$

28.

29.

30.

31.

32.

33.

34.

35.

36.

38.

39.

40.

31. Solve.

$$x^2 + 6 = 5x$$

32. One more than a number times one less than the number gives 24. What is the number?

Chapter 6

33. In which quadrant is the graph of the point $(5, -2)$?

34. Solve.

$$x^2 - 7x = -6$$

35. What is the x-intercept of

$$3x + 2y = 12?$$

36. What is the y-intercept of

$$3x + 2y = 12?$$

37. Graph the equation

$$y = -x + 2.$$

Attach your graph paper to this sheet.

38. Solve.

$$x + y = 13$$
$$x - y = 1$$

39. Solve.

$$2x + 3y = -1$$
$$3x - 2y = 18$$

40. The sum of two numbers is 23. The first number minus twice the second is 8. Find the numbers.

Note: Pretest covers only six chapters. This is because students who do well on this much should proceed to *Intermediate Algebra*, the next book in this series.

1 THE NUMBERS OF ORDINARY ARITHMETIC AND THEIR PROPERTIES

1.1 NUMBERS AND ALGEBRA

Numbers are abstract ideas. The idea that comes to mind when we answer the question "how many?" is a number, for example. The symbols we write to represent, or *name* numbers are called *numerals*. Thus symbols such as 4, 8, XVI, and 3/4 are numerals. In algebra as in arithmetic, we use symbols which represent numbers to calculate and to solve problems. The basic properties of numbers are needed so that we will know what kinds of procedures are correct. As you learn better how to handle algebraic symbols, you will be able to solve problems that you could not have solved before.

NUMBERS OF ORDINARY ARITHMETIC

There are several kinds of numbers. The numbers we use for counting are called *natural numbers*. They are

1, 2, 3, 4, 5, 6, 7, 8, and so on.

The *whole numbers* include the natural numbers and zero. The whole numbers are

0, 1, 2, 3, 4, 5, 6, 7, 8, and so on.

The *numbers of ordinary arithmetic* include the whole numbers and the fractions, such as 2/3 and 9/5. These numbers are sometimes called simply *the numbers of arithmetic*. These numbers are used in algebra as well as arithmetic.

All of the numbers of arithmetic can be named by fractional symbols. Such symbols are called *fractional numerals*, or *fractional notation*.

Examples

a) The number *three-fourths* can be named by the fractional numerals

$\frac{3}{4}, \frac{6}{8}, \frac{300}{400}$ and many more.

b) The number *one-third* can be named by the fractional numerals

$\frac{1}{3}, \frac{2}{6}, \frac{10}{30}$, and so on.

Any fractional numeral in which the denominator is three times the numerator names the number one-third.

The whole numbers can also be named with fractional notation.

Examples

a) $0 = \frac{0}{1} = \frac{0}{4}$ etc. b) $2 = \frac{2}{1} = \frac{6}{3} = \frac{200}{100}$ etc. c) $1 = \frac{2}{2} = \frac{5}{5}$ etc.

Do exercises 1 through 4 at the right.

You should be able to:

a) Write several numerals for any number of arithmetic (whole number or fraction) by multiplying by 1.

b) Find the simplest fractional numeral for a number of arithmetic.

c) Add, subtract, and multiply numbers of arithmetic using fractional numerals.

1. Write three fractional numerals for $\frac{2}{3}$.

2. Write three fractional numerals for $\frac{1}{2}$.

3. Write three fractional numerals for 1.

4. Write three fractional numerals for 4.

3

5. Multiply by 1 to find three different names for $\frac{4}{5}$.

6. Multiply by 1 to find three different names for $\frac{8}{7}$.

Simplify.

7. $\frac{18}{27}$

8. $\frac{38}{18}$

Every number of arithmetic has many fractional numerals, or fractional *names*. To find different names we can use the notion of multiplying by 1.

Examples

a) $\frac{2}{3} = \frac{2}{3} \times 1 = \frac{2}{3} \times \frac{5}{5} = \frac{10}{15}$ (Multiplying a number by 1 does not change the number)

b) $\frac{7}{5} = \frac{7}{5} \times 1 = \frac{7}{5} \times \frac{11}{11} = \frac{77}{55}$

Do exercises 5 and 6 at the left.

The simplest fractional numeral for a number has the smallest possible numerator and denominator. To simplify we can reverse the above process.

Examples

a) $\frac{10}{15} = \frac{2 \times 5}{3 \times 5} = \frac{2}{3} \times \frac{5}{5} = \frac{2}{3}$ (We factor numerator and denominator and then "remove" a factor of 1, in this case $\frac{5}{5}$)

b) $\frac{36}{24} = \frac{6 \times 6}{4 \times 6} = \frac{3 \times 2 \times 6}{2 \times 2 \times 6} = \frac{3}{2} \times \frac{2 \times 6}{2 \times 6} = \frac{3}{2}$

You may be in the habit of canceling in cases like this. For example, you might have done this example as follows.

$$\begin{array}{c} 2 \\ \cancel{12} \\ \cancel{24} \\ \cancel{36} \\ \cancel{18} \\ 3 \end{array} \qquad \text{or} \qquad \frac{24}{36} = \frac{2 \times \cancel{12}}{3 \times \cancel{12}} = \frac{2}{3}$$

Canceling causes many errors, so you should *avoid doing it*. Rather, use the method of removing factors of one, at least for now.

Do exercises 7 and 8 at the left.

It is sometimes helpful to include a factor of 1 in the numerator or denominator.

Examples

a) $\dfrac{18}{72} = \dfrac{2 \times 9}{8 \times 9} = \dfrac{1 \times 2 \times 9}{4 \times 2 \times 9} = \dfrac{1}{4} \times \dfrac{2 \times 9}{2 \times 9} = \dfrac{1}{4}$ b) $\dfrac{9}{72} = \dfrac{1 \times 9}{8 \times 9} = \dfrac{1}{8}$

Do exercises 9 and 10 at the right.

Example 1. Multiply and simplify.

$$\frac{5}{6} \cdot \frac{9}{25} = \frac{5 \cdot 9}{6 \cdot 25} = \frac{1 \cdot 3 \cdot 3 \cdot 5}{2 \cdot 3 \cdot 5 \cdot 5} = \frac{3 \cdot 5 \cdot 1 \cdot 3}{3 \cdot 5 \cdot 2 \cdot 5} = \frac{3 \cdot 5}{3 \cdot 5} = \frac{1 \cdot 3}{2 \cdot 5} = \frac{3}{10}$$

In the preceding example, we have used a dot · for a multiplication sign. It means exactly the same thing as ×.

Do exercises 11 and 12 at the right.

We can use multiplying by 1 to find common denominators for addition and subtraction.

Example 2. Add and simplify.

$$\frac{2}{8} + \frac{3}{5} = \frac{2}{8} \cdot \frac{5}{5} + \frac{3}{5} \cdot \frac{8}{8} = \frac{10}{40} + \frac{24}{40} = \frac{34}{40} = \frac{17}{20} \cdot \frac{2}{2} = \frac{17}{20}$$

Example 3. Add and simplify.

$$\frac{3}{8} + \frac{5}{12} = \frac{3}{8} \cdot \frac{3}{3} + \frac{5}{12} \cdot \frac{2}{2} = \frac{9}{24} + \frac{10}{24} = \frac{19}{24}$$

Do exercises 13 and 14 at the right.

Example 4. Subtract and simplify.

$$\frac{9}{8} - \frac{4}{5} = \frac{9}{8} \cdot \frac{5}{5} - \frac{4}{5} \cdot \frac{8}{8} = \frac{45}{40} - \frac{32}{40} = \frac{13}{40}$$

Do exercises 15 and 16 at the right.

Do exercise set 1.1, p. 35.

Simplify.

9. $\dfrac{27}{54}$

10. $\dfrac{48}{12}$

Multiply and simplify.

11. $\dfrac{6}{5} \cdot \dfrac{9}{25}$

12. $\dfrac{12}{11} \cdot \dfrac{14}{3} \cdot \dfrac{22}{35}$

Add and simplify.

13. $\dfrac{4}{5} + \dfrac{2}{3}$

14. $\dfrac{5}{12} + \dfrac{7}{18}$

Subtract and simplify.

15. $\dfrac{4}{5} - \dfrac{2}{3}$

16. $\dfrac{5}{12} - \dfrac{2}{9}$

You should be able to:

a) Write expanded numerals for numbers of arithmetic with or without exponents.

b) Given exponential notation such as x^3, write xxx and conversely.

c) Convert from fractional numerals to decimal numerals and conversely.

17. Write three kinds of expanded numerals for 52,374.

18. Write three kinds of expanded numerals for 234,809.

19. Write an expanded numeral for 752,398 using exponents.

20. Write an expanded numeral for 20,347 using exponents.

1.2 DECIMAL NOTATION

The numbers of arithmetic can all be named by fractional numerals. Another familiar and useful notation for these numbers is called *decimal notation*. It is called decimal because it is based on the number ten.* Our ordinary numerals for whole numbers are decimal numerals ("decimals" for short) based on ten. To show the meaning of decimal numerals we can write *expanded numerals*.

Example

An expanded numeral for 3542 is

$$3000 + 500 + 40 + 2$$

Note that the value of a place is ten times that to its right. We can emphasize this further with other expanded numerals.

Example

$$3542 = 3 \times 1000 + 5 \times 100 + 4 \times 10 + 2$$
$$= 3 \times 10 \times 10 \times 10 + 5 \times 10 \times 10 + 4 \times 10 + 2$$

Do exercises 17 and 18 at the left.

EXPONENTIAL NOTATION

We can use shorthand notation for $10 \times 10 \times 10$ and similar expressions. It is called *exponential notation*. For

$10 \times 10 \times 10$ we write 10^3

This is read "ten cubed" or "ten to the third power." The number 3 is called an *exponent* and we call 10 the *base*. Similarly, for 10×10 we write 10^2, read "ten squared" or "ten to the second power."

Example

An expanded numeral with exponents for 82,653 is

$$8 \times 10^4 + 2 \times 10^3 + 6 \times 10^2 + 5 \times 10 + 3$$

Do exercises 19 and 20 at the left.

An exponent tells how many times the base is used as a factor. The base can be a number other than ten.

---

* The Latin word for ten is *decem*.

Examples

a) 3^5 means $3 \times 3 \times 3 \times 3 \times 3$ b) 7^4 means $7 \cdot 7 \cdot 7 \cdot 7$

c) If we use n to stand for a number, n^4 means $n \cdot n \cdot n \cdot n$

Do exercises 21 and 22 at the right.

VARIABLES

A letter can often represent various numbers. In this case, we call it a *variable*. If we write two or more variables together, such as

nnn or xyz

the agreement is that it means multiplication.

Examples

a) nnn means $n \cdot n \cdot n$. It also means n^3

b) If n stands for 10, then nnn means $10 \times 10 \times 10$, or 1000

c) xyz means $x \cdot y \cdot z$

Do exercises 23 through 26 at the right.

The numbers 1 and 0 are also given meaning as exponents. To see how this is done we can look at an expanded numeral.

$3795 = 3 \times 10^3 + 7 \times 10^2 + 9 \times 10 + 5$

If we are to keep the decreasing pattern of exponents we would write

$3 \times 10^3 + 7 \times 10^2 + 9 \times 10^1 + 5 \times 10^0$

Accordingly, 10^1 should mean 10 and 10^0 should mean 1. These are the meanings we assign to 1 and 0 as exponents.

For any number n, n^1 means n.

For any number n, other than zero,* n^0 means 1.

Examples

a) $5^1 = 5$,   b) $8^1 = 8$,   c) $3^0 = 1$,   d) $7^0 = 1$,   e) $5^0 = 1$

Do exercises 27 and 28 at the right.

DECIMAL NUMERALS FOR FRACTIONS

Decimal numerals for fractions contain *decimal points*. The place values to the right of a decimal point are tenths, hundredths, and so on. We can show this with expanded numerals.

---

* We shall see later why 0 is excluded, i.e. 0^0 is meaningless.

21. What is the meaning of 5^4?

22. What is the meaning of x^5?

23. Write exponential notation for $nnnnn$.

24. What is the meaning of y^3? Do not use $\times$ or $\cdot$.

25. What number does xxx represent if x stands for 2?

26. What does n^4 represent if n stands for 10?

27. Write an expanded numeral for 3562 using exponents. Include exponents of 1 and 0.

28. What is 5^1? What is 7^0?

Write an expanded numeral for each of the following numbers.

29. 23.678

30. .27

31. 4.0067

Write fractional numerals. You need not simplify.

32. 1.62

33. 35.431

Write decimal numerals.

34. $\dfrac{7}{8}$

35. $\dfrac{4}{5}$

36. $\dfrac{9}{11}$

37. $\dfrac{23}{9}$

a) $34.243 = 3 \times 10 + 4 + 2 \times \dfrac{1}{10} + 4 \times \dfrac{1}{100} + 3 \times \dfrac{1}{1000}$

b) $.8 = 8 \times \dfrac{1}{10}$

Do exercises 29 through 31 at the left.

To convert from a decimal numeral to a fractional numeral, we can proceed as in the following examples.

Examples

a) $34.2 = \dfrac{342}{10}$ (34.2 is 34 and two-tenths, or 342 tenths)

b) $16.563 = \dfrac{16563}{1000}$ (16.563 is 16,563 thousandths)

Do exercises 32 and 33 at the left.

To convert from a fractional numeral to a decimal numeral we can divide.

Example 1

$\dfrac{3}{8}$ means $3 \div 8$, so we divide

$$
\begin{array}{r}
.375 \\
8)\overline{3.000} \\
2\,400 \\
\hline
600 \\
560 \\
\hline
40 \\
40 \\
\hline
\end{array}
$$

Sometimes we get a repeating decimal when we divide.

Example 2

$\dfrac{4}{11}$ means $4 \div 11$, so we divide

$$
\begin{array}{r}
.3636\ldots \\
11)\overline{4.00} \\
3\,3 \\
\hline
70 \\
66 \\
\hline
40 \\
33 \\
\hline
70 \\
66 \\
\hline
4 \\
\end{array}
$$

Thus $\frac{4}{11} = .363636\ldots$. Such decimals are often abbreviated by putting a bar over the repeating part, as follows.

$\frac{4}{11} = .36\overline{36}$

Do exercises 34 through 37 at the left.

Do exercise set 1.2, p. 37.

1.3 PROPERTIES OF THE NUMBERS OF ARITHMETIC

The numbers of arithmetic can be named in various ways. These numbers have certain properties which do not depend upon the kind of notation used for them. Some of the basic properties of these numbers are so simple and obvious that they may seem unimportant. This is deceiving, however. They are very important, especially in algebra.

ORDER IN ADDITION

One of the basic properties of numbers is that they can be added in any order. For example, $3 + 2$ and $2 + 3$ are the same. A similar thing is true for any two numbers. This simple fact goes by the hifalutin name of *commutative law (or property) of addition.**

For any numbers a and b, $a + b = b + a$.

This is the *commutative law of addition*.

Do exercises 38 and 39 at the right.

PARENTHESES, SYMBOLS OF GROUPING

Let us consider $5 \times 2 + 4$. What does it mean? If we multiply 5 by 2 and add 4 we get 14. If we add 2 and 4 and multiply by 5 we get 30. To tell which operation to do first, we use parentheses. In other words, parentheses show groupings.

Example. $(3 \times 5) + 6$ means $15 + 6$, or 21

$\qquad 3 \times (5 + 6)$ means 3×11, or 33

Do exercises 40 through 44 at the right.

GROUPING IN ADDITION

If we write $3 + 5 + 4$, what does it mean? Does it mean $(3 + 5) + 4$ or $3 + (5 + 4)$? Either way, we get 12, so it doesn't matter. In fact, if we are doing addition only, we can group numbers in any manner. This means we really don't need parentheses if we are doing only addition. This is another basic property.

---

* In learning to spell *commutative*, note that there is no n in it.

OBJECTIVES

You should be able to:

a) Do calculations as shown by parentheses.

b) Tell which laws are illustrated by certain sentences.

c) State the distributive law.

d) Use the distributive law to factor expressions like $3x + 3y$.

e) Evaluate an expression like $3x + 3y$ when numbers are specified for the letters.

Do these calculations.

38. $17 + 10$

39. $10 + 17$

Do these calculations.

40. $(5 \times 4) + 2$

41. $5 \times (4 + 2)$

42. $(4 \times 6) + 2$

43. $5 \times (2 \times 3)$

44. $(6 \times 2) + (3 \times 5)$

Add. Look for combinations that make ten.

45. $5 + 2 + 3 + 5 + 8$

46. $1 + 5 + 6 + 9 + 4$

Which laws are illustrated by these sentences?

47. $61 \times 56 = 56 \times 61$

48. $(3 + 5) + 2 = 3 + (5 + 2)$

49. $(4 + 2) + 5 = 5 + (2 + 4)$

50. $7 \cdot (9 \cdot 8) = 9 \cdot (8 \cdot 7)$

For any numbers a, b, and c, $(a + b) + c = a + (b + c)$.

This is the *associative law of addition*.

The commutative and associative laws together guarantee that we can change the order and grouping in any way when we are adding. This often helps make addition easier.

Example. To do the addition $3 + 4 + 7$, add 7 and 3 to make 10, then add 4, to get 14.

Do exercises 45 and 46 at the left.

COMMUTATIVE AND ASSOCIATIVE PROPERTIES OF MULTIPLICATION

For multiplication, does the order matter? For example, is $2 \times 5 = 5 \times 2$? Does the grouping matter? For example, is $2 \times (5 \times 3) = (2 \times 5) \times 3$? Our experience with arithmetic tells us that multiplication is both commutative and associative.

For any numbers a, b, and c,

$a \cdot b = b \cdot a$	(the *commutative law of multiplication*)
$a \cdot (b \cdot c) = (a \cdot b) \cdot c$	(the *associative law of multiplication*)

These laws together tell us that we can change the order in grouping in any way when we are multiplying.

Example. To do the multiplication $5 \cdot 7 \cdot 2$, multiply 5 by 2 to get 10, then multiply by 7, to get 70.

Examples. Which laws are illustrated by these sentences?

a) $3 + 5 = 5 + 3$ commutative law of addition (order changed)

b) $(2 + 3) + 5 = 2 + (3 + 5)$ associative law of addition (grouping changed)

c) $(3 \cdot 5) \cdot 2 = (2 \cdot 5) \cdot 3$ commutative and associative laws of multiplication (both order and grouping changed)

Do exercises 47 through 50 at the left.

THE DISTRIBUTIVE LAW

Another important and basic property, both in arithmetic and algebra, relates addition and multiplication. It is called the *distributive law of multiplication over addition.* In arithmetic, for example, it is used in multiplications like this:

$$\begin{array}{r} 25 \\ \times 7 \\ \hline \end{array}$$

Since $25 = 20 + 5$, what we want to find can be described like this:

$$(20 + 5)\cdot 7$$

But what do we actually do? We multiply the 5 by 7 first. Then we multiply the 20 by 7 and finally add the results. In other words, we do this:

$$(20\cdot 7) + (5\cdot 7)$$

This gives us $140 + 35$, or 175. Whenever we wish to multiply a number by a sum of several numbers, we can either add and then multiply or we we can multiply and then add.

Example. $(4 + 8)\cdot 5 = 12\cdot 5 = 60$

$$(4\cdot 5) + (8\cdot 5) = 20 + 40 = 60$$

Do exercises 51 through 53 at the right.

The property we are investigating is sometimes called the *distributive law* for short. It could be described by saying "For any numbers a, b, and c, $a\cdot (b + c) = (a\cdot b) + (a\cdot c)$." It is to be understood that there could be more than two numbers inside the parentheses on the left. Also, the a can be written on the right, as follows:

$$(b + c)\cdot a = (b\cdot a) + (c\cdot a).$$

AN AGREEMENT ABOUT PARENTHESES

Before we state the distributive law formally, let us make an agreement about parentheses, to shorten our writing. We agree that in an expression like $(4\cdot 5) + (3\cdot 7)$, we can omit the parentheses. Thus $4\cdot 5 + 3\cdot 7$ means $(4\cdot 5) + (3\cdot 7)$. In other words, do the multiplications first.

Do exercises 54 and 55 at the right.

Do the calculations as shown.

51. a) $(2 + 5)\cdot 4$

   b) $(2\cdot 4) + (5\cdot 4)$

52. a) $(7 + 4)\cdot 7$

   b) $(7\cdot 7) + (4\cdot 7)$

53. a) $6\cdot (3 + 2 + 4)$

   b) $(6\cdot 3) + (6\cdot 2) + (6\cdot 4)$

Evaluate and simplify.

54. $3\cdot 5 + 2\cdot 4$

55. $4\cdot 2 + 7\cdot 1$

Evaluate and simplify.

56. $ab + cd$, where $a = 2$, $b = 3$, $c = 5$, and $d = 2$.

57. $abc + ac$, where $a = 2$, $b = 3$, and $c = 4$.

Factor.

58. $4x + 4y$

59. $5a + 5b$

60. $7p + 7q + 7r$

Factor. Then evaluate both expressions when $x = 4$ and $y = 3$.

61. $5x + 5y$

62. $7x + 7y$

In an expression such as $ab + cd$, it is understood that parentheses belong around ab and also around cd. In other words, the multiplications are to be done first.

Do exercises 56 and 57 at the left.

Now, using our agreement about parentheses, we state the distributive law.

The distributive law:

For any numbers a, b, and c, $a(b + c) = ab + ac$.

Notice that we cannot omit the parentheses on the left. If we did we would have $ab + c$, which by our agreement means $(ab) + c$.

Notice also that the distributive law involves *two operations*. The commutative and associative laws involve just one.

FACTORING

Any equation can be reversed without changing its meaning. Thus for the statement of the distributive law we could also write $ab + ac = a(b + c)$. You should think of this law both ways. The distributive law is the basis of a process called *factoring*.

Example. $3x + 3y = 3(x + y)$ by the distributive law.
When we write $3(x + y)$ we say that we have *factored* $3x + 3y$.

Do exercises 58 through 60 at the left.

It is important to realize that when we factor an expression like $3x + 3y$ the factored expression represents the same number as the original one, *no matter what numbers x and y represent*!

Examples

a) Factor $4x + 4y$. Then evaluate both expressions when $x = 2$ and $y = 3$.

$4x + 4y = 4(x + y)$ $4 \cdot 2 + 4 \cdot 3 = 8 + 12 = 20$

$4(2 + 3) = 4 \cdot 5 = 20$

b) Factor $6x + 6y$. Then evaluate both expressions when $x = 3$ and $y = 2$.

$6x + 6y = 6(x + y)$ $6 \cdot 3 + 6 \cdot 2 = 18 + 12 = 30$

$6(3 + 2) = 6 \cdot 5 = 30$

Do exercises 61 and 62 at the left.

Do exercise set 1.3, p. 39.

1.4 USING THE DISTRIBUTIVE LAW

The distributive law is used as the basis of many procedures in both arithmetic and algebra. We have just seen how it serves as a basis of factoring, for example. Below are some further examples of factoring and some examples of other processes based on this property.

FACTORING

Examples. Factor:

a) $3x + 6 = 3x + 3 \cdot 2 = 3(x + 2)$

b) $ax + ay + az = a(x + y + z)$

The parts of an expression such as $5x + 2y + 3z$, separated by plus signs, are called *terms* of the expression. In the preceding examples, to see how to factor we look for a factor which is common to all the terms. Then by the distributive law we "remove" it, so to speak, or take it outside the parentheses.

Examples. Factor by "removing a common factor."

a) $7y + 14 + 21z = 7(y + 2 + 3z)$ (The common factor is 7)

b) $9x + 27y + 9 = 9x + 9 \cdot 3 \cdot y + 9 \cdot 1$

$\qquad\qquad\quad = 9(x + 3y + 1)$ (The common factor is 9)

Do exercises 63 through 66 at the right.

MULTIPLYING

We can reverse the above process. When we do, we say that we are "multiplying."

Examples. Multiply.

a) $3(x + 2) = 3x + 3 \cdot 2 = 3x + 6$

b) $(s + t + w)b = sb + tb + wb$

In the preceding examples, we used the distributive law. What we do amounts to this: Multiply each term inside the parentheses by the factor outside.

Do exercises 67 through 69 at the right.

Factor.

63. $5x + 10$

64. $12 + 3x$

65. $6x + 12 + 9y$

66. $5x + 10y + 5$

Multiply.

67. $5(y + 3)$

68. $4(x + 2y + 5)$

69. $a(m + n + p)$

Collect like terms.

70. a) $6y + 2y$

b) $7x + x$

c) $x + .03x$

71. $10p + 8p + 4q + 5q$

72. $7x + 3y + 4x + 5y$

Combine like terms.

73. $4y + 12y$

74. $3s + 4s + 6w + 7w$

75. $5x + 4y + 4x + 6y$

COLLECTING LIKE TERMS

If two terms have the same letters, we say that they are *similar* terms, or *like* terms. We can often simplify expressions by a process called *collecting like terms.**

Examples. Collect like terms

a) $3x + 4x = (3 + 4)x = 7x$

b) $2x + 3y + 5x + 2y = 2x + 5x + 3y + 2y$ (Don't forget the commutative and associative laws)

$$= (2 + 5)x + (3 + 2)y$$

$$= 7x + 5y$$

c) $5x + x = 5x + 1 \cdot x = (5 + 1)x = 6x$

d) $x + .05x = (1 + .05)x = 1.05x$

Collecting like terms uses the distributive property and is similar to factoring. In the above examples the main difference is that it is the letters that are factored out.

Do exercises 70 through 72 at the left.

It is not necessary to write all the steps as in the above examples.

Examples. Combine like terms.

a) $5y + 2y + 4y = 11y$

b) $3x + 7x + 2y = 10x + 2y$

c) $3s + 5s + 8t + 2t = 8s + 10t$

d) $3s + 4t + 7s + 5t = 10s + 9t$

Do exercises 73 through 75 at the left.

Do exercise set 1.4, p. 41.

---

* Collecting like terms is also called "combining like terms."

1.5 THE NUMBER 1 AND RECIPROCALS

The number 1 has some very special properties. These properties, like those mentioned earlier, are important in both arithmetic and algebra. The first of these properties is the simplest and most important. When we multiply any number by 1 we get that same number. Because of this we say that 1 is the *multiplicative identity*.

The number 1 is the *multiplicative identity*.

For any number *n*, *n*·1 = *n*.

What happens when we divide a number by 1? We again get the same number with which we started. This is a second property, one that you have relied upon many times.

For any number *n*, $\frac{n}{1} = n$.

When we divide a number by itself the result is the number 1. This is true for any number except zero. For reasons which we shall see later, we do not divide by zero.

For any number *n*, except zero, $\frac{n}{n} = 1$.

Examples

a) $\frac{\frac{3}{5}}{1} = \frac{3}{5}$,   b) $\frac{\frac{4}{3}}{\frac{4}{3}} = 1$

Do exercises 76 through 79 at the right.

RECIPROCALS

Two numbers whose product is 1 are called *reciprocals* of each other. They are also called *multiplicative inverses* of each other. All of the numbers of arithmetic, except zero, have reciprocals.

Examples

a) The reciprocal of $\frac{2}{3}$ is $\frac{3}{2}$ because $\frac{2}{3} \cdot \frac{3}{2} = \frac{6}{6} = 1$

b) The reciprocal of 9 is $\frac{1}{9}$ because $9 \cdot \frac{1}{9} = \frac{9}{9} = 1$

c) The reciprocal of $\frac{1}{4}$ is 4 because $4 \cdot \frac{1}{4} = 1$.

Do exercises 80 through 83 at the right.

RECIPROCALS AND DIVISION

The number 1 and reciprocals can be used to explain division of numbers of arithmetic. To divide, we can multiply by 1, choosing carefully the name that we use for 1.

You should be able to:

a) Name the reciprocal of any arithmetic number.

b) Divide with fractional notation by multiplying by 1.

c) Divide with fractional notation by multiplying by a reciprocal.

d) Graph numbers of arithmetic on a number line.

e) Insert the proper symbol $>$, $<$, or $=$ between two fractional numerals.

Simplify.

76. $\frac{7}{1}$ **77.** $\frac{67}{67}$

78. $\frac{\frac{2}{3}}{1}$ **79.** $\frac{\frac{7}{5}}{\frac{7}{5}}$

What is the reciprocal of each number?

80. $\frac{4}{11}$ **81.** $\frac{15}{7}$

82. 5   **83.** $\frac{1}{3}$

Divide, by multiplying by 1.

84. $\dfrac{\frac{3}{5}}{\frac{4}{7}}$

85. $\dfrac{\frac{5}{4}}{\frac{3}{2}}$

86. $\dfrac{\frac{9}{7}}{\frac{4}{5}}$

Divide, by multiplying by the reciprocal of the divisor.

87. $\dfrac{\frac{4}{3}}{\frac{7}{2}}$

88. $\dfrac{3}{5} \div \dfrac{7}{4}$

89. $\dfrac{2}{9} \div \dfrac{5}{7}$

Example. Divide $\frac{2}{3}$ by $\frac{7}{5}$

$$\dfrac{\frac{2}{3}}{\frac{7}{5}} = \dfrac{\frac{2}{3}}{\frac{7}{5}} \times \dfrac{\frac{5}{7}}{\frac{5}{7}} \qquad \text{because } \dfrac{\frac{5}{7}}{\frac{5}{7}} = 1$$

$$\dfrac{\frac{2}{3}}{\frac{7}{5}} \times \dfrac{\frac{5}{7}}{\frac{5}{7}} = \dfrac{\frac{2}{3} \times \frac{5}{7}}{\frac{7}{5} \times \frac{5}{7}} = \dfrac{\frac{10}{21}}{1} = \dfrac{10}{21}$$

In this example, we first wrote fractional notation to express the division. Then we multiplied by 1. Notice that after we multiplied we obtained 1 for a denominator. This was because we chose the name for 1 cleverly. We used the reciprocal of the divisor, $\frac{7}{5}$, for both numerator and denominator.

Do exercises 84 through 86 at the left.

When we multiply by 1 to divide we get a denominator of 1. What do we get in the numerator? In the above example we got $\frac{2}{3} \times \frac{5}{7}$. This is the product of $\frac{2}{3}$, the dividend, and $\frac{5}{7}$, the reciprocal of the divisor. A similar thing always happens. Thus we can always divide by multiplying by the reciprocal of the divisor.

To do a division $\dfrac{a}{b} \div \dfrac{c}{d}$, we can do the multiplication

$\dfrac{a}{b} \times \dfrac{d}{c}.$ **(We can multiply by the reciprocal of the divisor.)**

Example 1. Divide, by multiplying by the reciprocal of the divisor.

$$\dfrac{1}{2} \div \dfrac{3}{5} = \dfrac{1}{2} \cdot \dfrac{5}{3} = \dfrac{5}{6}$$

After a division, simplification is often possible and should be done.

Example 2.

$$\dfrac{2}{3} \div \dfrac{4}{9} = \dfrac{2}{3} \times \dfrac{9}{4} = \dfrac{18}{12} = \dfrac{3 \cdot 6}{2 \cdot 6} = \dfrac{3}{2} \cdot \dfrac{6}{6} = \dfrac{3}{2}$$

Example 3.

$$\dfrac{3}{5} \div \dfrac{4}{15} = \dfrac{3}{5} \times \dfrac{15}{4} = \dfrac{3 \cdot 15}{5 \cdot 4} = \dfrac{3 \cdot 5 \cdot 3}{5 \cdot 4} = \dfrac{9}{4} \times \dfrac{5}{5} = \dfrac{9}{4}$$

Do exercises 87 through 89 at the left.

THE NUMBER LINE

The order of the numbers of arithmetic can be shown on a line as in the figure below.

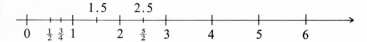

Do exercises 90 and 91 at the right.

Note that $\frac{1}{2}$ is less than 1, and $\frac{1}{2}$ is to the left of 1 on the number line.

For any numbers a and b, $a < b$ (read "a is less than b") means that a is to the left of b on the number line.

Examples. a) $\frac{1}{4} < \frac{1}{2}$,   b) $1 < 1.5$,   c) $3 < 6$

Also, $b > a$ means $a < b$. In other words, if a is less than b, then b is greater than a.

Examples. a) $\frac{1}{2} > \frac{1}{4}$ means $\frac{1}{4} < \frac{1}{2}$
            b) $1.5 > 1$ means $1 < 1.5$
            c) $3 < 6$ means $6 > 3$

Sentences such as $3 < 6$ and $x > 2$ are called *inequalities*.

For any numbers of arithmetic a and b (they might be the same) one and only one of the following must hold

$a > b$ or $a = b$ or $a < b$

Do exercises 92 through 94 at the right.

Consider $\frac{4}{5}$ and $\frac{3}{5}$. Checking the number line we see that $\frac{4}{5} > \frac{3}{5}$.

$$\begin{array}{cccc} + & + & + & + \rightarrow \\ 0 & \frac{3}{5} & \frac{4}{5} & 1 \end{array}$$

Note also that $4 > 3$. Thus when denominators are the same we need only compare numerators.

For any numbers of arithmetic $\frac{a}{b}$ and $\frac{c}{b}$,

$\frac{a}{b} > \frac{c}{b}$ when $a > c$

            and $\frac{a}{b} = \frac{c}{b}$ when $a = c$

$\frac{a}{b} < \frac{c}{b}$ when $a < c$

Do exercises 95 through 97 at the right.

Graph each number using a number line.

90. $\frac{6}{5}$

91. $\frac{17}{18}$

Give another meaning for each of the following.

92. $\frac{1}{4} < \frac{1}{2}$

93. $4.5 > 2.7$

94. $1 > \frac{99}{100}$

Insert the proper symbol $>$, $<$, or $=$ between each pair of numerals.

95. $\frac{3}{4}$ $\frac{4}{4}$

96. $\frac{1}{2}$ $\frac{1}{2}$

97. $\frac{22}{19}$ $\frac{21}{19}$

Find a common denominator and tell which number is larger.

98. $\dfrac{5}{9}$ or $\dfrac{7}{11}$

Insert the proper symbol $>$, $<$, or $=$ between each pair of numerals.

You may use either the cross products or the common denominator method.

99. $\dfrac{9}{5}$ $\dfrac{11}{7}$

100. $\dfrac{16}{12}$ $\dfrac{13}{8}$

101. $\dfrac{23}{7}$ $\dfrac{29}{9}$

102. $\dfrac{100}{197}$ $\dfrac{99}{198}$

It is not so easy to decide which of $\frac{2}{3}$ and $\frac{5}{7}$ is larger. Instead of using the number line let's use the quicker method of finding a common denominator and comparing numerators.

$$\frac{2}{3} = \frac{2}{3} \cdot \frac{7}{7} = \frac{14}{21} \quad \text{and} \quad \frac{5}{7} = \frac{5}{7} \cdot \frac{3}{3} = \frac{15}{21}$$

Since $15 > 14$, it follows that $\dfrac{5}{7} > \dfrac{2}{3}$.

Do exercise 98 at the left.

To compare $\dfrac{a}{b}$ and $\dfrac{c}{d}$ we can first find a common denominator:

$$\frac{a}{b} = \frac{a}{b} \cdot \frac{d}{d} = \frac{a \cdot d}{b \cdot d}; \quad \frac{c}{d} = \frac{c}{d} \cdot \frac{b}{b} = \frac{c \cdot b}{d \cdot b}, \quad \text{or} \quad \frac{b \cdot c}{b \cdot d}$$

When the common denominator is obtained we then look at the numerators to decide which is larger. This justifies this general principle.

For any numbers of arithmetic $\dfrac{a}{b}$ and $\dfrac{c}{d}$,

$$\frac{a}{b} > \frac{c}{d} \textbf{ when } a \cdot d > b \cdot c$$

$$\frac{a}{b} = \frac{c}{d} \textbf{ when } a \cdot d = b \cdot c$$

$$\frac{a}{b} < \frac{c}{d} \textbf{ when } a \cdot d < b \cdot c$$

(The products $a \cdot d$ and $b \cdot c$ are called *cross products*.)

Example 1. Insert the proper symbol $>$, $<$, or $=$ between $\frac{8}{9}$ and $\frac{7}{10}$.

$$\frac{8}{9} \diagdown\!\!\!\!\diagup \frac{7}{10}$$

$8 \cdot 10 \quad 9 \cdot 7$

$\qquad 80 > 63 \qquad$ Thus $\dfrac{8}{9} > \dfrac{7}{10}$

Example 2. Insert the proper symbol $>$, $<$, or $=$ between $\frac{6}{27}$ and $\frac{2}{9}$.

$$\frac{6}{27} \diagdown\!\!\!\!\diagup \frac{2}{9}$$

$6 \cdot 9 \quad 27 \cdot 2$

$\qquad 54 = 54 \qquad$ Thus $\dfrac{6}{27} = \dfrac{2}{9}$

Do exercises 99 through 102 at the left.

Do exercise set 1.5, p. 43.

1.6 EQUATIONS, OPPOSITE OPERATIONS, AND RELATED SENTENCES

There are several kinds of number sentences used in arithmetic and algebra. The most common kind is called an *equation*. By an equation we simply mean a number sentence with = for its verb. An example is $3 + 4 = 7$. Some equations are true. Some are false. Some are neither true nor false.

Examples

a) The equation $3 + 2 = 5$ is true.

b) The equation $5 - 4 = 3 + 7$ is false.

c) The equation $x + 2 = 8$ is neither true nor false because we don't know what x represents.

Do exercises 103 through 105 at the right.

An equation may be true or it may be false, but what does it mean? It simply says that the symbols on either side of the equals sign represent the same number. For example, $5 - 4 = 3 + 7$ says that $5 - 4$ represents the same number as $3 + 7$.

Do exercises 106 and 107 at the right.

SOLVING EQUATIONS

If an equation contains a variable it may be neither true nor false. Some replacements for the variable make it true. Some may make it false. The replacements making an equation true are called its *solutions*. When we find them we say we have *solved* the equation.

Example. Solve $x + 4 = 9$ by trial.

If we replace x by 2 we get a false sentence: $2 + 4 = 9$

If we replace x by 7 we get a false sentence: $7 + 4 = 9$

If we replace x by 5 we get a true sentence: $5 + 4 = 9$

No other replacement makes the equation true, so the only solution is the number 5.

Do exercises 108 through 112 at the right.

OBJECTIVES

You should be able to:

a) Define an equation and tell what an equation means.

b) Find replacements to make an equation false; to make it true.

c) Solve simple equations by trial.

d) Given a simple sentence, write two related sentences.

e) Point out impossible divisions, (divisor zero).

103. Write three true equations.

104. Write three false equations.

105. Write three equations that are neither true nor false.

Tell what each equation means.

106. $78 + 9 = 80 + 7$

107. $5 + \dfrac{1}{2} = 9 - 2$

108. Find three replacements that make $x + 5 = 12$ false.

109. Find the replacement that makes $x + 5 = 12$ true.

Solve these equations by trial.

110. $x + 4 = 10$

111. $3x = 12$

112. $3y + 1 = 4$

Squaring with a unit's digit of five.

For example, to square 65, add 1 to 6, and multiply by 6, obtaining 42. Place "42" to the left of "25," and you have

$$65^2 = 4225$$

To square 35, add 1 to 3, and multiply by 3, obtaining 12. Place "12" to the left of "25" and you have
$$35^2 = 1225$$

Use the above method to square each of the following.

a) 15   b) 45

c) 55   d) 25

e) 75   f) 85

g) 95

h) Using the distributive law, try to show why this works.

OPPOSITE OPERATIONS AND RELATED SENTENCES

We say that addition and subtraction are opposite operations. Let us see why. Suppose we add 3 and 2 and then subtract 2.

$$(3 + 2) - 2$$

The answer is 3, the number we started with. Subtraction "undoes" addition. Similarly, if we subtract 2 from a number and then add 2 we will get that number. Addition "undoes" subtraction.

RELATED SENTENCES

Consider the sentence $7 - 4 = 3$. It is true, so we have one number (three) named in two different ways. That is, $7 - 4$ and 3 represent the same number. Suppose we add 4. We can think of it in two ways because $7 - 4$ and 3 represent the same number.

$$(7 - 4) + 4 \quad \text{or} \quad 3 + 4$$

Addition "undoes" subtraction, so $(7 - 4) + 4$ is 7. Thus we have another equation, which is also true.

$$7 = 3 + 4 \quad (\text{or, } 7 = 4 + 3)*$$

Now 7 and $3 + 4$ represent the same number. Let us subtract 3. Again we can think of this in two ways.

$$7 - 3 \quad \text{or} \quad (4 + 3) - 3$$

Now $(4 + 3) - 3 = 4$, so we have the following equation. It is also true.

$$7 - 3 = 4$$

These three equations are called *related sentences*.

(a) $7 - 4 = 3$ ————————(We added 4 to get (b))

(b) $7 = 4 + 3$

(c) $7 - 3 = 4$ ————————(We subtracted 3 to get (c))

Any simple addition sentence has two related subtraction sentences. Any simple subtraction sentence has a related addition sentence and a related subtraction sentence. If a sentence is true, both of its related sentences are true.

* Because addition is commutative we regard $7 = 4 + 3$ and $7 = 3 + 4$ as the same.

Examples. For each sentence, write the related sentences.

a) $8 - 3 = 5$.   Related sentences $8 - 5 = 3$ and $8 = 5 + 3$

b) $9 = 2 + 5$.   Related sentences $9 - 5 = 2$ and $9 - 2 = 5$

c) $x - 4 = 12$. Related sentences $x - 12 = 4$ and $x = 12 + 4$

From the second example above you may suspect that if a sentence is false, then its related sentences are false. This is the case.

Sentences like the following are *related sentences.* **A sentence and its related sentences have the same solutions.**

① $a - b = c$

       ② $a = c + b$

③ $a - c = b$

Do exercises 113 through 116 at the right.

MULTIPLICATION AND DIVISION ARE OPPOSITES

Multiplication and division "undo" each other. For a multiplication or a division sentence there are two related sentences.

Sentences like the following are *related sentences.*

① $\dfrac{a}{b} = c$ ——————(We can multiply by b to get ②.)

       ② $a = bc$

③ $\dfrac{a}{c} = b$ ——————(We can divide by c to get ③.)

Examples. For each sentence, write the related sentences.

a) $\dfrac{12}{4} = 3$ Related sentences $\dfrac{12}{3} = 4$ and $12 = 4 \cdot 3$

b) $\dfrac{15}{5} = 3$ Related sentences $\dfrac{15}{3} = 5$ and $15 = 5 \cdot 3$

c) $\dfrac{17}{x} = 19$ Related sentences $\dfrac{17}{19} = x$ and $17 = 19x$

d) $10 = 19x$ Related sentences $\dfrac{10}{19} = x$ and $\dfrac{10}{x} = 19$

Do exercises 117 through 121 at the right.

For each sentence, write two related sentences.

113. $17 - 5 = 12$

114. $10 = 4 + 5$

115. $3 - x = 1$

116. $10.5 = x + 2$

For each sentence, write two related sentences.

117. $\dfrac{18}{6} = 3$

118. $35 = 5 \cdot 7$

119. $5 \cdot 7 = 35$

120. $\dfrac{39}{x} = 13$

121. $5x = 27$

Which of these divisions are possible?

122. $\dfrac{80}{4}$

123. $\dfrac{5}{0}$

124. $\dfrac{0}{5}$

125. $\dfrac{0}{0}$

126. $\dfrac{11}{10 - 10}$

127. $\dfrac{9}{12 - (4 \cdot 3)}$

128. $\dfrac{P}{x - x}$

SUBTRACTION AND DIVISION

The ideas of related sentences and opposite operations actually come from the definitions of subtraction and division.

The difference $a - b$ is the number (if it exists) which when added to b gives a.

Thus,

a) $7 - 5 = 2$ because $7 = 2 + 5$ (when 2 is added to 5 it gives 7)

b) $3.5 - 2 = 1.5$ because $3.5 = 1.5 + 2$

The quotient $\dfrac{a}{b}$ is the number (if it exists) which when multiplied by b gives a.

Thus,

a) $\dfrac{39}{3} = 13$ because $39 = 3 \cdot 13$ (when 13 is multiplied by 3 it gives 39)

b) $\dfrac{\frac{1}{4}}{\frac{1}{2}} = \dfrac{1}{2}$ because $\dfrac{1}{4} = \dfrac{1}{2} \cdot \dfrac{1}{2}$

DIVISION BY ZERO

Why do we not divide by zero?

$\dfrac{n}{0}$ would be some number r such that $r \cdot 0 = n$. But $r \cdot 0 = 0$, so the only possible number n which could be divided by 0 is 0.
Look for a pattern.

a) $\dfrac{0}{0} = 5$ because $0 = 0 \cdot 5$ b) $\dfrac{0}{0} = 789$ because $0 = 0 \cdot 789$

c) $\dfrac{0}{0} = 17$ because $0 = 0 \cdot 17$ d) $\dfrac{0}{0} = \dfrac{1}{2}$ because $0 = 0 \cdot \dfrac{1}{2}$

It looks as if $\frac{0}{0}$ could be any number at all. This would be very confusing, getting any answer we want when we divide 0 by 0. Thus we agree to exclude division by zero.

We never divide by zero.

Do exercises 122 through 128 at the left.

Do exercise set 1.6, p. 45.

1.7 SOLVING EQUATIONS

One of the most important skills of elementary algebra is the ability to solve equations. Simple equations can be solved using the ideas of related sentences and opposite operations. Later other techniques will be developed for solving equations. The idea is to look for an equation which has the variable alone on one side. That equation will tell you which calculations to do to find the solution.

Example 1. Solve: $17 - x = 14$

Related sentence: $17 - 14 = x$ (see page 21)

We do the calculation $17 - 14$ and get $3 = x$.

In this example we get the equation $3 = x$. Its solution is obviously the number 3. Is 3 also a solution of the original equation $17 - x = 14$? We check to find out. To do this we replace x by 3 in the equation and see.

Check: $17 - x = 14$

$$17 - \boxed{3} \,\bigg|\, 14$$
$$14 \,\bigg|$$

When we simplify we get 14 on both sides. So 3 is a solution.

Example 2. Solve: $3x = 39$ Now we check:

related sentence $x = \dfrac{39}{3}$

$$x = 13$$

$$3x = 39$$
$$3 \cdot \boxed{13} \,\bigg|\, 39$$
$$39 \,\bigg|$$

Do exercises 129 through 132 at the right.

Example 3. Solve: $3(x + 2) = 15$ Check: $3(x + 2) = 15$

related sentence $x + 2 = \dfrac{15}{3}$

$$x = \frac{15}{3} - 2 = 3$$

$$3(\boxed{3} + 2) \,\bigg|\, 15$$
$$3 \cdot 5$$
$$15$$

Do exercises 133 through 135 at the right.

You should be able to solve and check simple equations using opposite operations and related sentences.

Solve and check.

129. $15 - x = 7$

130. $5x = 35$

131. $11y = 44$

132. $x - 56 = 101$

Solve and check.

133. $4(x + 1) = 12$

134. $40 = 5(3 + y)$

135. $3(x - 2) = 18$

Solve and check.

136. $1.5x = 4.2$

137. $20y = 50$

138. Solve and check.

$z - \dfrac{5}{6} = \dfrac{3}{8}$

139. Solve and check.

$54 = \dfrac{9}{8}y$

Example 4

Solve: $3.5y = 14.7$

$$y = \frac{14.7}{3.5}$$

$$y = \frac{14.7}{3.5} \cdot \frac{10}{10}$$

$$y = \frac{147}{35}, \text{ or } 4.2$$

Check: $3.5y = 14.7$

$$3.5 \times \boxed{4.2} \mid 14.7$$
$$14.7 \mid$$

Do exercises 136 and 137 at the left.

Example 5

Solve: $p - \dfrac{3}{5} = \dfrac{1}{4}$

$$p = \frac{1}{4} + \frac{3}{5}$$

$$p = \frac{5}{20} + \frac{12}{20}$$

$$p = \frac{17}{20}$$

Check: $p - \dfrac{3}{5} = \dfrac{1}{4}$

$$\boxed{\frac{17}{20}} - \frac{3}{5} \ \Big| \ \frac{1}{4}$$

$$\frac{17}{20} - \frac{12}{20}$$

$$\frac{5}{20}$$

$$\frac{1}{4}$$

Do exercise 138 at the left.

Example 6

Solve: $33 = \dfrac{2}{3}t$

$$\frac{33}{\frac{2}{3}} = t$$

$$33 \cdot \frac{3}{2} = t$$

$$\frac{99}{2} = t$$

Check: $33 = \dfrac{2}{3}t$

$$33 \ \Big| \ \frac{2}{3} \cdot \boxed{\frac{99}{2}}$$

$$\frac{99}{3}$$

$$33$$

Do exercise 139 at the left.

Do exercise set 1.7, p. 47.

1.8 SOLVING PROBLEMS

Why should one learn to solve equations? The answer is that applied problems can be solved using equations. To do this, we first translate the problem situation to an equation and then solve the equation. The problem situation may be explained in words (as in a textbook) or may come from an actual situation in the real world.

Example 1. Solve this problem.

Thirty-five is three-fourths of what number?

$$35 \quad = \quad \tfrac{3}{4} \quad \cdot \quad x$$

The translation gives us the equation $35 = \tfrac{3}{4}x$. We solve it:

$$x = \frac{35}{\dfrac{3}{4}} = 35 \cdot \frac{4}{3} = \frac{140}{3}.$$

To check we find out if $\tfrac{3}{4}$ of this number is 35.

$$\frac{3}{4} \times \frac{140}{3} = \frac{3 \cdot 140}{4 \cdot 3} = \frac{3 \cdot 35 \cdot 4}{4 \cdot 3} = 35.$$

Notice that in translating, *is* translates to $=$. The word *of* translates to $\cdot$, and the unknown number translates to a variable.

Do exercises 140 and 141 at the right.

Translating a problem to a number sentence is usually the most important part of solving a problem, and may take the most time. Sometimes it is helpful to reword the problem before translating.

Example 2

Hugh Diddy's salary is $12,000. This is 1.5 times John Jones' salary. What is Jones' salary?

Rewording: $12,000 is 1.5 times Jones' salary.

Translation: $12,000 = 1.5 \cdot y$

Do exercises 142 and 143 at the right.

Do exercise set 1.8, p. 49.

Do exercise set 1.8, p. 49.

OBJECTIVES

You should be able to:

a) Translate problem situations to equations.

b) Solve applied problems by translating to equations and solving the equations.

Translate to equations. Then solve and check.

140. Thirty-seven plus what number is seventy-three?

141. Forty-four is two-thirds of what number?

Translate to equations. Do not solve.

142. There were 224 washing machines in a warehouse. When some of them were removed, three-fourths of them were left. How many were removed?

143. Bob R. Ooky's batting average is one and a half times that of George Hurler. Ooky's average is .320. What is Hurler's average.

OBJECTIVES

You should be able to:

a) Convert percent notation to decimal notation, and conversely.

b) Convert percent notation to fractional notation, and conversely.

c) Solve applied problems involving percents.

The symbol % was derived from the numeral 100.

Find a decimal numeral for each of the following.

144. 46.2%

145. 100%

Find a fractional numeral for each of the following. You need not simplify.

146. 67%

147. 45.6%

148. $\frac{1}{4}\%$

1.9 PERCENT NOTATION

THE PERCENT SYMBOL

There are other ways to name numbers of arithmetic besides using fractional and decimal numerals. Among these are numerals which contain the percent symbol %, which means "per hundred."

We can regard the percent symbol as part of a numeral. For example

37%

is defined to mean

$$37 \times .01 \quad \text{or} \quad 37 \times \frac{1}{100}$$

In general,

$$n\% \quad \text{means} \quad n \times .01 \quad \text{or} \quad n \times \frac{1}{100}$$

CONVERTING TO DECIMAL NOTATION

Example 1. Find a decimal numeral for 78.5%.

$$78.5\% = 78.5 \times .01 \quad \text{(by definition of \%)}$$
$$= .785$$

Do exercises 144 and 145 at the left.

CONVERTING TO FRACTIONAL NOTATION

Example 2. Find a fractional numeral for 89%.

$$89\% = 89 \times \frac{1}{100} \quad \text{(by definition of \%)}$$
$$= \frac{89}{100}$$

Example 3. Find a fractional numeral for 34.7%.

$$34.7\% = 34.7 \times \frac{1}{100} \quad \text{(by definition of \%)}$$
$$= \frac{34.7}{100}$$
$$= \frac{34.7}{100} \cdot \frac{10}{10}$$
$$= \frac{347}{1000}$$

Do exercises 146 through 148 at the left.

CONVERTING FROM DECIMAL TO PERCENT NOTATION

By applying the definition of percent in reverse we can convert from decimals to percents. We multiply by 1, naming it $100 \times .01$ (that is, $100 \times .01 = 1$).

Example 1. Name 4.93 using a percent symbol.

$$4.93 = 4.93 \times (100 \times .01)$$
$$= (4.93 \times 100) \times .01 \qquad \text{(using associativity)}$$
$$= 493 \times .01$$
$$= 493\%$$

Example 2. Name .002 using a percent symbol.

$$.002 = .002 \times (100 \times .01) = (.002 \times 100) \times .01$$
$$= .2 \times .01$$
$$= .2\%$$

Do exercises 149 and 150 at the right.

CONVERTING FROM FRACTIONAL TO PERCENT NOTATION

We can also convert from fractional numerals to percents. Again, we multiply by 1, but this time we name it $100 \cdot \frac{1}{100}$.

Example 3. Name $\frac{5}{8}$ using a percent symbol.

$$\frac{5}{8} = \frac{5}{8} \cdot \left(100 \cdot \frac{1}{100}\right)$$
$$= \left(\frac{5}{8} \cdot 100\right) \cdot \frac{1}{100}$$
$$= \frac{500}{8} \cdot \frac{1}{100}$$
$$= \frac{500}{8}\%, \text{ or } 62.5\%$$

Do exercises 151 through 153 at the right.

SOLVING PROBLEMS INVOLVING PERCENTS

Let's solve some applied problems involving percents. Again, it is helpful to translate the problem situation to an equation and then solve the equation.

Name each of the following using a percent symbol.

149. 6.77

150. .9944

Name each of the following using a percent symbol.

151. $\frac{1}{4}$

152. $\frac{3}{8}$

153. $\frac{2}{3}$

Translate and solve.

154. What is 23% of 48?

155. 25% of 40 is what?

Translate and solve.

156. 15 is what percent of 60?

157. What percent of 50 is 16?

Example 1. What is 12% of 59?

We first translate to a number sentence.

$$\begin{array}{ccccc} \text{What is} & 12\% & \text{of} & 59 \\ \downarrow & \downarrow\ \downarrow & \downarrow & \downarrow \\ x & =\ 12\% & \cdot & 59 \end{array}$$

Solve: $x = 12 \times .01 \times 59$

$x = .12 \times 59$

$x = 7.08$

The solution is 7.08, so 7.08 is 12% of 59.

Do exercises 154 and 155 at the left.

Example 2. 15 is what percent of 45?

We translate to a number sentence.

$$\begin{array}{cccccc} 15 & \text{is} & \text{what} & \text{percent} & \text{of} & 45? \\ \downarrow & \downarrow & \downarrow & \downarrow & \downarrow & \downarrow \\ 15 & = & x & \% & \cdot & 45 \end{array}$$

Solve: $15 = x \times .01 \times 45$

$15 = x \times .45$

$\dfrac{15}{.45} = x$

Now $\dfrac{15}{.45} = \dfrac{15}{.45} \cdot \dfrac{100}{100} = \dfrac{1500}{45} = 33\frac{1}{3}$

The solution is $33\frac{1}{3}$, so 15 is $33\frac{1}{3}$% of 45.

Example 3. 7.5 is what percent of 1.5?

We translate to a number sentence.

$$\begin{array}{cccccc} 7.5 & \text{is} & \text{what} & \text{percent} & \text{of} & 1.5? \\ \downarrow & \downarrow & \downarrow & \downarrow & \downarrow & \downarrow \\ 7.5 & = & x & \% & \cdot & 1.5 \end{array}$$

Solve: $7.5 = x \cdot .01 \cdot 1.5$
$7.5 = .015x$

$x = \dfrac{7.5}{.015} = \dfrac{7.5}{.015} \cdot \dfrac{1000}{1000} = \dfrac{7500}{15} = 500$

The solution is 500, so 7.5 is 500% of 1.5.

Do exercises 156 and 157 at the left.

Example 4. 3 is 16 percent of what?

Translate: $3 = 16\% \cdot y$

Solve: $3 = 16 \times .01 \times y$

$3 = .16 \times y$

$\dfrac{3}{.16} = y$

Now $\dfrac{3}{.16} = \dfrac{3}{.16} \cdot \dfrac{100}{100} = \dfrac{300}{16} = 18.75$

The number 18.75 checks, so it is the solution.

Do exercises 158 and 159 at the right.

Sometimes it is helpful to reword the problem before translating.

Example 5. Blood is 90% water. The average adult has 5 quarts of blood. How much water is in the average adult's blood?

Rewording: 90% of 5 is what?

Translate: $90\% \cdot 5 = x$

Solve: $90 \times .01 \times 5 = x$

$.90 \times 5 = x$

$4.5 = x$

The number 4.5 checks, so it is the solution. Thus there are 4.5 quarts of water in the average adult's blood.

Example 6. Jim Brown gets a 6% raise of $.85 in his daily wage as a butcher. What is his old wage? His new wage?

Rewording: 6% of what is .85

Translate: $6\% \cdot W = .85$

Solve: $6 \times .01 \times W = .85$

$.06 \times W = .85$

$W = \dfrac{.85}{.06} = \dfrac{.85}{.06} \cdot \dfrac{100}{100} = \dfrac{85}{6} = 14.16\overline{6} \approx 14.17$

("$\approx$" means "approximately equal to")

So his old daily wage is about $14.17 and his new wage is $14.17 + $.85, or $15.02.

Do exercises 160 and 161 at the right.

Do exercise set 1.9, p. 51.

Translate and solve.

158. 45 is 20 percent of what?

159. 120 percent of what is 60?

Translate to equations. Then solve.

160. The area of Greenland is 25% of the area of the USA. If the area of the USA is 3,615,000 sq mi, what is the area of Greenland?

161. The price of an item increased from 35¢ to 40¢. What was the percent of increase? (Hint: 5 is what percent of 35?)

You should be able to:

a) Find the area of a given rectangle, square, parallelogram, trapezoid, or circle.

b) Given the length of a radius of a circle, find the length of a diameter, and conversely.

c) Solve applied problems involving geometric formulas.

d) Find perimeters of a rectangle or square.

e) Find the volume of a rectangular solid.

162. Find the area and perimeter of this rectangle.

1.10 GEOMETRIC FORMULAS

In this section we shall review some formulas from geometry. We first consider rectangles. The area of a rectangle is the number of *unit* squares it takes to fill it up. Unit squares look like these.

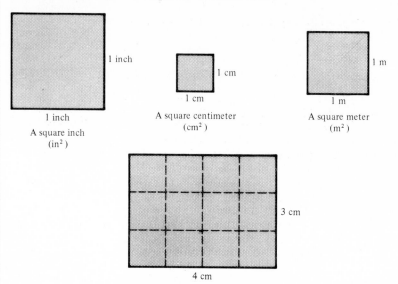

1 inch

A square inch
(in²)

A square centimeter
(cm²)

A square meter
(m²)

This rectangle has length 4 cm and width 3 cm. It takes 12 unit squares to fill it up. Therefore the area is 12 cm² (or 12 sq cm). In any rectangle we can find the area by multiplying the length by the width. The *perimeter* of a rectangle is the distance around it. We can find the perimeter by adding the lengths of the four sides. Or we could double the length and double the width and then add.

If a rectangle has length l and width w, the area is given by $A = l \cdot w$ (area is length times width).

The perimeter is given by $P = 2l + 2w$ [or $2(l + w)$].

When we make computations with dimension symbols, such as inches (in.) or centimeters (cm), we can treat them as if they were numerals or variables.

Compare

a) $3x \cdot 4x = 3 \cdot 4 \cdot x \cdot x = 12x^2$ with 3 in. · 4 in. = 3 · 4 in. · in. = 12 in²
b) $2x + 5x = (2 + 5)x = 7x$ with 2 cm + 5 cm = (2 + 5) cm = 7 cm

Example 1. Find the area and perimeter of this rectangle.

$$A = lw$$
$$= (4.7 \text{ cm}) \cdot (2.4 \text{ cm})$$
$$= 4.7 \times 2.4 \text{ cm} \cdot \text{cm}$$
$$= 11.28 \text{ cm}^2 \text{ (or 11.28 square centimeters)}$$

2.4 cm

4.7 cm

$P = 2(l + w)$
$= 2(4.7 \text{ cm} + 2.4 \text{ cm})$
$= 2 \times 7.1 \text{ cm} = 14.2 \text{ cm}$

Do exercise 162 at the left.

If the length and width of a rectangle are the same, then the rectangle is a square. In a square, all four sides are the same length. Suppose a square has sides of length s. Then its area is $s \cdot s$ or s^2 and its perimeter is $s + s + s + s$, or $4s$.

If a square has sides of length s, then the area is given by $A = s^2$ and the perimeter is given by $P = 4s$.

PARALLELOGRAMS, TRAPEZOIDS, AND TRIANGLES

A parallelogram is a four-sided figure with two pairs of parallel sides. To find the area of a parallelogram, we can think of cutting off a part of it as shown below. We can then place the part as shown to form a rectangle. The area of the rectangle is $b \cdot h$ (length of the base times height). This is also the area of the parallelogram.

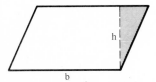 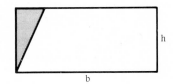

The area of a parallelogram is given by $A = b \cdot h$, where b is the length of the base and h is the height.

Do exercises 163 and 164 at the right.

A trapezoid is a four-sided figure with at least one pair of parallel sides. To find the area of a trapezoid we can think of cutting out another one just like the given one and placing the two of them together, as shown below. This forms a parallelogram, whose area is $h \cdot (a + b)$.

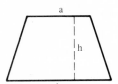 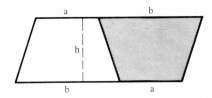

The trapezoid is half of this parallelogram, so its area is $\frac{1}{2}h \cdot (a + b)$. We call the parallel sides of the trapezoid the *bases*, so the area is half the product of the height and the sum of the bases.

If a trapezoid has bases of lengths a and b and has height h, its area is given by $A = \frac{1}{2} \cdot h \cdot (a + b)$.

163. Find the area and perimeter of this square.

5 m

5 m

164. Find the area of this parallelogram.

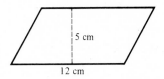

5 cm

12 cm

165. Find the area of this trapezoid.

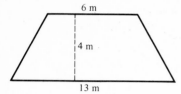

166. Find the area of this triangle.

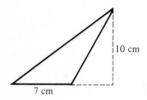

167. Find the missing angle measure.

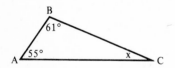

Example 2. Find the area of this trapezoid.

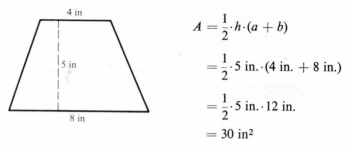

$$A = \frac{1}{2} \cdot h \cdot (a + b)$$

$$= \frac{1}{2} \cdot 5 \text{ in.} \cdot (4 \text{ in.} + 8 \text{ in.})$$

$$= \frac{1}{2} \cdot 5 \text{ in.} \cdot 12 \text{ in.}$$

$$= 30 \text{ in}^2$$

Do exercise 165 at the left.

To find the area of a triangle, we can think of cutting out another one like the one given and placing the two of them together, as shown below.

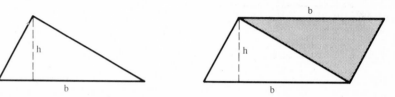

This forms a parallelogram, whose area is $b \cdot h$. The triangle is half of the parallelogram, so the area is half of $b \cdot h$.

If a triangle has a base of length b and has height h, then the area is given by $A = \frac{1}{2}b \cdot h$.

The measures of the angles of a triangle add up to 180°. Thus if we know the measures of two angles of a triangle we can calculate the third.

In any triangle the sum of the measures of the angles is 180°.
$$m(\angle A) + m(\angle B) + m(\angle C) = 180°.$$

Example 3. Find the area of this triangle.

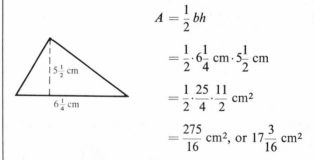

$$A = \frac{1}{2} bh$$

$$= \frac{1}{2} \cdot 6\frac{1}{4} \text{ cm} \cdot 5\frac{1}{2} \text{ cm}$$

$$= \frac{1}{2} \cdot \frac{25}{4} \cdot \frac{11}{2} \text{ cm}^2$$

$$= \frac{275}{16} \text{ cm}^2, \text{ or } 17\frac{3}{16} \text{ cm}^2$$

Example 4. Find the missing angle measure.

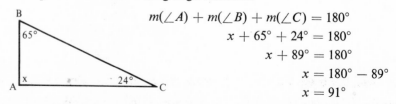

$$m(\angle A) + m(\angle B) + m(\angle C) = 180°$$
$$x + 65° + 24° = 180°$$
$$x + 89° = 180°$$
$$x = 180° - 89°$$
$$x = 91°$$

Do exercises 166 and 167 at the left.

VOLUMES OF RECTANGULAR SOLIDS

The volume of a rectangular solid is the number of *unit* cubes it takes to fill it up. Unit cubes look like these.

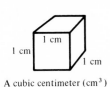

A cubic centimeter (cm³)

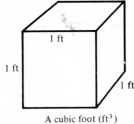

A cubic foot (ft³)

This rectangular solid has length 4 cm, width 3 cm, and height 2 cm. There are thus two layers of cubes, each having 3 × 4, or 12 cubes. The total volume is thus 24 cm³ (cubic centimeters).

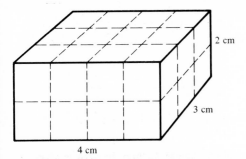

In any rectangular solid we can find the volume by multiplying the length by the width by the height.

If a rectangular solid has length *l*, width *w*, and height *h*, then the volume is given by $V = l \cdot w \cdot h.$

Example 5. Find the volume of this solid.

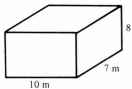

$$V = lwh$$
$$= 10\,\text{m} \cdot 7\,\text{m} \cdot 8\,\text{m}$$
$$= 10 \cdot 56\,\text{m}^3 \text{ (cubic meters)}$$
$$= 560\,\text{m}^3$$

Do exercise 168 at the right.

CIRCLES

Here is a circle, with center O. Also shown are a diameter d and a radius r. A diameter is twice as long as a radius.

In any circle, if *d* is the diameter and *r* is the radius, then $d = 2 \cdot r.$

The *circumference* of a circle is the distance around it. If we divide the circumference of a circle by the diameter we always get the same number.

---

168. Find the volume of this solid.

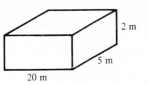

169. Find the length of a diameter of a circle whose radius is 9 meters.

This number is not a number of arithmetic, so we use the Greek letter π to name it. This number is approximately 3.14 or 22/7. We use the letter C for circumference. Then $C/d = \pi$ in any circle, or $C = \pi d$.

If C is the circumference of a circle and r is the radius, then $C = \pi d$, or since $d = 2r$, $C = 2\pi r$.

Example 1. Find the length of a diameter of a circle whose radius is 14 ft.

$d = 2r$

$\quad = 2 \cdot 14\,\text{ft} = 28\,\text{ft}$

Example 2. Find the circumference of a circle whose radius is 14 ft. Use 22/7 for π.

$C = 2 \cdot \pi \cdot r$

$\quad = 2 \cdot \dfrac{22}{7} \cdot 14\,\text{ft}$

$\quad = 88\,\text{ft}$

170. Find the circumference of a circle whose radius is 9 cm. Use 3.14 for π.

Do exercises 169 and 170 at the left.

AREAS OF CIRCLES

Suppose we take half a circular region, cut it into small wedges, and arrange the wedges as shown below. If the wedges are very small the

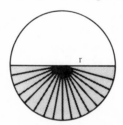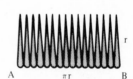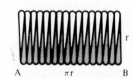

distance from A to B will be very nearly πr (half the circumference of the circle). Now if we cut the other half of the circle the same way and place the wedges together, we get what is almost a rectangle, with area $\pi r \cdot r$.

The area of a circle of radius r is given by $A = \pi r^2$.

171. Find the area of a circle with a 9 ft radius. Use 3.14 for π.

Example 3. Find the area of this circle. Use $\frac{22}{7}$ for π.

$A = \pi \cdot r \cdot r$

$A \approx \dfrac{22}{7} \cdot 14\,\text{cm} \cdot 14\,\text{cm}$

$A \approx \dfrac{22}{7} \cdot 196\,\text{cm}^2$

$A \approx 616\,\text{cm}^2$

The area is about 616 cm².

Do exercise 171 at the left.

Do exercise set 1.10, p. 53.

EXERCISE SET 1.1 (pp. 3–5)

Review the objectives for this lesson before doing these exercises.

Write four different fractional numerals for each number.

1. $\dfrac{4}{3}$　　　　**2.** $\dfrac{5}{9}$　　　　**3.** $\dfrac{6}{11}$　　　　**4.** $\dfrac{15}{7}$

5. $\dfrac{2}{11}$　　　　**6.** 1　　　　**7.** 5　　　　**8.** 0

Write the simplest fractional numeral for each number.

9. $\dfrac{8}{6}$　　　　**10.** $\dfrac{15}{25}$　　　　**11.** $\dfrac{17}{34}$　　　　**12.** $\dfrac{35}{25}$

13. $\dfrac{100}{50}$　　　　**14.** $\dfrac{13}{39}$　　　　**15.** $\dfrac{250}{75}$　　　　**16.** $\dfrac{12}{18}$

Multiply and simplify.

17. $\dfrac{1}{4} \cdot \dfrac{1}{2}$　　　**18.** $\dfrac{11}{10} \cdot \dfrac{8}{5}$　　　**19.** $\dfrac{17}{2} \cdot \dfrac{3}{4}$　　　**20.** $\dfrac{11}{12} \cdot \dfrac{12}{11}$

1. _____

2. _____

3. _____

4. _____

5. _____

6. _____

7. _____

8. _____

9. _____

10. _____

11. _____

12. _____

13. _____

14. _____

15. _____

16. _____

17. _____

18. _____

19. _____

20. _____

ANSWERS

21. _____

22. _____

23. _____

24. _____

25. _____

26. _____

27. _____

28. _____

29. _____

30. _____

31. _____

32. _____

Add and simplify.

21. $\dfrac{1}{2} + \dfrac{1}{2}$

22. $\dfrac{1}{2} + \dfrac{1}{4}$

23. $\dfrac{4}{9} + \dfrac{13}{18}$

24. $\dfrac{5}{12} + \dfrac{7}{18}$

25. $\dfrac{12}{15} + \dfrac{8}{35}$

26. $\dfrac{9}{8} + \dfrac{7}{12}$

Subtract and simplify.

27. $\dfrac{1}{4} - \dfrac{1}{4}$

28. $\dfrac{4}{5} - \dfrac{2}{5}$

29. $\dfrac{13}{18} - \dfrac{4}{9}$

30. $\dfrac{13}{15} - \dfrac{8}{45}$

31. $\dfrac{11}{12} - \dfrac{4}{10}$

32. $\dfrac{73}{30} - \dfrac{21}{35}$

NAME

CLASS

ANSWERS

EXERCISE SET 1.2 (pp. 6–8)

Review the objectives for this lesson.

Write three kinds of expanded numerals, without exponents, for each number.

1. 5677 **2.** 34,678 **3.** 908,563

Write an expanded numeral, with exponents, for each number.

4. 432,123 **5.** 3466 **6.** 589

What is the meaning of each of the following?

7. 4^5 **8.** 9^2 **9.** 45^0 **10.** 1^3

Write exponential notation for each of the following.

11. *nnn* **12.** *yyyy* **13.** *rr*

What is the meaning of each of the following? Do not use $\times$ or $\cdot$.

14. x^3 **15.** y^4 **16.** m^2

What does each of the following represent if *n* stands for 10?

17. n^1 **18.** n^2 **19.** n^3

What does each of the following represent if *x* stands for 3?

20. x^2 **21.** x^3 **22.** x^4

1.
2.
3.
4.
5.
6.
7.
8.
9.
10.
11.
12.
13.
14.
15.
16.
17.
18.
19.
20.
21.
22.

ANSWERS

23.

_____

24.

_____

25.

_____

26.

_____

27.

_____

28.

_____

29.

_____

30.

_____

31.

_____

32.

_____

33.

_____

34.

_____

35.

_____

36.

_____

37.

_____

38.

_____

What does each of the following represent if y stands for 2?

23. y^0 **24.** y^1 **25.** y^3

Write an expanded numeral for each of the following numbers.

26. 29.1 **27.** 16.33 **28.** 16.789

Write fractional numerals. You need not simplify.

29. 4.67 **30.** 3.09 **31.** 3.1415 **32.** .001

Write decimal numerals.

33. $\dfrac{1}{4}$ **34.** $\dfrac{1}{2}$ **35.** $\dfrac{3}{5}$

36. $\dfrac{2}{9}$ **37.** $\dfrac{4}{9}$ **38.** $\dfrac{75}{100}$

NAME

CLASS

ANSWERS

EXERCISE SET 1.3 (pp. 9–12)

Review the objectives for this lesson.

Do these calculations.

1. $(10 + 4) + 8$ **2.** $10 \times (9 + 4)$ **3.** $(10 \times 7) + 19$

Do these calculations, choosing your grouping and ordering so as to make the work easy.

4. $8 + 4 + 5 + 2 + 6 + 15 + 1$ **5.** $2 \times 2 \times 2 \times 5 \times 5 \times 5$

Which laws are illustrated by these sentences?

6. $67 + 3 = 3 + 67$ **7.** $5 \times (9 \times 10) = (5 \times 9) \times 10$

8. $6 \cdot (9 + 5) = (6 \cdot 9) + (6 \cdot 5)$ **9.** $8 \cdot (7 \cdot 6) = (8 \cdot 7) \cdot 6$

Simplify.

10. $5 \cdot 4 + 8 \cdot 10$ **11.** $10 \cdot 9 + 8 \cdot 4$ **12.** $\dfrac{1}{2} \cdot \dfrac{1}{2} + \dfrac{1}{4} \cdot \dfrac{1}{4}$

Evaluate and simplify.

13. $ab + cd$, where $a = 3$, $b = 5$, $c = 10$ and $d = 0$ **14.** $xy + mp$, where $x = 4$, and $y = 10$, $m = 3$, and $p = 1$

15. $abc + c^2$, where $a = 2$, $b = 4$ and $c = 10$ **16.** $x^2 - y^2$, where $x = 4$ and $y = 2$

17. $(x - y) \cdot (x + y)$, where $x = 4$ and $y = 2$ **18.** $2(l + w)$, where $l = 15$ and $w = 7$

1. _____

2. _____

3. _____

4. _____

5. _____

6. _____

7. _____

8. _____

9. _____

10. _____

11. _____

12. _____

13. _____

14. _____

15. _____

16. _____

17. _____

18. _____

ANSWERS

19.

_____

20.

_____

21.

_____

22.

_____

23.

_____

24.

_____

25.

_____

26.

_____

27.

_____

28.

_____

29.

_____

30.

_____

31.

_____

32.

_____

33.

_____

Factor.

19. $6x + 6y$

20. $7w + 7u$

21. $9t + 9e$

22. $4x + 4y + 4z$

23. $10a + 10b + 10c$

24. $87x + 87y + 87z$

Factor. Then evaluate both expressions when $x = 5$ and $y = 10$.

25. $4x + 4y$

26. $8x + 8y$

27. $10x + 10y$

Factor. Then evaluate both expressions when $a = 2$ and $b = 7$.

28. $2a + 2b$

29. $3a + 3b$

30. $14a + 14b$

Factor. Then evaluate both expressions when $p = 0$ and $q = 9$.

31. $5p + 5q$

32. $7p + 7q$

33. $10p + 10q$

EXERCISE SET 1.4 (pp. 13–14)

Review the objectives for this lesson.

Factor.

1. $2x + 4$

2. $2x + 4y$

3. $3x + 6y$

4. $9x + 27$

5. $6x + 24$

6. $30 + 5y$

7. $5x + 10 + 15y$

8. $8a + 16b + 64$

9. $7 + 14b + 56w$

10. $100x + 1000y$

11. $db + dg + dh$

12. $30 + 40y + 5$

Multiply.

13. $3(x + 1)$

14. $2(x + 2)$

15. $5(x + 5)$

16. $7(x + 8)$

17. $4(1 + y)$

18. $7(9 + t)$

19. $7(x + 4 + 6y)$

20. $9(4t + 3z)$

21. $8(9x + 5y + 80)$

22. $a(x + y)$

23. $a(t + u + v)$

24. $10(10 + 10y + 9)$

1.

2.

3.

4.

5.

6.

7.

8.

9.

10.

11.

12.

13.

14.

15.

16.

17.

18.

19.

20.

21.

22.

23.

24.

ANSWERS

Collect like terms.

25. _____

26. _____

27. _____

28. _____

29. _____

30. _____

31. _____

32. _____

33. _____

34. _____

35. _____

36. _____

37. _____

38. _____

39. _____

40. _____

41. _____

42. _____

25. $2x + 3x$ **26.** $7y + 9y$ **27.** $10a + a$

28. $15x + 7x + 5y + 12y$ **29.** $2x + 9z + 6x$ **30.** $50a + 40a + 9b + 2b$

31. $41a + 90c + 60c + 2a$ **32.** $9a + a$ **33.** $42x + 6b + 4x + 2b$

34. $8a + 8b + 3a + 3b$ **35.** $8u + 3t + 10u + 6u + 9u + 2t$

36. $100y + 200z + 190y + 400z$ **37.** $5t + 6h + t + 8t + 10t + h$

38. $45 + 90d + 87 + 9d + d$ **39.** $23 + 5t + 7y + t + y + 27$

40. $\frac{1}{2}x + \frac{1}{2}x$ **41.** $2y + \frac{1}{4}y + y$ **42.** $\frac{1}{2}a + a + 2b + \frac{1}{2}b$

EXERCISE SET 1.5 (pp. 15–18)

Review the objectives for this lesson.

What is the reciprocal of each number?

1. $\dfrac{1}{4}$ **2.** $\dfrac{3}{4}$ **3.** $\dfrac{1}{2}$ **4.** 1   **5.** 8   **6.** 9

Divide, by multiplying by 1.

7. $\dfrac{\frac{6}{7}}{\frac{3}{5}}$ **8.** $\dfrac{\frac{1}{4}}{\frac{7}{9}}$ **9.** $\dfrac{\frac{11}{7}}{\frac{8}{5}}$

Divide, by multiplying by the reciprocal of the divisor.

10. $\dfrac{\frac{7}{6}}{\frac{3}{5}}$ **11.** $\dfrac{1}{4} \div \dfrac{1}{2}$ **12.** $\dfrac{7}{5} \div \dfrac{3}{4}$ **13.** $\dfrac{9}{10} \div \dfrac{14}{11}$

14. $\dfrac{13}{12} \div \dfrac{39}{5}$ **15.** $\dfrac{17}{6} \div \dfrac{3}{8}$ **16.** $100 \div \dfrac{1}{5}$ **17.** $\dfrac{3}{4} \div 10$

Graph each number using a number line. Use separate paper.

18. $\dfrac{5}{4}$ **19.** $\dfrac{15}{16}$ **20.** $\dfrac{123}{15}$

NAME

CLASS

ANSWERS

1.

2.

3.

4.

5.

6.

7.

8.

9.

10.

11.

12.

13.

14.

15.

16.

17.

ANSWERS

Insert the proper symbol $<$, $>$, or $=$ between each pair of numerals.

21. _____

_____

22. _____

21. $\dfrac{1}{2}$ $\dfrac{2}{4}$

22. $\dfrac{11}{31}$ $\dfrac{41}{13}$

23. $\dfrac{5}{9}$ $\dfrac{11}{7}$

23. _____

_____

24. _____

_____

24. $\dfrac{1}{4}$ $\dfrac{6}{24}$

25. $\dfrac{4}{5}$ $\dfrac{8}{10}$

26. $\dfrac{9}{12}$ $\dfrac{3}{4}$

25. _____

_____

26. _____

_____

27. $\dfrac{11}{15}$ $\dfrac{13}{24}$

28. $\dfrac{7}{3}$ $\dfrac{77}{33}$

29. $\dfrac{9}{7}$ $\dfrac{13}{10}$

27. _____

_____

28. _____

_____

29. _____

_____

30. $\dfrac{13}{8}$ $\dfrac{8}{5}$

31. $\dfrac{21}{16}$ $\dfrac{5}{4}$

32. $\dfrac{19}{16}$ $\dfrac{5}{4}$

30. _____

_____

31. _____

_____

32. _____

_____

NAME

CLASS

EXERCISE SET 1.6 (pp. 19–22)

Review the objectives for this lesson.

1. Write two true equations

2. Write two false equations

3. Write two equations which are neither true nor false.

Tell what each equation means.

4. $5 = 15 - 10$

5. $7 - 2 = 1 + 4$

6. Find three replacements that make $7x = 28$ false.

7. Find the replacement that makes $7x = 28$ true.

Solve these equations by trial.

8. $x + 8 = 10$

9. $5 + x = 13$

10. $x - 4 = 5$

11. $5x = 25$

12. $7y = 35$

13. $10x = 100$

14. $2y + 1 = 9$

15. $7x - 1 = 48$

16. $5y + 7 = 107$

ANSWERS

1.

2.

3.

4.

5.

6.

7.

8.

9.

10.

11.

12.

13.

14.

15.

16.

ANSWERS

17. _____

18. _____

19. _____

20. _____

21. _____

22. _____

23. _____

24. _____

25. _____

26. _____

27. _____

28. _____

29. _____

30. _____

31. _____

32. _____

33. _____

34. _____

35. _____

36. _____

37. _____

For each sentence, write two related sentences.

17. $15 - 7 = 8$ **18.** $19 = 6 + 13$ **19.** $8 - x = 7$

20. $11 = x + 5$ **21.** $9 = 4 + x$ **22.** $x = 6 + 7$

23. $\dfrac{21}{7} = 3$ **24.** $45 = 5 \cdot 9$ **25.** $56 = 7 \cdot 8$

26. $\dfrac{x}{5} = 20$ **27.** $\dfrac{5}{1} = x$ **28.** $4x = 18$

29. $17 = x - 2$ **30.** $4.3 = \dfrac{x}{7.5}$ **31.** $45 = \dfrac{3}{4}x$

Which of these divisions are possible?

32. $\dfrac{7}{0}$ **33.** $\dfrac{0}{9}$ **34.** $\dfrac{0}{0}$

35. $\dfrac{98}{0}$ **36.** $\dfrac{7}{5-5}$ **37.** $\dfrac{12}{y-y}$

NAME

CLASS

ANSWERS

EXERCISE SET 1.7 (pp. 23–24)

Review the objectives for this lesson.

Solve and check.

1. $19 - x = 11$ **2.** $5x = 45$ **3.** $x + 17 = 22$

4. $x - 34 = 10$ **5.** $\dfrac{50}{x} = 25$ **6.** $56 + x = 75$

7. $\dfrac{x}{2.5} = 14$ **8.** $15y = 42$ **9.** $45.8 + x = 79.3$

10. $5(x + 2) = 45$ **11.** $49 = 7(x + 3)$ **12.** $8(x - 2) = 40$

13. $72 = 8(x - 7)$ **14.** $90 = 2(x + 24)$ **15.** $7(x + 4) = 700$

1. _____

2. _____

3. _____

4. _____

5. _____

6. _____

7. _____

8. _____

9. _____

10. _____

11. _____

12. _____

13. _____

14. _____

15. _____

ANSWERS

16. _____

17. _____

18. _____

19. _____

20. _____

21. _____

22. _____

23. _____

24. _____

25. _____

26. _____

27. _____

16. $4x = 5$

17. $10y = 2.4$

18. $\frac{1}{2}x = 23$

19. $x + \frac{4}{5} = \frac{9}{8}$

20. $\frac{17}{19} = x - \frac{7}{38}$

21. $\frac{10}{27} = \frac{2}{3}t$

22. $x - 45.89 = 12.7$

23. $56.78 + x = 72.87$

24. $\frac{x}{\frac{1}{4}} = 25$

25. $\frac{x}{\frac{2}{3}} = \frac{4}{5}$

26. $\frac{x}{10.5} = 24.6$

27. $\frac{22.4}{x} = 7$

NAME

CLASS

ANSWERS

EXERCISE SET 1.8 (p. 25)

Review the objectives for this lesson.

Translate to equations. Then solve and check.

1. Forty-eight is two-thirds of what number?

2. Twenty-two is what number plus five?

3. Fifteen minus what number is seven?

4. Four-fifths of what number is twelve?

5. What number is 4 more than 5?

6. Twice some number is 3 more than five. What is the number?

Translate to equations. Do not solve.

7. George S. Patton, highly regarded and controversial general of World War II, died in 1945 at the age of 60. When was he born?

8. The boiling point of water is 212° F. This is 180° F more than its freezing point. What is its freezing point?

9. The area of Lake Superior is four times the area of Lake Ontario. The area of Lake Superior is 30,160 sq. mi. What is the area of Lake Ontario?

10. Normally your heart beats 4320 times every hour. How many times does it beat each minute?

1. _____

2. _____

3. _____

4. _____

5. _____

6. _____

7. _____

8. _____

9. _____

10. _____

11. Two-thirds of the human body is water. If your body contains 125 pounds of water, how much is your total weight?

12. After spending $27 for books and $43 for clothes you have $16 left. How much did you have to begin with?

13. Izzi Zlow's typing speed is 35 words per minute. This is two-fifths of Ty Preitter's speed. What is Preitter's speed?

14. The area of Alaska is 586,400 sq. mi. This is 483 times the area of Rhode Island. What is the area of Rhode Island?

15–22. Solve the equations in Exercises 7–14.

EXERCISE SET 1.9 (pp. 26–29)

Note: If you need extra practice on percent, do exercise set 1.11, on page 55.

Find a decimal numeral for each of the following.

1. 76%   **2.** 54.7%   **3.** 100%   **4.** 1%

Name each of the following using a percent symbol.

5. 4.54   **6.** .998   **7.** .75   **8.** 1

Find a fractional numeral for each of the following. You need not simplify.

9. 20%   **10.** 78.6%   **11.** 12.5%   **12.** $\frac{1}{2}$%

Name each of the following using a percent symbol.

13. $\frac{1}{4}$ **14.** $\frac{3}{20}$ **15.** $\frac{1}{3}$ **16.** $\frac{99}{100}$

Translate and solve.

17. What is 65% of 840?   **18.** 56.2% of 48 is what?

19. What percent of 80 is 100?   **20.** 45 is 20 percent of what?

1. _____

2. _____

3. _____

4. _____

5. _____

6. _____

7. _____

8. _____

9. _____

10. _____

11. _____

12. _____

13. _____

14. _____

15. _____

16. _____

17. _____

18. _____

19. _____

20. _____

21. _____

22. _____

23. _____

24. _____

25. _____

26. _____

27. _____

28. _____

Translate to equations. Then solve.

21. On a test of 88 items a student got 76 correct.
What percent did he get correct?

22. A baseball player made 13 hits in 25 times at bat. What was his percent of hits?

23. A family spent $85 a month for food. This was 20% of its monthly income. What was their monthly income?

24. A man's brain is 2.7% of his body weight. His brain weighs 2.1 kilograms. How much does the man weigh?

25. The sales tax rate in New York City is 5%. How much tax would be charged on a purchase of $428.86?

26. The volume of water increases 9% when it freezes. If you froze 400 cubic centimeters of water, how much would its volume increase? What would be the volume of the ice?

27. A woman earned $9600 one year. She received a 6% raise in salary, but the cost of living rose 5.4%. How much additional earning power did she actually get?

28. Due to inflation the price of an item rose 4%, which was 6¢. What was the old price? The new price?

EXERCISE SET 1.10 (pp. 30–34)

Review the objectives for this lesson.

Find the area and perimeter of each figure.

1.

44 cm

25 cm

2.

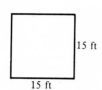

15 ft

15 ft

3.

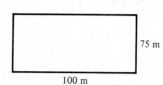

100 m

75 m

Find the area of each figure.

4.

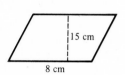

8 cm

15 cm

5.

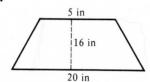

5 in

16 in

20 in

6.

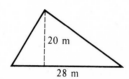

20 m

28 m

7.

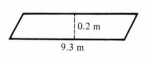

9.3 m

0.2 m

8.

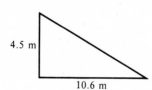

4.5 m

10.6 m

9.

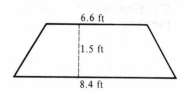

6.6 ft

1.5 ft

8.4 ft

10. Use 3.14 for π.

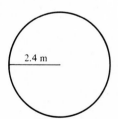

2.4 m

11. Use 3.14 for π.

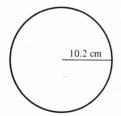

10.2 cm

12. Use $\frac{22}{7}$ for π.

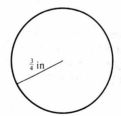

$\frac{3}{4}$ in

13–15. Find the length of a diameter of each of the above circles.

16–18. Find the circumference of each of the above circles.

1. A =
P =

2. A =
P =

3. A =
P =

4. _____

5. _____

6. _____

7. _____

8. _____

9. _____

10. _____

11. _____

12. _____

13. _____

14. _____

15. _____

16. _____

17. _____

18. _____

ANSWERS

19. _____

20. _____

21. _____

22. _____

23. _____

24. _____

25. _____

26. _____

27. _____

28. _____

Find the volume of each solid.

19.

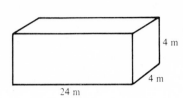

24 m, 4 m, 4 m

20.

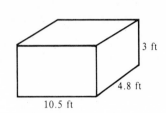

3 ft, 4.8 ft, 10.5 ft

21.

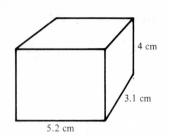

4 cm, 3.1 cm, 5.2 cm

Find the missing angle measures.

22.

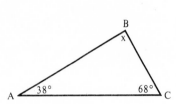

23.

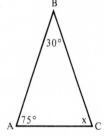

24.

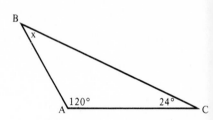

Applied problems

25. The length of a diameter of a circle is 24.6 cm. What is the length of a radius?

26. The area of a rectangle is 49.2 ft² and the length is 10 ft. What is the width?

27. The circumference of a circle is 22.608. What is the length of a diameter? Use 3.14 for π.

28. The area of a parallelogram is 254.8 square meters and the base is 52 meters long. What is the height?

EXERCISE SET 1.11

Find a decimal numeral for each of the following.

1. 38%   **2.** 35%   **3.** 72.1%   **4.** 35.6%

5. 65.4%   **6.** 82.6%   **7.** 3.25%   **8.** 5.32%

9. 8.24%   **10.** 9.45%   **11.** .61%   **12.** .83%

13. .43%   **14.** .73%   **15.** .012%   **16.** .023%

17. .045%   **18.** .053%   **19.** .0035%   **20.** .0041%

21. 125%   **22.** 135%   **23.** 240%   **24.** 320%

Name each of the following using a percent symbol.

25. .62   **26.** .73   **27.** .85   **28.** .91

ANSWERS

29. _____

30. _____

31. _____

32. _____

33. _____

34. _____

35. _____

36. _____

37. _____

38. _____

39. _____

40. _____

41. _____

42. _____

43. _____

44. _____

45. _____

46. _____

47. _____

48. _____

49. _____

50. _____

51. _____

52. _____

53. _____

54. _____

55. _____

56. _____

29. .623   **30.** .741   **31.** .812   **32.** .732

33. 7.2   **34.** 8.3   **35.** 3.5   **36.** 2.6

37. 2   **38.** 3   **39.** 4   **40.** 5

41. .072   **42.** .085   **43.** .013   **44.** .045

45. .0013   **46.** .0057   **47.** .0073   **48.** .0068

Find a fractional numeral for each of the following. You need not simplify.

49. 30%   **50.** 40%   **51.** 70%   **52.** 80%

53. 13.6%   **54.** 17.8%   **55.** 73.4%   **56.** 82.5%

NAME

CLASS

ANSWERS

57. 3.2% **58.** 4.8% **59.** 8.4% **60.** 7.6%

61. 120% **62.** 140% **63.** 250% **64.** 370%

65. $.35\%$ **66.** $.53\%$ **67.** $.48\%$ **68.** $.59\%$

69. $.042\%$ **70.** $.035\%$ **71.** $.083\%$ **72.** $.074\%$

Name each of the following using a percent symbol.

73. $\dfrac{17}{100}$ **74.** $\dfrac{35}{100}$ **75.** $\dfrac{119}{100}$ **76.** $\dfrac{173}{100}$

77. $\dfrac{7}{10}$ **78.** $\dfrac{3}{10}$ **79.** $\dfrac{8}{10}$ **80.** $\dfrac{9}{10}$

81. $\dfrac{7}{20}$ **82.** $\dfrac{11}{20}$ **83.** $\dfrac{7}{25}$ **84.** $\dfrac{12}{25}$

57. _____
58. _____
59. _____
60. _____
61. _____
62. _____
63. _____
64. _____
65. _____
66. _____
67. _____
68. _____
69. _____
70. _____
71. _____
72. _____
73. _____
74. _____
75. _____
76. _____
77. _____
78. _____
79. _____
80. _____
81. _____
82. _____
83. _____
84. _____

ANSWERS

85. _____

86. _____

87. _____

88. _____

89. _____

90. _____

91. _____

92. _____

93. _____

94. _____

95. _____

96. _____

97. _____

98. _____

99. _____

100. _____

101. _____

102. _____

103. _____

104. _____

85. $\dfrac{1}{2}$ 86. $\dfrac{3}{2}$ 87. $\dfrac{1}{4}$ 88. $\dfrac{3}{4}$

89. $\dfrac{3}{5}$ 90. $\dfrac{4}{5}$ 91. $\dfrac{17}{50}$ 92. $\dfrac{31}{50}$

93. $\dfrac{1}{3}$ 94. $\dfrac{2}{3}$ 95. $\dfrac{3}{8}$ 96. $\dfrac{5}{8}$

Translate and solve.

97. What is 38% of 250?   **98.** What is 47% of 320?

99. 37.2% of 85 is what?   **100.** 17.6% of 70 is what?

101. What percent of 80 is 20?   **102.** What percent of 60 is 20?

103. 35 is 20 percent of what?   **104.** 16 is 25 percent of what?

NAME_____

CHAPTER 1 TEST

CLASS_____SCORE_____GRADE_____

Before taking the test *be sure* to allow yourself a day or so for review. Use the objectives listed in the margins to guide your study. The test will evaluate your progress and aid your preparation for a possible classroom test. Allow about an hour for the test. Remove the test from the book. When you finish read the test analysis on the answer page at the end of the book.

1. Write three different fractional numerals for $\frac{5}{6}$.

2. Write the simplest fractional numeral for $\frac{24}{84}$.

3. Multiply and simplify.

$\frac{5}{8} \cdot \frac{24}{32}$

4. Add and simplify.

$\frac{5}{12} + \frac{3}{8}$

5. Subtract and simplify.

$\frac{13}{24} - \frac{3}{8}$

6. Write an expanded numeral, with exponents, for 34,809.

7. What is the meaning of 9^4?

8. Write exponential notation for $xxxx$.

9. What is the meaning of y^3? Do not use $\times$ or $\cdot$

10. What does x^3 represent if x stands for 5?

11. Write a fractional numeral for 24.7.

12. Write a decimal numeral for $\frac{1}{8}$

13. Calculate. $10 \times (9 + 4)$.

14. What law is illustrated by this sentence? $16 + (9 + 8) = (16 + 9) + 8$.

15. Evaluate and simplify $ab + bc$, where $a = 10$, $b = 7$ and $c = 8$.

Factor.

16. $15y + 5$

17. $24x + 16 + 8y$

18. $ax + ay + ab$

Multiply.

19. $10(9x + 3y)$

20. $7(9m + 2x + 1)$

Collect like terms.

21. $21x + 9b + 5x + b$

22. What is the reciprocal of $\frac{8}{7}$?

ANSWERS

1. _____

2. _____

3. _____

4. _____

5. _____

6. _____

7. _____

8. _____

9. _____

10. _____

11. _____

12. _____

13. _____

14. _____

15. _____

16. _____

17. _____

18. _____

19. _____

20. _____

21. _____

22. _____

ANSWERS

23. _____

24. _____

25. _____

26. _____

27. _____

28. _____

29. _____

30. _____

31. _____

32. _____

33. _____

34. _____

35. _____

36. _____

37. _____

38. _____

39. _____

23. Divide. $\dfrac{5}{6} \div \dfrac{24}{45}$

24. Graph $\dfrac{8}{3}$ on a number line.

25. Insert the proper symbol $>$, $<$, or $=$ between the numerals.

$\dfrac{4}{11}$ $\dfrac{3}{8}$

26. Write a true equation.

Solve and check.

27. $14 - x = 11$

28. $4(x + 5) = 44$

Translate to equations. Then solve and check.

29. What percent of 75 is 5?

30. Five-fourths of what number is 75?

31. Find a decimal numeral for 45.8%.

32. Name $\dfrac{14}{25}$ using a percent symbol.

Find the area of each figure.

33.

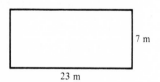
23 m, 7 m

34.

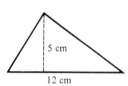
5 cm, 12 cm

35.

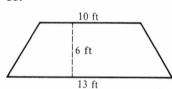
10 ft, 6 ft, 13 ft

Applied Problems.

36. A man wagered $12.50 and lost $10. What percent of his wager did he lose?

37. The area of a rectangular lot is 5717.8 ft², and the length is 46 ft. What is the width?

38. Find the area. Use 3.14 for π.

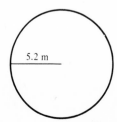
5.2 m

39. Find the volume.

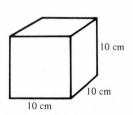
10 cm, 10 cm, 10 cm

2 INTEGERS AND RATIONAL NUMBERS

2.1 INTEGERS AND THE NUMBER LINE

The set of numbers we call integers includes the set of whole numbers. We extend the set of whole numbers to the integers by including some *negative* numbers. This is shown below on a number line. The new numbers

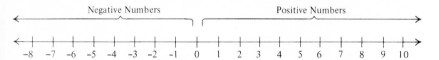

are those pictured to the left of 0. We read -2 as "negative two".
We say that one integer is greater than another if it occurs to the right of the first.

Examples

$-2 < 1, \quad -8 < -3, \quad -5 < 0, \quad 5 > -5$

Do exercises 1 through 3 at the right.

Notice that all of the negative integers are less than zero. All of the positive numbers are greater than zero.

The *absolute value* of an integer can be thought of as its distance from 0 on the number line. The absolute value of an integer *n* can be named $|n|$.

Examples

$|3| = 3, \quad |-5| = 5, \quad |0| = 0$

Notice that the absolute value of an integer is never negative.

Do exercises 4 and 5 at the right.

ADDITION OF INTEGERS

To explain addition of integers we can use the number line. To do the addition $a + b$, we start at a and then move a distance b. If b is negative we move to the left.

Examples

$$2 + (-5) = -3$$

$$-6 + 4 = -2$$

OBJECTIVES

You should be able to:

a) Tell which of two integers is greater, using $<$ or $>$.

b) Name the additive inverse of any integer, and simplify expressions like $^-(-5)$.

c) Name the absolute value of any integer, and simplify expressions such as $|-3|$.

d) Add several integers, positive or negative.

Insert $<$ or $>$ to make true sentences.

1. $-5 \quad -7$ **2.** $0 \quad -3$

3. $-5 \quad -2$

4. What is the absolute value of each integer?

a) 5   b) -3

c) -7 d) 120

5. Simplify

a) $|2|$ b) $|-7|$

c) $|-3|$ d) $|0|$

Add these integers

6. $3 + (-2)$ **7.** $5 + (-7)$

8. $4 + (-4)$ **9.** $-7 + 3$

10. $-5 + (-2)$ **11.** $(-4) + 4$

12. Name the additive inverse of each integer.

a) 5   b) -13

c) -12 d) 0

13. Simplify:

a) $^-(-2)$ b. $^-(-7)$

c) $^-0$ d) $^-4$

14. Find ^-x when x is

a) 3   b) -12

c) -5 d) 0

Add. Do not use a number line, except as a check.

15. $-4 + (-5)$ **16.** $-7 + (-8)$

17. $-5 + (-10)$ **18.** $\begin{array}{r} -7 \\ -3 \\ -2 \\ \hline \end{array}$

19. $\begin{array}{r} -8 \\ -5 \\ -4 \\ \hline \end{array}$ **20.** $\begin{array}{r} -9 \\ -3 \\ -7 \\ \hline \end{array}$

Do exercises 6 through 11 at the left.

When we add 0 to any integer, the result is that same integer. Thus we say that 0 is the *additive identity*. Suppose we add 5 and -5. The result is 0, the identity. Thus we say that the numbers 5 and -5 are *additive inverses* of each other. In the set of integers, every number has an additive inverse. We use a symbol that looks like a raised minus sign to denote an additive inverse. The additive inverse of a number n is denoted ^-n. Additive inverses are sometimes called "opposites."

Examples

$^-3$ (the additive inverse of 3) is -3 because $3 + (-3) = 0$

$^-(-7)$ (the additive inverse of negative 7) is 7 because $-7 + 7 = 0$

$^-0$ (the additive inverse of 0) is 0 because $0 + 0 = 0$

Do exercises 12 through 14 at the left.

In reading a symbol like ^-n we often omit the word "additive" and simply say "the inverse of n." Notice that $^-3 = -3$. Thus $^-3$ is sometimes read "negative three." However, $^-(-4)$ should be read "the inverse of negative four."

When we add positive integers and zero the additions are of course additions of whole numbers. When we add 0 to any integer we get that same integer. When we add an integer and its additive inverse we get 0. What happens when we add two negative integers?

Examples

$-5 + (-7) = -12$

$-3 + (-5) = -8$

When we add two negative integers, the result is negative. We can generalize as follows.

To add two negative integers, we can add their absolute values and then take the additive inverse. The result is negative.

We can add more than two negative integers, of course, and it may help to use a vertical form.

Example
$\begin{array}{r} -12 \\ -5 \\ -8 \\ \hline -25 \end{array}$

Do exercises 15 through 20 at the left.

What happens when we add a positive and a negative integer with different absolute values?

Examples

a) $3 + (-5) = -2$ b) $-7 + 4 = -3$

c) $7 + (-3) = 4$ d) $-6 + 10 = 4$

e) $5 + (-10) = -5$ f) $5 + (-2) = 3$

Sometimes the answer is positive and sometimes negative, depending on which number has the greater absolute value. We can generalize as follows.

To add a positive and a negative integer find the difference of their absolute values. If the negative integer has the greater absolute value the answer is negative. If the positive integer has the greater absolute value the answer is positive.

Do exercises 21 through 23 at the right.

We now have rules for adding any two integers, positive, negative or zero. Thus we can add integers without using the number line. You should practice addition without the number line.
Suppose we wish to add several integers, some of them positive and some negative. It is best to pick out the positive ones and add them, then add the negative ones. Then we add the results.

Example

25	First, add the	Next, add the
-14	positives	negatives
-127	25	-14
45	45	-127
32	32	-118
-118	87	$\overline{-259}$
87	$\overline{189}$	

Now add the results
$$-259$$
$$189$$
$$\overline{-70}$$

Do exercise 24 at the right.

Review the objectives for this lesson.

Do exercise set 2.1, p. 79.

Add.

21. $-4 + 6$ **22.** $-7 + 3$

23. $5 + (-7)$

Add.

24. -35
        17
        14
        -27
        31
        -12
        $\overline{}$

OBJECTIVES

You should be able to:

a) Add and subtract rational numbers, positive, negative or zero.

b) Simplify expressions containing additive inverses.

c) Simplify expressions containing absolute value notation.

d) Tell which of two numbers is greater, using $<$ or $>$.

e) State the properties of rational numbers under addition (commutative, associative, etc.).

f) Tell which properties are illustrated by certain sentences.

25. Draw a number line. Locate points for these rational numbers.

a) 4.1   b) −3.5

c) $\frac{5}{2}$ d) $-\frac{8}{3}$

Add.

26. $-8.6 + 2.4$ **27.** $\frac{5}{9} + \left(-\frac{7}{9}\right)$

28.   −3.2
            7.5
          −13.1
            8.2
          _____

2.2 RATIONAL NUMBERS

The set of rational numbers consists of all of the numbers of ordinary arithmetic and their additive inverses. The rational numbers include all of the integers, and it is these numbers that we use mostly in elementary algebra. There are positive and negative rational numbers, the same as for integers, but between any two integers there are many other rational numbers.

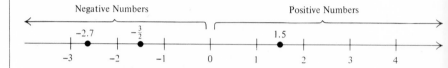

Everything we have said about integers so far also holds for the rational numbers. For example, all rational numbers have additive inverses, and absolute value means the same as before.

Examples

a) $^-4.2 = -4.2$ b) $^-\left(-\frac{3}{2}\right) = \frac{3}{2}$

c) $|-9.35| = 9.35$ d) $\left|-\frac{3}{16}\right| = \frac{3}{16}$

Addition is done just as it is with integers.

Examples

a) $-9.2 + 3.1 = -6.1$ b) $-\frac{3}{2} + \frac{9}{2} = \frac{6}{2} = 3$

Do exercises 25 through 28 at the left.

SUBTRACTION

Subtraction is defined for rational numbers as it was for the numbers of arithmetic. That is, $a - b$ is the number (if it exists) which when added to b gives a. Remembering this definition, let us look at subtraction on a number line.

Example

$-2 - 3 = -5$

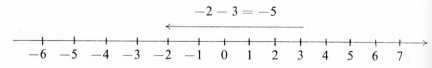

To subtract 3 from -2 we find the number which when added to 3 gives -2. Thus we start at 3. We find we must move 5 units to the left (negative direction) to get to -2. Thus adding -5 to 3 gives -2. Now we know that $-2 - 3 = -5$.

Do exercise 29 at the right.

To see how to subtract without a number line, we look for a pattern.

Examples

Compare:

a) $3 - (-2)$ with $3 + 2$

b) $-5 - 4$ with $-5 + (-4)$

Do exercise 30 at the right.

The preceding exercises indicate that we can always subtract by adding an inverse. This can be proved to hold in all cases.

For any integers a and b, $a - b = a + {}^-b$.
(We can subtract by adding the additive inverse of the subtrahend.)

Examples

a) $3 - 5 = 3 + {}^-5 = 3 + (-5) = -2$

b) $-5 - 7 = -5 + {}^-7 = -5 + (-7) = -12$

c) $3 - (-7) = 3 + 7 = 10$

d) $-4 - (-5) = -4 + 5 = 1$

Do exercise 31 at the right.

We speak of the "sign" of a rational number. The sign of a positive number is positive. The sign of a negative number is negative. When we replace a number by its additive inverse, we sometimes say that we "change the sign." In this sense it is correct to describe subtraction as follows: "change the sign of the subtrahend and then add."

A PROOF (optional)

We said we could prove that we can subtract by adding an inverse. Here is a proof. We want to prove that if a and b are rational numbers then $a - b$ is the same as $a + {}^-b$. Let's add b to $a - b$. We get a by definition of subtraction. Let's add b to $a + {}^-b$. We get $a + {}^-b + b$, which equals $a + 0$ (Why?), or a. We got a both times. Thus $a - b$ and $a + {}^-b$ must be the same number.

Do exercise 32 at the right.

29. Draw a number line. Then do these subtractions. Check by adding.

a) $5 - 2$ b) $3 - (-2)$

c) $-2 - 4$

30. Do these additions and subtractions and compare

a) $5 - 2$ $5 + (-2)$

b) $6 - (-4)$ $6 + 4$

c) $-3 - 5$ $-3 + (-5)$

d) $-4 - (-2)$ $-4 + 2$

31. Subtract by adding an inverse

a) $5 - 8$ b) $-3 - 6$

c) $4 - (-3)$

32. (optional)

a) Add b to $a - b$ and simplify. How do you know your answer is correct?

b) Add b to $a + {}^-b$ and simplify. How do you know your answer is correct?

Which property is illustrated by each sentence?

33. $5 + (-10.3) = -10.3 + 5$

34. $\frac{3}{5} + \left(-\frac{4}{3} + \frac{2}{3}\right) = \left(\frac{3}{5} + \left(-\frac{4}{3}\right)\right) + \frac{2}{3}$

35. $-\frac{9}{5} + \frac{9}{5} = 0$

36. $0 + (-9.2) = -9.2$

PROPERTIES OF RATIONAL NUMBERS UNDER ADDITION

Rational numbers, with respect to addition, have properties similar to those of the numbers of ordinary arithmetic. There is one new property, too, that the numbers of arithmetic do not have.

Addition is commutative. For any rational numbers a and b,

$$a + b = b + a.$$

The commutative property, you will remember, says that we can change the order in addition without changing the sum.

Addition is associative. For any rational numbers a, b and c,

$$a + (b + c) = (a + b) + c.$$

This property tells us that parentheses are not really necessary when only addition is considered. Both properties together tell us that we can change the grouping and ordering as we wish.

The number 0 is the additive identity. For any rational number a,

$$a + 0 = a.$$

The new property is the following.

Every rational number a has an additive inverse ^-a, for which

$$a + {}^-a = 0.$$

This new property tells us several things. One is that we can always subtract any two rational numbers.

Examples

Which property of rational numbers is illustrated by each sentence?

a) $3.2 + 5.7 = 5.7 + 3.2$ (commutative law of addition)

b) $\frac{2}{3} + \left(\frac{4}{5} + \frac{5}{9}\right) = \left(\frac{2}{3} + \frac{4}{5}\right) + \frac{5}{9}$ (associative law of addition)

c) $3 + (-3) = 0$ (property of inverses)

d) $-17.3 + 0 = -17.3$ (property of zero)

Do exercises 33 through 36 at the left.

Always review the objectives before working the exercise sets.

Do exercise set 2.2, p. 81.

2.3 MULTIPLICATION AND DIVISION OF RATIONAL NUMBERS

Multiplication of rational numbers is very much like multiplication of numbers of ordinary arithmetic. The only difference is that we must determine whether the answer is positive or negative. To see how this is done, consider the pattern in the following.

This number decreases ⟶ | | ⟵ This number decreases
by 1 each time

$$
\begin{array}{rcr}
4\cdot 5 = & 20 \\
3\cdot 5 = & 15 \\
2\cdot 5 = & 10 \\
1\cdot 5 = & 5 \\
0\cdot 5 = & 0 \\
-1\cdot 5 = & -5 \\
-2\cdot 5 = & -10 \\
-3\cdot 5 = & -15 \\
\end{array}
$$

by 5 each time

Do exercise 37 at the right.

According to this pattern, it looks as if the product of a positive number and a negative number should be negative. Let us look at this another way. As you probably suspect, the distributive law holds for rational numbers. Now consider the following example.

$$5(-2 + 4) = 5\cdot 2 = 10$$

But

$$5(-2 + 4) = 5(-2) + 5\cdot 4 \qquad \text{by the distributive law}$$
$$= 5(-2) + 20$$
$$\text{So } 5(-2) + 20 = 10. \qquad \text{Therefore } 5(-2) \text{ must be } -10.$$

Do exercise 38 at the right.

To multiply a positive and a negative number, multiply their absolute values. The answer is negative.

Do exercises 39 through 44 at the right.

How do we multiply two negative numbers? To see this we can again look at a pattern.

OBJECTIVES

You should be able to:

a) Multiply integers or rational numbers using standard notation or fractional notation.

b) Multiply rational numbers using decimal notation.

c) Find the reciprocal of a negative number (fractional notation).

d) Divide rational numbers using the definition of division.

e) Divide rational numbers by multiplying by 1 (fractional notation).

37. Complete, as in the example

$$
\begin{array}{rcl}
4\cdot 10 & = & 40 \\
3\cdot 10 & = & 30 \\
2\cdot 10 & & \\
1\cdot 10 & & \\
0\cdot 10 & & \\
-1\cdot 10 & & \\
-2\cdot 10 & & \\
-3\cdot 10 & & \\
\end{array}
$$

38. Complete two ways as in the example.

$$4(-3 + 5) = 4(\quad) = \underline{\hspace{3cm}}$$

$$4(-3) + 4(5) = \underline{\hspace{2cm}} + \underline{\hspace{2.5cm}}$$

Multiply

39. $3\cdot(-5)$ **40.** $-7\cdot 4$

41. $6(-5)$ **42.** $\dfrac{2}{3}\left(-\dfrac{5}{9}\right)$

43. $-\dfrac{4}{5}\cdot\dfrac{7}{3}$ **44.** -4.23×7.1

Complete, as in the example

45. $3(-10) = -30$
   $2(-10)$
   $1(-10)$
   $0(-10)$
   $-1(-10)$
   $-2(-10)$
   $-3(-10)$

This number decreases ———↓   ↓——— This number increases
by 1 each time $4(-5) = -20$ by 5 each time
                                         $3(-5) = -15$
                                         $2(-5) = -10$
                                         $1(-5) = -5$
                                         $0(-5) = 0$
                                        $-1(-5) = 5$
                                        $-2(-5) = 10$
                                        $-3(-5) = 15$

Do exercise 45 at the left.

We can also use the distributive law to see how to multiply two negative numbers, as in this example.

$$-5(-2 + 4) = -5\cdot2 = -10$$

But

$$-5(-2 + 4) = -5(-2) + (-5)\cdot4 \qquad \text{by the distributive law.}$$
$$= -5(-2) + (-20)$$
So $-5(-2) + (-20) = -10$.   Therefore $-5(-2)$ must be 10.

Multiply

46. $-3(-4)$ **47.** $-5(-10)$

To multiply two negative numbers, multiply their absolute values. The answer is positive.

Do exercises 46 through 51 at the left.

DIVISION AND RECIPROCALS

48. $-16(-4)$ **49.** $-\dfrac{4}{7}\left(-\dfrac{5}{9}\right)$

The number 1 is the multiplicative identity for rational numbers, as it is for numbers of arithmetic. The number 1 can be named in many ways.

Examples

$$1 = \frac{-3}{-3} = \frac{\frac{2}{3}}{\frac{2}{3}} = \frac{-\frac{3}{5}}{-\frac{3}{5}}$$

To see that this is so, remember the definition of division (see page 22), $\dfrac{a}{b}$ is the number which when multiplied by b gives a.

50. $-\dfrac{2}{3}\left(-\dfrac{4}{5}\right)$ **51.** $-3.25(-4.14)$

Examples

a) $\dfrac{3}{3} = 1$ because $1\cdot3 = 3$

b) $\dfrac{-5}{-5} = 1$ because $1\cdot(-5) = -5$.

The number 1 is the multiplicative identity for rational numbers. That is, $n\cdot1 = n$ for any rational number n.

For any non-zero rational number m, $\dfrac{m}{m} = 1$.

Now let us see how to handle signs when dividing rational numbers.

Examples

a) $\dfrac{8}{-2} = -4$ because $-4(-2) = 8$

b) $\dfrac{-10}{-2} = 5$ because $5(-2) = -10$

Do exercises 52 through 54 at the right.

From these examples we can see how to handle signs in division.

When we divide a positive number by a negative or a negative by a positive, the answer is negative. When we divide two negative numbers, the answer is positive.

RECIPROCALS

Two numbers are reciprocals of each other if their product is 1. What is the reciprocal of a negative number?

Examples

a) The reciprocal of $-\dfrac{2}{3}$ is $-\dfrac{3}{2}$ because $-\dfrac{2}{3}\left(-\dfrac{3}{2}\right) = 1$.

b) The reciprocal of a number n is $\dfrac{1}{n}$ because

$$n \cdot \frac{1}{n} = \frac{n}{n} = 1.$$

Do exercises 55 through 58 at the right.

Any nonzero rational number n has a reciprocal $\dfrac{1}{n}$. The reciprocal of a negative number is negative. The reciprocal of a positive number is positive.

As with numbers of arithmetic we can use the idea of multiplying by 1 to divide.

Example

Divide: $\dfrac{2}{3} \div \left(-\dfrac{5}{4}\right)$ or $\dfrac{\frac{2}{3}}{-\frac{5}{4}}$

$$\frac{\frac{2}{3}}{-\frac{5}{4}} \times \frac{-\frac{4}{5}}{-\frac{4}{5}} = \frac{\frac{2}{3}\left(-\frac{4}{5}\right)}{1} = -\frac{8}{15}$$

Do exercises 59 through 61 at the right.

Do exercise set 2.3, p. 83.

Use the definition of division (page 22) to divide.

52. $\dfrac{10}{-5}$ **53.** $\dfrac{-16}{4}$

54. $\dfrac{-20}{-10}$

Find the reciprocal of each number.

55. $-\dfrac{5}{4}$ **56.** $-\dfrac{3}{4}$

57. -3 **58.** $-\dfrac{1}{5}$

Divide by multiplying by 1.

59. $\dfrac{4}{7} \div \left(-\dfrac{3}{5}\right)$ **60.** $-\dfrac{8}{5} \div \dfrac{2}{3}$

61. $-\dfrac{13}{7} \div \left(-\dfrac{3}{4}\right)$

OBJECTIVES

You should be able to:

a) State the distributive law of multiplication over subtraction.

b) Factor by removing a common factor.

c) Multiply, where one expression has several terms.

d) Collect like terms.

Compute:

62. a) $4(5 - 3)$

b) $4 \cdot 5 - 4 \cdot 3$

63. a) $-2(5 - 2)$

b) $-2 \cdot 5 - (-2) \cdot 2$

64. a) $5(2 - 7)$

b) $5 \cdot 2 - 5 \cdot 7$

65. a) $-4(3 - (-2))$

b) $-4 \cdot 3 - (-4)(-2)$

Name the terms of each

66. $5x - 4y + 3$

67. $-4y - 2x + 3z$

2.4 DISTRIBUTIVE LAWS AND THEIR USE

For rational numbers, multiplication is commutative and associative. There is a multiplicative identity, 1, and every non-zero number has a multiplicative inverse, or reciprocal. Also, the distributive law of multiplication over addition holds. But there is another distributive law that holds for rational numbers. It is a distributive law of multiplication over subtraction.

Example

Compare

$3(4 - 2)$	$3 \cdot 4 - 3 \cdot 2$
$3 \cdot 2$	$12 - 6$
6	6

When we multiply a number by a difference we can either subtract and then multiply or multiply and then subtract.

Do exercises 62 through 65 at the left.

The distributive law of multiplication over subtraction:
For any rational numbers *a*, *b* and *c*, $a(b - c) = ab - ac$.

In an expression like $(ab) - (ac)$ we can omit the parentheses. In other words, in $ab - ac$ we *agree* that the multiplications are to be done first. The distributive laws are used in various ways in algebra. The most fundamental ones are factoring, multiplying and collecting similar terms.

FACTORS AND TERMS

Factors are numbers to be multiplied. The factors of $3xy$ are 3, *x*, and *y*. The factors of 6 are 1, 2, 3, and 6. Each is a factor because we can *multiply* it by some number to get 6.

What do we mean by the *terms* of an expression? When they are all separated by addition signs it is easy to tell, but if there are some subtraction signs, we might have reason to wonder. If there are subtraction signs, we can always rewrite using addition signs.

Examples
What are the terms of $3x - 4y + 2z$?
$3x - 4y + 2z = 3x + {}^-4y + 2z$.
Thus the terms are: $3x$, $-4y$ and $2z$.

Do exercises 66 and 67 at the left.

FACTORING

When all the terms of an expression contain a common factor, we can "remove" it. This procedure is based upon the distributive laws.

Examples

Factor:

a) $5x - 10 = 5x - 5 \cdot 2 = 5(x - 2)$

b) $ax - ay + az = a(x - y + z)$

c) $9x + 27y - 9 = 9x + 27y - 9 \cdot 1 = 9(x + 3y - 1)$

Do exercises 68 through 71 at the right.

MULTIPLYING

One kind of multiplying is the reverse of the factoring process above.

Examples

Multiply

a) $4(x - 2) = 4x - 4 \cdot 2 = 4x - 8$

b) $b(s - t + w) = bs - bt + bw$

Do exercises 72 through 75 at the right.

COLLECTING LIKE TERMS

The process of "collecting like terms" is also based upon the distributive laws.

Examples

Collect like terms

a) $4x - 2x = (4 - 2)x = 2x$

b) $2x + 3y - 5x - 2y = -3x + y$

c) $3x - x = 3x - 1 \cdot x = (3 - 1)x = 2x$

d) $x - 5x = 1 \cdot x - 5x = (1 - 5)x = -4x$

Do exercises 76 through 79 at the right.

Don't forget to review the objectives.

Do exercise set 2.4, p. 85.

Factor

68. $4x - 8$

69. $3x - 6y + 9$

70. $bx + by - bz$

71. $7p + 7q - 14t$

Multiply

72. $3(x - 5)$

73. $5(x - y + 4)$

74. $-2(x - 3)$

75. $b(x - 2y + z)$

Collect like terms

76. a) $6x - 3x$

    b) $7x - x$

    c) $x - 6x$

77. $5x + 4y - 2x - y$

78. $3x - 7x + 2y - 3z$

79. $3x - 2y - 5x - 3y$

OBJECTIVES

You should be able to:

a) Rename an additive inverse without parentheses, where an expression has several terms.

b) Name the terms of an expression.

c) Simplify expressions by removing parentheses and collecting like terms.

Rename each additive inverse without parentheses.

80. $^-(x + 2)$

81. $^-(4x - 5y + 2)$

2.5 MULTIPLYING BY −1 AND SIMPLIFYING

What happens when we multiply a rational number by −1? The following examples illustrate.

Examples

a) $-1 \cdot 7 = -7$, b) $-1 \cdot (-5) = 5$, c) $-1 \cdot 0 = 0$

When we multiply a number by −1 the result is the additive inverse of that number.

For any rational number *n*, $-1 \cdot n = {}^-n$. That is, negative one times *n* is the additive inverse of *n*.

This fact enables us to rename an additive inverse when an expression has several terms.

Examples

a) $^-(3 + x) = -1 \cdot (3 + x)$ [Taking the additive inverse is the same as multiplying by −1]

$= -1 \cdot 3 + -1 \cdot x$ [Using the distributive law]

$= -3 + {}^-x$ [Here, we replaced $-1 \cdot x$ by ^-x, again using the fact that $-1 \cdot n = {}^-n$]

or $-3 \div x$ [Since adding an inverse is the same as subtracting]

b) $^-(3x - 2y + 4) = -1 \cdot (3x - 2y + 4) = -1 \cdot 3x - (-1) \cdot 2y + (-1) \cdot 4$

$= -3x + 2y - 4$

Do exercises 80 and 81 at the left.

The above examples show that we can rename an additive inverse by multiplying every term by −1. We could also say that we "change the sign" of every term inside the parentheses.

REMOVING CERTAIN PARENTHESES

In some expressions commonly encountered in algebra there are parentheses preceded by subtraction signs. These parentheses can be removed.

Examples

Remove parentheses and simplify:

a) $3x - (4x + 2)$ $= 3x + {}^-(4x + 2)$ [subtracting is adding the inverse]

$= 3x + (-4x) + (-2)$ [taking the inverse is the same as multiplying by −1]

$= 3x - 4x - 2 = -x - 2$

b) $3y - 2 - (2y - 4) = 3y - 2 - 2y + 4$

$= y + 2$

When removing parentheses preceded by a subtraction sign or an additive inverse sign, the sign of each term inside the parentheses is changed. If parentheses are preceded by an addition sign, no signs are changed.

Do exercises 82 and 83 at the right.

PARENTHESES WITHIN PARENTHESES

It often happens that parentheses occur within parentheses. When this happens, we may use parentheses of different shapes, such as [ ] (also called "brackets") and { } (usually called "braces"). These all have the same meaning. The different shapes help us keep track of them.

When parentheses occur within parentheses, the computations in the innermost ones are to be done first.

Example

Simplify: (Note that we work from inside out)

$4(2 + 3) - \{6 - [3 - (7 + 3)]\}$

$4 \cdot 5 - \{6 - [3 - 10]\}$

$20 - \{6 - (-7)\}$

$20 - \{13\}$ or 7

Do exercises 84 and 85 at the right.

When expressions with parentheses contain variables we may still sometimes simplify by removing parentheses. We still work from the inside out.

Example

Simplify:

$[5(x + 2) - 3x] - [3(y + 2) - 4(y + 2)]$

$[5x + 10 - 3x] - [3y + 6 - 4y - 8]$

$[2x + 10] - [-y - 2]$

$2x + 10 + y + 2$

$2x + y + 12$

Do exercises 86 and 87 at the right.

Do exercise set 2.5, p. 87.

Remove parentheses and simplify

82. $5x - (3x + 9)$

83. $5y - 2 - (2y - 4)$

Simplify

84. $3(4 + 2) - \{7 - [4 - (6 + 5)]\}$

85. $\{-4 - [-5 - (3 - 5)]\} - \dfrac{3}{2}(5 + 1)$

86. $[3(x + 2) + 2x] - [4(y + 2) - 3(y - 2)]$

87. $[3(x + 2) - 2(x - 3)] + 2[y + 2(y - 5)]$

OBJECTIVES

You should be able to:

a) Explain the meaning of exponential notation when exponents are negative integers.

b) Rename a number without negative exponents.

c) Use exponents in multiplying (adding the exponents).

d) Use exponents in dividing (subtracting the exponents).

e) Use exponents in raising a power to a power (multiplying the exponents).

Explain the meaning of the following.

88. 4^{-3} **89.** 5^{-2} **90.** 2^{-4}

Rename, using a negative exponent.

91. $\dfrac{1}{3^2}$ **92.** $\dfrac{1}{5^4}$ **93.** $\dfrac{1}{7^3}$

Rename, without a negative exponent.

94. 5^{-3} **95.** 7^{-5} **96.** 10^{-4}

2.6 INTEGERS AS EXPONENTS

The use of positive integers and zero as exponents is familiar. Negative integers can also be used as exponents. One way to see how this is done is by considering expanded notation. Let us write expanded notation for 4256.387.

$$4256.387 = 4 \times 10^3 + 2 \times 10^2 + 5 \times 10^1 + 6 \times 10^0$$
$$+ \; 3 \times 10^{-1} + 8 \times 10^{-2} + 7 \times 10^{-3}$$

If we are to continue the descending pattern of exponents past the decimal point the exponents become negative. So, we do this and hence agree that 10^{-1} means 1/10, and that 10^{-2} means 1/100, and so on. We adopt a similar agreement even though the base involved is not 10.

If n is any positive integer, b^{-n} is given the meaning $\dfrac{1}{b^n}$.

In other words, b^n and b^{-n} are reciprocals.

Examples

a) Explain the meaning of 3^{-4} without using negative exponents.

3^{-4} means $\dfrac{1}{3^4}$ or $\dfrac{1}{3 \cdot 3 \cdot 3 \cdot 3}$, or $\dfrac{1}{81}$

b) Rename $\dfrac{1}{5^2}$ using a negative exponent.

$\dfrac{1}{5^2} = 5^{-2}$

c) Rename 4^{-3} without using a negative exponent.

$4^{-3} = \dfrac{1}{4^3}$

Do exercises 88 through 96 at the left.

PROPERTIES OF EXPONENTS

Let us consider an expression with exponents, such as $a^3 \cdot a^2$, and see how we could simplify it. We recall the definition of exponents.

$a^3 \cdot a^2$ means $(a \cdot a \cdot a)(a \cdot a)$,

and $(a \cdot a \cdot a)(a \cdot a) = a^5$. Note that the exponent in a^5 is the sum of those in $a^3 a^2$. In general, when we multiply like this the exponents are added. But notice that the base must be the same in both parts.

Now let us consider a case where one exponent is positive and one is negative.

$a^5 \cdot a^{-2}$ means $(a \cdot a \cdot a \cdot a \cdot a) \cdot \left(\dfrac{1}{a \cdot a}\right)$, which simplifies as follows:

$a \cdot a \cdot a \cdot \left(\dfrac{a \cdot a}{1}\right)\left(\dfrac{1}{a \cdot a}\right) = a \cdot a \cdot a \cdot \dfrac{a \cdot a}{a \cdot a}$, and we have

$a \cdot a \cdot a$ or a^3

In this case if we add the exponents we again get the correct result. Next we consider a case in which both exponents are negative.

$a^{-3} \cdot a^{-2}$ means $\dfrac{1}{a \cdot a \cdot a} \cdot \dfrac{1}{a \cdot a}$. This is equal to $\dfrac{1}{a \cdot a \cdot a \cdot a \cdot a}$ or $\dfrac{1}{a^5}$ or a^{-5}

Again, adding the exponents gives the correct result. The same is true if one or both exponents are zero.

In multiplication with exponential notation, we can add exponents if the bases are the same.

$a^m \cdot a^n = a^{m+n}$

Do exercises 97 through 102 at the right.

Let us consider an example of division with exponential notation.

$\dfrac{5^4}{5^2}$ means $\dfrac{5 \cdot 5 \cdot 5 \cdot 5}{5 \cdot 5}$, which is equal to $5 \cdot 5 \cdot \dfrac{5 \cdot 5}{5 \cdot 5}$ and we have

$5 \cdot 5$ or 5^2

Note that in division we subtracted exponents. This is true whether the exponents are positive, negative or zero.

Examples

a) $\dfrac{5^4}{5^{-2}} = (5 \cdot 5 \cdot 5 \cdot 5) \cdot (5 \cdot 5) = 5^{4-(-2)} = 5^6$

b) $\dfrac{7^{-2}}{7^3} = 7^{-2-3} = 7^{-5}$

c) $\dfrac{3^{-4}}{3^{-5}} = 3^{-4-(-5)} = 3^1 = 3$

Multiply and simplify

97. $3^5 \cdot 3^3$ **98.** $5^{-2} \cdot 5^4$

99. $6^{-3} \cdot 6^{-4}$ **100.** $5^0 \cdot 5^{-5}$

101. $7^4 \cdot 7^0$ **102.** $4^2 \cdot 3^2$

Divide and simplify

103. $\dfrac{4^5}{4^2}$ **104.** $\dfrac{5^4}{5^{-2}}$

In a division $\dfrac{a^m}{a^n}$ we can subtract exponents, to obtain a^{m-n}.

Do exercises 103 through 108 at the left.

Now we consider raising a power to a power. Look at this example.

Example

$(3^2)^4$ means $3^2 \cdot 3^2 \cdot 3^2 \cdot 3^2$. This is equal to $3 \cdot 3 \cdot 3 \cdot 3 \cdot 3 \cdot 3 \cdot 3 \cdot 3$, or 3^8.

In this example, we could have multiplied the exponents in $(3^2)^4$. Suppose the exponents are not both positive, as in the following example.

Example

105. $\dfrac{3^2}{3^{-5}}$ **106.** $\dfrac{7^{-2}}{7^{-3}}$

$(5^{-2})^3$ means $\dfrac{1}{5^2} \cdot \dfrac{1}{5^2} \cdot \dfrac{1}{5^2}$. This is equal to $\dfrac{1}{5 \cdot 5} \cdot \dfrac{1}{5 \cdot 5} \cdot \dfrac{1}{5 \cdot 5}$, which is $\dfrac{1}{5^6}$ or 5^{-6}.

Again, we could have multiplied the exponents. Actually, this works for any integer exponents.

To raise a power to a power we can multiply the exponents.
For any exponents m and n, $(a^m)^n = a^{mn}$.

107. $\dfrac{6^0}{6^{-3}}$ **108.** $\dfrac{8^{-7}}{8^0}$

Examples

a) $(3^5)^4 = 3^{5 \cdot 4}$ $= 3^{20}$

b) $(y^{-5})^7 = y^{-5 \cdot 7}$ $= y^{-35}$

c) $(x^4)^{-2} = x^{4(-2)}$ $= x^{-8}$

d) $(a^{-4})^{-6} = a^{(-4)(-6)} = a^{24}$

Do exercises 109 through 111 at the left.

There may be several factors inside parentheses as in the following.

Simplify

109. $(3^4)^5$ **110.** $(x^{-3})^4$

Examples

a) $(3x^2y^{-2})^3 = 3^3x^6y^{-6}$

b) $(5x^3y^{-5}z^2)^4 = 5^4x^{12}y^{-20}z^8$

111. $(y^{-5})^{-3}$

Are you remembering to review the objectives?

Do exercise set 2.6, p. 89.

2.7 SOLVING EQUATIONS AND PROBLEMS

In chapter 1 (page 23) it was explained how to solve equations and then how to use them to solve applied problems. The same principles and methods apply whether we use rational numbers or only the numbers of arithmetic.

SOLVING EQUATIONS

We can solve simple equations by remembering that addition and subtraction are opposites. Some we can solve by remembering that multiplication and division are opposites. In either case, we write the proper related sentence and solve it.

Examples

a) Solve $\frac{3}{4} - x = \frac{3}{2}$.

Related sentence: $\frac{3}{4} - \frac{3}{2} = x$

We do the calculation $\frac{3}{4} - \frac{3}{2}$ and get $-\frac{3}{4} = x$.

Check: $\frac{3}{4} - x = \frac{3}{2}$

$$
\begin{array}{c|c}
\frac{3}{4} - \left(-\frac{3}{4}\right) & \frac{3}{2} \\
\frac{3}{4} + \frac{3}{4} & \\
\frac{6}{4} & \\
\frac{3}{2} &
\end{array}
$$
Thus $-\frac{3}{4}$ is the solution

b) Solve $\frac{4}{3}y = -\frac{5}{6}$

Related sentence: $y = \dfrac{-\frac{5}{6}}{\frac{4}{3}}$

We do the calculation $-\frac{5}{6} \div \frac{4}{3}$ or $-\frac{5}{6} \cdot \frac{3}{4}$

and get $-\frac{5}{8}$

Check: $\frac{4}{3}y = -\frac{5}{6}$

$$
\begin{array}{c|c}
\frac{4}{3}\left(-\frac{5}{8}\right) & -\frac{5}{6} \\
-\frac{5}{6} &
\end{array}
$$
Thus $-\frac{5}{8}$ is the solution

Do exercises 112 and 113 at the right.

OJBECTIVES

You should be able to:

a) Translate problems to equations.
b) Solve simple equations in rational numbers.
c) Solve applied problems.

Solve and check.

112. $\dfrac{4}{5} - y = \dfrac{6}{5}$

113. $\dfrac{3}{4}x = -\dfrac{5}{9}$

114. Eggs at one time sold for 48¢ per dozen. This is $\frac{3}{4}$ of the price four years later. What was the price four years later?

SOLVING PROBLEMS

To solve an applied problem we first translate it to an equation and then solve the equation.

Example

Solve this problem.

Negative 17 is three-fourths of what number

$$-17 = \frac{3}{4} \cdot x$$

The translation gives us the equation $-17 = \frac{3}{4} \cdot x$. We solve it.

$\dfrac{-17}{\frac{3}{4}} = x$. We do the calculation: $-17 \cdot \frac{4}{3} = -\frac{68}{3}$

The number $-\frac{68}{3}$ checks, so it is the answer.
Remember that translating a problem to an equation is the most important part of solving the problem and may take the most time and thought. Sometimes it is helpful to reword a problem before translating.

Example

George Hamil's salary is $46 per week. This is two thirds of Fred Perkins' salary. What is Perkins' salary?

Rewording: 46 is two-thirds of Perkins' salary.

Translation: $46 = \frac{2}{3} \cdot x$.

We get the equation $46 = \frac{2}{3}x$.

We solve: $\dfrac{46}{\frac{2}{3}} = x$

$$46 \cdot \frac{3}{2} = x$$

$$69 = x$$

The number 69 checks, so the answer is $69.

Do exercise 114 at the left.

Do exercise set 2.7, p. 91.

ANSWERS

EXERCISE SET 2.1 (pp. 61–63)

Always review the objectives before working exercises.

Copy these pairs of numerals in the answer spaces. Insert $<$, $>$, or $=$ to make true sentences.

1. 5   0   **2.** -9 0   **3.** -9 5   **4.** 8 -3

5. $^-6$ 6   **6.** 5   8   **7.** -5 -8 **8.** -4 -4

Find the absolute value of each integer.

9. 17   **10.** 0   **11.** -5 **12.** -13 **13.** 10

14. -9 **15.** 9   **16.** 4   **17.** -6 **18.** 5

Simplify.

19. $|3|$ **20.** $|-6|$ **21.** $|43|$ **22.** $|0|$ **23.** $|6 + 5|$

24. $|-8|$ **25.** $|33|$ **26.** $|-14|$ **27.** $|-5|$ **28.** $|10 - 7|$

Name the additive inverse of each integer.

29. 12   **30.** -23 **31.** -8 **32.** 0   **33.** 41

Simplify.

34. $^-5$ **35.** $^-(-11)$ **36.** $^-(-5)$ **37.** $^-18$ **38.** $^-(-9)$

Find ^-x when x is

39. 6   **40.** -10

1.
2.
3.
4.
5.
6.
7.
8.
9.
10.
11.
12.
13.
14.
15.
16.
17.
18.
19.
20.
21.
22.
23.
24.
25.
26.
27.
28.
29.
30.
31.
32.
33.
34.
35.
36.
37.
38.
39.
40.

ANSWERS

41. _____

42. _____

43. _____

44. _____

45. _____

46. _____

47. _____

48. _____

49. _____

50. _____

51. _____

52. _____

53. _____

54. _____

55. _____

56. _____

57. _____

58. _____

59. _____

60. _____

Add.

41. $17 + 14$

42. $-7 + 0$

43. $-3 + (-9)$

44. $-11 + (-15)$

45. $-8 + (-7)$

46. $-13 + (-8)$

47. $16 + (-5)$

48. $6 + (-19)$

49. $35 + (-27)$

50. $(-4) + (-8) + (-10)$

51. $18 + (-33) + 15$

52.
```
 -35
 -63
 -27
 -14
 -59
```

53.
```
  24
  37
  19
 -45
 -35
```

54.
```
 -296
 -384
 -709
 -157
  -83
```

55.
```
  27
 -54
 -32
  65
  46
```

56.
```
 -62
  53
 -87
  14
 -28
```

57.
```
  403
 -790
  527
 -135
  246
```

58.
```
 -45
 -93
 -58
 -37
 -16
 -28
```

59.
```
 494
 328
 641
 209
 350
 717
```

60.
```
 -643
 -271
  536
  390
  416
 -119
```

EXERCISE SET 2.2 (pp. 64–66)

Are you remembering to review the objectives?

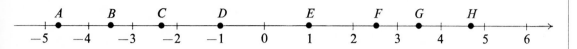

On the number line above, each of the lettered points represents a rational number.
Write the letter of the point which corresponds to each of these numbers.

1. 4.7 **2.** -3.5 **3.** $\dfrac{5}{2}$ **4.** $-\dfrac{7}{2}$ **5.** $-\dfrac{7}{3}$ **6.** $-\dfrac{8}{8}$

Add.

7. $-6.5 + 4.7$ **8.** $\dfrac{7}{3} + \dfrac{5}{3}$ **9.** $-\dfrac{3}{5} + \dfrac{7}{5}$

10. $-\dfrac{3}{7} + \left(-\dfrac{5}{7}\right)$ **11.** $-\dfrac{2}{3} + \dfrac{2}{3}$ **12.** $\dfrac{4}{9} + \left(-\dfrac{6}{9}\right)$

13. $3.6 + (-1.9)$ **14.** $\dfrac{3}{4} + \left(-\dfrac{2}{3}\right)$ **15.** $-\dfrac{5}{8} + \dfrac{1}{4}$

16. $-\dfrac{3}{5} + \dfrac{2}{3}$ **17.** $-2.8 + (-5.3)$ **18.** $-\dfrac{3}{7} + \left(-\dfrac{2}{5}\right)$

19. $5.7 + (-7.2) + 6.6$ **20.** $\dfrac{1}{3} + \left(-\dfrac{11}{6}\right) + \dfrac{1}{2}$

NAME

CLASS

ANSWERS

1. _____

2. _____

3. _____

4. _____

5. _____

6. _____

7. _____

8. _____

9. _____

10. _____

11. _____

12. _____

13. _____

14. _____

15. _____

16. _____

17. _____

18. _____

19. _____

20. _____

ANSWERS

21. _____

22. _____

23. _____

24. _____

25. _____

26. _____

27. _____

28. _____

29. _____

30. _____

31. _____

32. _____

33. _____

34. _____

35. _____

36. _____

37. _____

38. _____

39. _____

40. _____

Subtract.

21. $3 - (-5)$ **22.** $6.1 - 3.8$ **23.** $1.5 - (-3.5)$

24. $-9 - (-2)$ **25.** $-9 - 5$ **26.** $-3.2 - 5.8$

27. $\dfrac{3}{4} - \dfrac{2}{3}$ **28.** $\dfrac{3}{4} - \left(-\dfrac{2}{3}\right)$ **29.** $-\dfrac{3}{4} - \dfrac{2}{3}$

Add or subtract as indicated.

30. $5 + (-8)$ **31.** $5 - (-8)$ **32.** $-5 + 8$

33. $2.7 - 5.9$ **34.** $-\dfrac{3}{5} + \left(-\dfrac{2}{3}\right)$ **35.** $-\dfrac{3}{5} - \left(-\dfrac{2}{3}\right)$

36. $10.3 + (-7.5) + 3.1 - 1.8$

Look at examples A through D.

A. $-7.5 + 0 = -7.5$ **B.** $\dfrac{3}{7} + \dfrac{5}{11} = \dfrac{5}{11} + \dfrac{3}{7}$

C. $\dfrac{3}{7} + \left(-\dfrac{3}{7}\right) = 0$ **D.** $\left(\dfrac{1}{2} + \dfrac{1}{3}\right) + \dfrac{1}{4} = \dfrac{1}{2} + \left(\dfrac{1}{3} + \dfrac{1}{4}\right)$

Which of them illustrates the

37. Commutative property of addition?

38. Associative property of addition?

39. Additive identity (property of 0)?

40. Additive inverse property?

EXERCISE SET 2.3 (pp. 67–69)

Did you review the objectives?

Multiply.

1. $-7 \cdot 4$

2. $-6.3 \cdot 2.7$

3. $-4.1 \cdot 9.5$

4. $-8 \cdot (-5)$

5. $-4.5 \cdot (-3.6)$

6. $-5.2 \cdot (-3.1)$

7. $\dfrac{2}{3} \cdot \left(-\dfrac{3}{5}\right)$

8. $-\dfrac{5}{7} \cdot \left(-\dfrac{1}{3}\right)$

9. $4 \cdot \left(-\dfrac{1}{8}\right)$

10. $9 \cdot (-7)$

11. $11 \cdot (-12)$

12. $-16 \cdot (-16)$

13. $\dfrac{3}{8} \cdot \left(-\dfrac{2}{9}\right)$

14. $\dfrac{2}{11} \cdot \left(-\dfrac{4}{3}\right)$

15. $-\dfrac{3}{8} \cdot \left(-\dfrac{8}{3}\right)$

16. $-17 \cdot (-9)$

17. $135 \cdot (-7)$

18. $\dfrac{a}{b} \cdot \left(-\dfrac{c}{d}\right)$

19. $6 \cdot (-5) \cdot (-4) \cdot 3$

20. $-\dfrac{2}{3} \cdot \dfrac{1}{2} \cdot \left(-\dfrac{6}{7}\right) \cdot \left(-\dfrac{1}{3}\right)$

ANSWERS	
1.	
2.	
3.	
4.	
5.	
6.	
7.	
8.	
9.	
10.	
11.	
12.	
13.	
14.	
15.	
16.	
17.	
18.	
19.	
20.	

ANSWERS

21. _____

22. _____

23. _____

24. _____

25. _____

26. _____

27. _____

28. _____

29. _____

30. _____

31. _____

32. _____

33. _____

34. _____

35. _____

36. _____

37. _____

38. _____

39. _____

40. _____

Name the reciprocal of each.

21. 5   **22.** -7 **23.** $\dfrac{1}{4}$ **24.** $-\dfrac{1}{8}$

25. $\dfrac{2}{3}$ **26.** $-\dfrac{3}{4}$ **27.** $\dfrac{5}{3}$ **28.** $-\dfrac{7}{5}$

Divide.

29. $18 \div (-6)$ **30.** $-35 \div 7$ **31.** $-12 \div (-3)$

32. $-4.5 \div 1.5$ **33.** $7.2 \div (-1.2)$ **34.** $-4.8 \div (-4)$

35. $-\dfrac{3}{4} \div \dfrac{2}{3}$ **36.** $\dfrac{1}{2} \div \left(-\dfrac{1}{2}\right)$ **37.** $-\dfrac{5}{4} \div \left(-\dfrac{3}{4}\right)$

38. $\dfrac{3}{5} \div \left(-\dfrac{5}{3}\right)$ **39.** $-\dfrac{2}{7} \div \dfrac{4}{9}$ **40.** $-\dfrac{5}{9} \div \left(-\dfrac{5}{6}\right)$

EXERCISE SET 2.4 (pp. 70–71)

Multiply.

1. $7(x - 3)$

2. $15(6 - x)$

3. $-3(t - 7)$

4. $-5(14 - n)$

5. $8(2x - 9)$

6. $6(13 - 5x)$

7. $-4(3x - 11)$

8. $3(2x + 5y - 9)$

9. $-9(12 - 4x)$

10. $-7(-2x - 4y + 3)$

Factor.

11. $5x - 15$

12. $28 - 7x$

13. $-8x + 24$

14. $-50 + 10x$

15. $6x - 10$

16. $25 - 15x$

17. $-12x + 28$

18. $9x + 6y - 15$

19. $ax - 7a$

20. $3bx - 12b$

1. _____
2. _____
3. _____
4. _____
5. _____
6. _____
7. _____
8. _____
9. _____
10. _____
11. _____
12. _____
13. _____
14. _____
15. _____
16. _____
17. _____
18. _____
19. _____
20. _____

ANSWERS

21. _____

22. _____

23. _____

24. _____

25. _____

26. _____

27. _____

28. _____

29. _____

30. _____

31. _____

32. _____

33. _____

34. _____

35. _____

36. _____

Simplify by collecting like terms.

21. $11x - 3x$

22. $17t - 9t$

23. $6n - n$

24. $y - 17y$

25. $9x + 2y - 5x$

26. $3x - 7y - 8x$

27. $11x + 2y - 4x - y$

28. $3a + 9b - a - 4b$

29. $7x - 3y - 9x + 8y$

30. $6.7a + 4.3b - 4.1a - 2.9b$

31. $6x + 9y - 8 - x + 3y$

32. $3a + 11b + c - 8b - 6c$

33. $5x + 7y - 4x - 6y$

34. $-4a - 6b - 7a + 10b$

35. $\frac{1}{5}x + \frac{4}{5}y + \frac{2}{5}x - \frac{1}{5}y$

36. $\frac{7}{8}x + \frac{5}{9}y - \frac{1}{2}x - \frac{2}{3}y$

EXERCISE SET 2.5 (pp. 72–73)

Did you review the objectives?

Rename each additive inverse without parentheses.

1. $^-(2x + 7)$

2. $^-(3x - 5)$

3. $^-(4a - 3b + 7c)$

4. $^-(-2a + 5b)$

5. $^-(6x + 8y - 5)$

6. $^-(-3x - 9y + 6)$

Name the terms of each expression separating them with commas.

7. $3x + 5y + 6$

8. $6a - 4b + 7c$

9. $-8x - 6y - 4$

10. $-2a + 9b - 5c$

Remove parentheses and simplify.

11. $9x - (4x + 3)$

12. $7y - (2y - 9)$

13. $2a + (5a - 9)$

14. $11n - (7 - 3n)$

15. $2x + 7x - (4x + 6)$

16. $3a - (7 - 4a) + 2a$

17. $2x - 4y - (7x - 2y)$

18. $a + 7b - (2c - 3a + 4b)$

1. _____

2. _____

3. _____

4. _____

5. _____

6. _____

7. _____

8. _____

9. _____

10. _____

11. _____

12. _____

13. _____

14. _____

15. _____

16. _____

17. _____

18. _____

ANSWERS

19.
_____

20.
_____

21.
_____

22.
_____

23.
_____

24.
_____

25.
_____

26.
_____

27.
_____

28.
_____

29.
_____

30.
_____

31.
_____

32.
_____

33.
_____

34.
_____

35.
_____

36.
_____

37.
_____

Simplify.

19. $(2 + 8) - (7 - 3)$

20. $5(6 - 4) + 2(8 - 5)$

21. $(7 + 3) \cdot 6 + 5$

22. $7 + (3 \cdot 6) + 5$

23. $2 \cdot [4 + 3(7 - 2)]$

24. $3[2 + 2(16 + 9)]$

25. $[4(9 - 6) + 11] - [14 - (6 + 4)]$

26. $[3(8 - 4) + 12] - [10 - (3 + 5)]$

27. $7(9 - 4) - [5(10 - 7) - 2(8 - 3)]$

28. $3(8 - 2) - [3(8 - 3) - 3(5 - 2)]$

29. $5(x + 2) - 3x$

30. $37(6x - 5x) - 35x$

31. $16x - 5(2x + 7)$

32. $12x - 3(4x - 5)$

33. $[10(x + 3) - 4] + [2(x - 17) + 6]$

34. $[4(2x - 5) + 7] + [3(x + 3) + 5x]$

35. $[7(x + 5) - 19] - [4(x - 6) + 10]$

36. $3\{[6(x - 2) + 4] - [2(2x - 5) + 6]\}$

37. $4\{5[(x - 3) + 2] - 3[2(x + 5) - 6]\}$

NAME

CLASS

EXERCISE SET 2.6 (pp. 74–76)

ANSWERS

1.

Explain the meaning of the following.

2.

1. 3^{-2} **2.** 5^{-1} **3.** 2^{-3}

3.

4.

4. 7^{-2} **5.** 10^{-1} **6.** 10^{-3}

5.

6.

7.

Rename, using a negative exponent.

8.

7. $\dfrac{1}{4^3}$ **8.** $\dfrac{1}{5^2}$ **9.** $\dfrac{1}{3^4}$ **10.** $\dfrac{1}{2^5}$

9.

10.

11. $\dfrac{1}{x^3}$ **12.** $\dfrac{1}{y^2}$ **13.** $\dfrac{1}{a^4}$ **14.** $\dfrac{1}{t^5}$

11.

12.

13.

14.

Rename, using a positive exponent.

15. 7^{-3} **16.** 5^{-2} **17.** 2^{-4} **18.** 10^{-7}

15.

16.

19. a^{-3} **20.** x^{-2} **21.** y^{-4} **22.** t^{-7}

17.

18.

19.

20.

21.

22.

Multiply and simplify.

23. $2^4 \cdot 2^3$ **24.** $3^{-5} \cdot 3^3$ **25.** $5^{-8} \cdot 5^9$

26. $6^0 \cdot 6^3$ **27.** $7^2 \cdot 7^3$ **28.** $7^2 \cdot 7^{-3}$

29. $10^1 \cdot 10^3$ **30.** $4^5 \cdot 4^{-3}$ **31.** $4^{-5} \cdot 4^3$

32. $x^4 \cdot x^3$ **33.** $x^9 \cdot x^{-6}$ **34.** $x^{-7} \cdot x^5$

Divide and simplify.

35. $\dfrac{7^5}{7^2}$ **36.** $\dfrac{4^5}{4^3}$ **37.** $\dfrac{10^7}{10^3}$

38. $\dfrac{3^{-5}}{3^{-2}}$ **39.** $\dfrac{2^{-3}}{2^4}$ **40.** $\dfrac{5^2}{5^{-3}}$

41. $\dfrac{x^3}{x^6}$ **42.** $\dfrac{t^6}{t^3}$ **43.** $\dfrac{n^3}{n^{-4}}$

Simplify.

44. $(2^3)^2$ **45.** $(3^4)^3$ **46.** $(5^2)^{-3}$

47. $(x^{-3})^2$ **48.** $(t^2)^2$ **49.** $(a^{-3})^{-3}$

EXERCISE SET 2.7 (pp. 77–78)

Solve each equation by writing a related sentence. Check the result.

1. $x + 5 = 8$ **2.** $x + 3 = 7$ **3.** $x + 9 = 4$

Check: $x + 5 = 8$ Check: Check:

4. $x + 7 = 3$ **5.** $x + 1.9 = 3.5$ **6.** $x + \dfrac{3}{4} = 1$

Check: Check: Check:

7. $x - 7 = 3$ **8.** $x - 10 = 1$ **9.** $x - 5 = 9$

Check: Check: Check:

10. $9 - x = 4$ **11.** $13 - x = 5$ **12.** $7 - x = 9$

Check: Check: Check:

13. $x - \dfrac{1}{2} = 2\dfrac{1}{2}$ **14.** $x - \dfrac{2}{3} = \dfrac{1}{3}$ **15.** $\dfrac{3}{4} - x = \dfrac{1}{2}$

Check: Check: Check:

NAME

CLASS

ANSWERS

1.

2.

3.

4.

5.

6.

7.

8.

9.

10.

11.

12.

13.

14.

15.

ANSWERS

16. _____

17. _____

·18. _____

19. _____

20. _____

21. _____

22. _____

23. _____

24. _____

25. _____

26. _____

27. _____

28. _____

Solve each equation by writing a related sentence. Check the result.

16. $3 \cdot x = 18$ **17.** $7 \cdot x = 28$ **18.** $9 \cdot x = 63$

Check: ——|——   Check: ——|——   Check: ——|——

19. $8x = 4$ **20.** $9x = 5$ **21.** $1.7x = 6.8$

Check: ——|——   Check: ——|——   Check: ——|——

22. $\dfrac{x}{7} = 3$ **23.** $\dfrac{x}{6} = 2$ **24.** $\dfrac{x}{10} = 30$

Check: ——|——   Check: ——|——   Check: ——|——

25. $\dfrac{x}{1.5} = 3$ **26.** $\dfrac{3}{5}x = 15$ **27.** $\dfrac{21}{x} = 7$

Check: ——|——   Check: ——|——   Check: ——|——

Write an equation from the problem. Solve the equation and check the answer.

28. If Bill earned 3 times as much as he does, he would be making $48 per day. How much does he make now?

NAME_____

CHAPTER 2   TEST   CLASS_____ SCORE_____ GRADE_____

Before taking the test *be sure* to allow yourself a day or so for review. Use the objectives listed in the margins to guide your study. The test will evaluate your progress and aid your preparation for a possible classroom test. Allow about an hour for the test. Remove the test from the book. When you finish, read the test analysis on the answer page at the end of the book.

1. _____

2. _____

3. _____

Which symbol, $<$ or $>$, makes a true statement.

4. _____

1. 8   5   **2.** -8 -5 **3.** 5 -8

5. _____

Simplify.

6. _____

4. $|7|$ **5.** $|-9|$ **6.** $^-(-6)$ **7.** $^-(0)$

7. _____

8. _____

Add.

9. _____

8. $(-5) + 14 + (-3) + 2$ **9.** $6 + (-9) + (-8) + 7$

10. _____

11. _____

12. _____

Add or subtract as indicated.

13. _____

10. $-3.9 + (-7.4)$ **11.** $-4.6 + 6.3$ **12.** $5.8 - 9.5$

14. _____

15. _____

13. $8.2 - (-11.4)$ **14.** $-\dfrac{5}{9} - \dfrac{2}{9}$ **15.** $\dfrac{3}{5} - \left(-\dfrac{2}{3}\right)$

16. _____

17. _____

Multiply or divide as indicated.

18. _____

16. $-9 \cdot (-6)$ **17.** $2.7 \cdot (-3.4)$ **18.** $-5.1 \div 1.7$

19. _____

20. _____

19. $\dfrac{2}{3} \cdot \left(-\dfrac{3}{7}\right)$ **20.** $-\dfrac{3}{5} \div \dfrac{4}{5}$ **21.** $\dfrac{2}{3} \div \dfrac{4}{9}$

21. _____

22. _____

22. $3 \cdot (-7) \cdot (-2) \cdot (-5)$ **23.** $-1 \cdot (-1) \cdot (-1) \cdot (-1)$

23. _____

24. _____

25. _____

26. _____

27. _____

28. _____

29. _____

30. _____

31. _____

32. _____

33. _____

34. _____

35. _____

36. . _____

37. _____

38. _____

39. _____

40. _____

41. _____

Multiply.

24. $5(3x - 7)$ **25.** $-2(4x - 5)$ **26.** $6(9 - 2x)$

Factor.

27. $2x - 14$ **28.** $6x - 9$ **29.** $-7x + 21$

Simplify by collecting like terms.

30. $11a + 2b - 4a - 5b$ **31.** $7x - 3y - 9x + 8y$

Remove parentheses and simplify.

32. $2a - (5a - 9)$ **33.** $7y - (9 - 2y)$

34. $4(x + 3) - 2x$ **35.** $10(x - 2) + 4$

Rename, using a negative exponent.

36. $\dfrac{1}{4^2}$ **37.** $\dfrac{1}{5^3}$

Multiply or divide as indicated.

38. $6^2 \cdot 6^3$ **39.** $x^{-6} \cdot x^2$ **40.** $\dfrac{7^2}{7^5}$

41. $\dfrac{x^3}{x^{-2}}$

3 SOLVING EQUATIONS

3.1 THE ADDITION PRINCIPLE

You have already learned to solve certain equations using the idea of opposite operations. In this chapter we shall develop some more useful principles for solving equations.

What does an equation actually say? An equation $a = b$ says that a and b represent the same number. Suppose this is true and then add a number c to the number a. We will get the same answer as if we add c to b, because a and b are the same number. Thus $a + c = b + c$ is true if $a = b$ is true. This is the addition principle.

The *addition principle*: If an equation $a = b$ is true, then $a + c = b + c$ is true for any number c.

When we use the addition principle, we sometimes say that we "add the same number to both sides of an equation". Let's see how this principle can help in solving equations.

Example

Solve $x + 5 = -7$

$$x + 5 + (-5) = -7 + (-5)$$ Here we used the addition principle. We added -5.

$$\left. \begin{array}{c} x + 0 = -7 + (-5) \\ x = -12 \end{array} \right\}$$ Here we simplified.

Check: $x + 5 = -7$

$$\begin{array}{c|c} \boxed{-12} + 5 & -7 \\ \hline -7 & \end{array}$$ The solution is -12.

In the above example, why did we choose to add -5? Because we wanted x alone on one side of the equation, we added the inverse of 5. This "got rid of" the 5 on the left.

Do exercises 1 through 4 at the right.

Always remember to review the objectives before doing the exercise sets.

Do exercise set 3.1, p. 105.

OBJECTIVES

You should be able to:

a) State the addition principle.

b) Solve simple equations using the addition principle.

Solve, using the addition principle.

1. $x + 7 = 2$ **2.** $3 = -5 + y$

3. $x + \dfrac{1}{2} = -\dfrac{3}{2}$

4. $y - 3.2 = 7.6$

You should be able to:

a) State the multiplication principle.
b) Solve simple equations using the multiplication principle.

Solve, using the multiplication principle.

5. $5x = 25$

6. $4 = \frac{1}{3}y$

7. $4x = -7$

8. $-2.1y = 6.3$

3.2 THE MULTIPLICATION PRINCIPLE

Using the addition principle we can simplify certain equations. That is, we can get a variable alone on one side of the equation. This should lead us to suspect that there is also a multiplication principle. This is true, and it is very much like the addition principle.

Remember what an equation says. An equation $a = b$ says that a and b name the same number. If this is true and if we multiply the number a by some number c, we will get the same answer as if we multiply b by c. This must be so because a and b are the same number.

When we use the multiplication principle we sometimes say that we "multiply both sides by the same number".

The *multiplication principle:* **If an equation** $a = b$ **is true, then** $a \cdot c = b \cdot c$ **is true for any number** c**.**

Example

Solve $\quad 3x = 9$

$\quad \frac{1}{3} \cdot 3x = \frac{1}{3} \cdot 9 \quad$ Here we used the multiplication principle. We multiplied by $\frac{1}{3}$.

$\left. \begin{array}{l} 1 \cdot x = 3 \\ x = 3 \end{array} \right\} \quad$ Here we simplified.

Check: $3x = 9$

$$\begin{array}{c|c} 3 \cdot 3 & 9 \\ \hline 9 & \end{array}$$ The solution is 3.

In the above example, why did we choose to multiply by $\frac{1}{3}$? Because we wanted x alone on one side of the equation, we multiplied by the reciprocal of 3. Then when we multiplied we got $1 \cdot x$, which simplified to x. In other words, this enabled us to "get rid of" the 3 on the left.

Do exercises 5 through 8 at the left.

Don't forget to review the objectives.

Do exercise set 3.2, p. 107.

3.3 USING THE PRINCIPLES TOGETHER

Now that we have established the addition and multiplication principles, let us see how to use them together.

Example 1

Solve $3x + 4 = 13$

$3x + 4 + (-4) = 13 + (-4)$ Here we used the addition principle, adding -4.

$3x = 9$ Here we simplified.

$\frac{1}{3} \cdot 3x = \frac{1}{3} \cdot 9$ Here we used the multiplication principle, multiplying by $\frac{1}{3}$.

$x = 3$ Here we simplified.

Check: $3x + 4 = 13$

$$
\begin{array}{c|c}
3 \cdot 3 + 4 & 13 \\
9 + 4 & \\
13 &
\end{array}
$$

The solution is 3.

Do exercises 9 and 10 at the right.

Note from example 1 that we used the addition principle first. This is usually best, although there might be equations for which we would wish to use the multiplication principle first. If there are like terms on one side of an equation, they should be combined. This should be done before using the principles.

Example 2

Solve $3x + 2x = 15$

$5x = 15$ Here we combined like terms.

$\frac{1}{5} \cdot 5x = \frac{1}{5} \cdot 15$ Here we multiplied by $\frac{1}{5}$.

$x = 3$

The number 3 checks, so the solution is 3.

If there are like terms on opposite sides of an equation, we can get them on the same side using the addition principle, and then "combine" them.

You should be able to:

Solve equations using both the addition and multiplication principles.

Solve.

9. $-4 + 3x = 8$

10. $-\frac{1}{2}x + 3 = 1$

Solve

11. $4x + 3x = 21$

12. $7y + 5 = 2y + 10$

13. $5 - 2y = 3y - 5$

14. $7x - 17 + 2x = 2 - 8x + 15$

Solve

15. $3x - 15 = 5x + 2 - 4x$

16. $1.2x - 6 = 3.1x - 3 - 2.2x$

Example 3

Solve $2x - 2 = -3x + 3$

$2x - 2 + 2 = -3x + 3 + 2$ Here we added 2.

$2x = -3x + 5$ Here we simplified.

$2x + 3x = -3x + 3x + 5$ Here we added $3x$.

$5x = 5$ Here we combined like terms and simplified.

$\frac{1}{5} \cdot 5x = \frac{1}{5} \cdot 5$ Here we multiplied by $\frac{1}{5}$.

$x = 1$ Here we simplified.

Check: $2x - 2 = -3x + 3$

$2 \cdot 1 - 2$	$-3 \cdot 1 + 3$
$2 - 2$	$-3 + 3$
0	0

The solution is 1.

Notice in the above example, that we use the addition principle to get all terms with the variable on one side of the equation and all other terms on the other side. Then we can combine like terms and proceed as before.

Do exercises 11 through 14 at the left.

Using the addition and multiplication principles and the combining of like terms, we can solve a great many equations. Let us look at another example.

Example 4

Solve $6x + 5 - 7x = 10 - 4x + 3$

$6x - 7x = 10 - 4x + 3 - 5$ (adding -5)

$4x + 6x - 7x = 10 + 3 - 5$ (adding $4x$)

$3x = 8$ (combining like terms and simplifying)

$\frac{1}{3} \cdot 3x = \frac{1}{3} \cdot 8$ (multiplying by $\frac{1}{3}$)

$x = \frac{8}{3}$ (simplifying)

The number $\frac{8}{3}$ checks, so it is the solution.

Do exercises 15 and 16 at the left.

Do exercise set 3.3, p. 109.

3.4 EQUATIONS CONTAINING PARENTHESES

Certain equations containing parentheses can be solved by first multiplying to remove parentheses and then proceeding as before.

Example 1

Solve $4x = 2(12 - 2x)$

$$4x = 24 - 4x \quad \text{We multiplied to remove parentheses.}$$

$$4x + 4x = 24 \quad \text{Here we added } 4x.$$

$$8x = 24 \quad \text{We combined like terms.}$$

$$x = 3 \quad \text{Here we multiplied by } \tfrac{1}{8}.$$

Check: $4x = 2(12 - 2x)$

$$
\begin{array}{c|l}
4 \cdot 3 & 2(12 - 2 \cdot 3) \\
 & 2(12 - 6) \\
12 & 2 \cdot 6 \\
 & 12
\end{array}
$$
        The solution is 3.

Do exercise 17 at the right.

Let us look at another example.

Example 2

Solve $3(x - 2) - 1 = 2 - 5(x + 5)$

$$3x - 6 - 1 = 2 - 5x - 25 \quad \text{We multiplied to remove parentheses.}$$

$$3x - 7 = -5x - 23 \quad \text{We simplified.}$$

$$3x + 5x = -23 + 7 \quad \text{Here we added } 5x \text{ and also } 7.$$

$$8x = -16 \quad \text{Here we combined like terms and simplified.}$$

$$x = -2 \quad \text{Here we multiplied by } \tfrac{1}{8}.$$

Check: $3(x - 2) - 1 = 2 - 5(x + 5)$

$$
\begin{array}{c|c}
3(-2 - 2) - 1 & 2 - 5(-2 + 5) \\
3 \cdot (-4) - 1 & 2 - 5(3) \\
-12 - 1 & 2 - 15 \\
-13 & -13
\end{array}
$$
      The solution is -2.

Do exercise 18 at the right.

Do exercise set 3.4, p. 113.

OBJECTIVES

You should be able to:

Solve equations with parentheses like those in the examples.

Solve

17. $2(2y + 3) = 14$

18. $3(7 + 2x) = 30 + 7(x - 1)$

You should be able to:

Solve equations (already factored) using the principle of zero products.

Solve, using the principle of zero products.

19. $(x - 3)(x + 4) = 0$

20. $x(3x - 17) = 0.$

3.5 THE PRINCIPLE OF ZERO PRODUCTS

When we multiply two numbers, the answer will be zero if one of the factors is zero. For example, $3 \cdot 0 = 0$. Furthermore, *if any product is zero, then a factor must be zero.* This gives us another principle for solving equations.

Example 1

Solve $(x + 3)(x - 2) = 0$.

Here we have a product which is zero. This equation will become true when either factor is zero. Hence it is true when

$x + 3 = 0$ or $x - 2 = 0$. Here we have two simple equations to solve.

There are two solutions, -3 and 2.

The above example illustrates the new principle, stated below.

The principle of zero products: **An equation with 0 on one side and with factors on the other can be solved by finding those numbers which make the factors 0. (Do not make the mistake of using this principle when there is not a zero on one side.)**

Example 2

Solve $(5x + 1)(x - 7) = 0$.

$5x + 1 = 0$ or $x - 7 = 0$ (using the principle of zero products)

$5x = -1$ or $x = 7$

$x = -\dfrac{1}{5}$ or $x = 7$

The solutions are $-\dfrac{1}{5}$ and 7.

Example 3

Solve $x(2x - 9) = 0$

$x = 0$ or $2x - 9 = 0$ (principle of zero products)

$x = 0$ or $x = \dfrac{9}{2}$

The solutions are 0 and $\dfrac{9}{2}$.

Do exercises 19 and 20 at the left.

Do exercise set 3.5, p. 117.

3.6 APPLIED PROBLEMS

The first step in solving an applied problem is to translate it to mathematical language. Very often this means translating to an equation. Then we solve the equation and check to see if we have a solution to the problem.

Example 1

A 6 ft board is cut into two pieces, one twice as long as the other. How long are the pieces?

One way to translate is this:

Length of one piece   plus   length of other piece = 6

x + $2x$ = 6

(We use x for the length of one piece and $2x$ for the length of the other because we know one is twice as long as the other.)

We have let x = length of one piece. Then $2x$ = length of other piece.

Now we solve $x + 2x = 6$

$$3x = 6 \quad \text{(collecting like terms)}$$

$$x = 2 \quad \text{(multiplying by } \tfrac{1}{3}\text{)}$$

Do we have an answer to the problem itself? If one piece is 2 ft long, the other, to be twice as long, must be 4 ft long, and the lengths of the pieces add up to 6 ft. This checks.

In the above example, notice that we did not need to check the solution of the equation. Rather, we checked in the problem itself. We might have made a mistake in translating, so we check back to the very beginning.

The steps in solving a problem are: 1. Translate to an equation. 2. Solve the equation. 3. Check the answer in the original problem.

Do exercise 21 at the right.

Example 2

If 6 is added to twice a certain number the result is 20. What is the number?

First we translate:

6 added to twice a certain number is 20

6   + $2x$ = 20

OBJECTIVES

You should be able to:

a) Translate a problem to an equation.
b) Solve applied problems like those in the examples.

21. An 8 ft board is cut into two pieces. One piece is 2 ft longer than the other. How long are the pieces?

22. If 5 is subtracted from three times a certain number, the result is 10. What is the number?

23. One angle of a triangle is 3 times as large as another. The third angle measures 30° more than the smallest angle. Find the measures of the angles.

We have let x = the number.

Now we solve. $6 + 2x = 20$

$$2x = 14$$

$$x = 7$$

Now we check. Twice 7 is 14. If we add 6 to 14 we get 20. Hence this checks and the answer is 7.

Do exercise 22 at the left.

Example 3

One angle of a triangle is twice as large as another.

The measure of the third angle is 20° greater than that of the smallest angle. How large are the angles?

To solve this problem, we must remember that the measures of the angles of a triangle add up to 180°. This gives us a translation.

First angle	+	second angle	+	third angle	=	180
x	+	$2x$	+	$(x + 20)$	=	180

We let x = the first angle. Then $2x$ = the second angle and $x + 20$ = the third angle.

Now we solve: $x + 2x + (x + 20) = 180$

$$4x + 20 = 180$$

$$4x = 160$$

$$x = 40$$

The measures of the angles are x, $2x$ and $x + 20$. If $x = 40°$, then $2x = 80°$ and $x + 20 = 60°$.

We check: These add up to 180°. One angle is twice another and the third is 20° larger than the smallest.

Hence these numbers check.

Do exercise 23 at the left.

Example 4

The sum of two consecutive integers is 29. (*Consecutive* integers are next to each other, such as 3 and 4. The larger is 1 plus the smaller.) What are the integers? We translate:

First integer   +   second integer   =   29

$\quad x \quad\quad + \quad\quad (x+1) \quad\quad = \quad 29$

We have let $x =$ the first integer. Then $x + 1 =$ the second.

We solve: $x + (x+1) = 29$

$$2x + 1 = 29$$
$$2x = 28$$
$$x = 14$$

We check: Our answers are 14 and 15. These are consecutive integers. Their sum is 29, so the answers check in the problem.

Do exercise 24 at the right.

Example 5

The perimeter of a rectangle is 150 cm. The length is 15 cm more than the width. Find the dimensions.

We first draw a picture.

We have let $x =$ the width. Then $x + 15 =$ the length. Now we translate:

width   +   width   +   length   +   length   =   150

$\quad x \quad\; + \quad\; x \quad\; + \quad (x+15) \quad + \quad (x+15) \quad = \quad 150$

Next, we solve, and get $x = 30$ (width) and $x + 15 = 45$ (length).

Finally we check:

The length is 15 cm more than the width. The perimeter is $30 + 30 + 45 + 45$, which is 150. The answer checks.

Do exercise 25 at the right.

Do exercise set 3.6, p. 119.

24. The sum of two consecutive even integers is 38. (Consecutive even integers are next to each other such as 4 and 6. The larger is 2 plus the smaller.) What are the integers?

25. The length of a rectangle is twice the width. The perimeter is 60 meters. Find the dimensions.

You should be able to:

Solve a formula for a specified letter.

26. Solve $E = IR$ for I.

27. Solve $d = rt$ for t.

3.7 FORMULAS

A formula is a kind of recipe for doing a certain kind of calculation. Formulas are very often given in the form of equations. A familiar formula from electricity is

$$E = IR.$$

This formula tells us that to calculate the voltage E in an electric circuit we multiply the current I by the resistance R. But suppose we know E and I and wish to calculate R. We would like to have R alone on one side to see how to do this calculation. Our knowledge of equations now allows us to get R alone on one side, or as we say "solve" the formula for R.

Example 1

Solve $E = IR$ for R.

We use the multiplication principle. We multiply by $\frac{1}{I}$.

$$\frac{1}{I} \cdot E = \frac{1}{I} \cdot IR$$

$$\frac{E}{I} = R$$

Notice that multiplying by $\frac{1}{I}$ is the same as dividing by I. Our new formula says we can find R by dividing E by I.

Do exercise 26 at the left.

Example 2

A formula for the circumference of a circle is $C = 2\pi r$. Solve it for r.

We multiply by $\frac{1}{2\pi}$:

$$\frac{1}{2\pi} \cdot C = \frac{1}{2\pi} \cdot 2\pi r = \frac{2\pi}{2\pi} \cdot r = 1 \cdot r$$

$$\frac{C}{2\pi} = r$$

Do exercise 27 at the left.

Do exercise set 3.7, p. 121.

NAME _____

CLASS _____

ANSWERS

EXERCISE SET 3.1 (p. 95)

Solve using the addition principle. Check.

1. $x + 2 = 6$ **2.** $x + 5 = 8$ **3.** $x + 15 = -5$

Check: ——————   Check: ——————   Check: ——————

4. $m + \dfrac{5}{2} = \dfrac{8}{2}$ **5.** $r + \dfrac{2}{3} = -\dfrac{5}{6}$ **6.** $t + \dfrac{3}{4} = 1$

Check: ——————   Check: ——————   Check: ——————

7. $x - 5 = -6$ **8.** $x - \dfrac{2}{3} = \dfrac{7}{3}$ **9.** $x - \dfrac{5}{6} = \dfrac{7}{8}$

Check: ——————   Check: ——————   Check: ——————

10. $5 + t = 7$ **11.** $\dfrac{1}{3} + a = \dfrac{5}{6}$ **12.** $-\dfrac{1}{5} + z = -\dfrac{1}{4}$

Check: ——————   Check: ——————   Check: ——————

13. $x - 2.3 = -7.4$ **14.** $x - 3.7 = 8.4$ **15.** $-2.6 + x = 8.3$

Check: ——————   Check: ——————   Check: ——————

1. _____

2. _____

3. _____

4. _____

5. _____

6. _____

7. _____

8. _____

9. _____

10. _____

11. _____

12. _____

13. _____

14. _____

15. _____

ANSWERS

16. _____

17. _____

18. _____

19. _____

20. _____

21. _____

22. _____

23. _____

24. _____

25. _____

26. _____

27. _____

28. _____

29. _____

30. _____

16. $5 = y + 7$

17. $8 = y - 8$

18. $8 = y + 8$

Check: ———————|————

Check: ———————|————

Check: ———————|————

19. $-6 = -2 + y$

20. $-8 = y - 2$

21. $10 = y - 6$

Check: ———————|————

Check: ———————|————

Check: ———————|————

22. $-\dfrac{2}{3} = y + \dfrac{1}{3}$

23. $-\dfrac{2}{3} = y - \dfrac{4}{3}$

24. $-7.8 = 2.8 + x$

Check: ———————|————

Check: ———————|————

Check: ———————|————

25. $5\dfrac{1}{6} + x = 7$

26. $4\dfrac{2}{3} + x = 5\dfrac{1}{4}$

27. $p + \dfrac{2}{3} = 7\dfrac{1}{3}$

Check: ———————|————

Check: ———————|————

Check: ———————|————

28. $q + \dfrac{1}{3} = -\dfrac{1}{7}$

29. $22\dfrac{1}{7} = 30 + t$

30. $84.6 + r = 83.9$

Check: ———————|————

Check: ———————|————

Check: ———————|————

EXERCISE SET 3.2 (p. 96)

Solve using the multiplication principle. Check.

1. $6x = 36$

2. $3x = 39$

3. $72 = 9x$

Check: ——————

Check: ——————

Check: ——————

4. $7x = -49$

5. $-12x = 72$

6. $-15x = -105$

Check: ——————

Check: ——————

Check: ——————

7. $\frac{1}{7}t = 9$

8. $7 = -\frac{1}{3}t$

9. $\frac{1}{5} = \frac{1}{3}t$

Check: ——————

Check: ——————

Check: ——————

10. $6.3x = 44.1$

11. $-2.7y = 54$

12. $-3.1y = 21.7$

Check: ——————

Check: ——————

Check: ——————

NAME ————————

CLASS

ANSWERS

1. ————————

2. ————————

3. ————————

4. ————————

5. ————————

6. ————————

7. ————————

8. ————————

9. ————————

10. ————————

11. ————————

12. ————————

ANSWERS

13. _____

14. _____

15. _____

16. _____

17. _____

18. _____

19. _____

20. _____

21. _____

22. _____

23. _____

24. _____

13. $-\dfrac{1}{5}r = 10$

14. $-\dfrac{1}{5} = -10y$

15. $-\dfrac{4}{3} = -\dfrac{2}{5}y$

Check: _____

Check: _____

Check: _____

16. $\dfrac{3}{2}r = 27$

17. $12 = \dfrac{6}{5}y$

18. $\dfrac{4}{5}y = 20$

Check: _____

Check: _____

Check: _____

19. $-3.3y = 6.6$

20. $4x = 85.2$

21. $38.7m = -309.6$

Check: _____

Check: _____

Check: _____

22. $-\dfrac{2}{3}y = -10.6$

23. $20.07 = \dfrac{3}{2}y$

24. $-\dfrac{9}{7}y = 12.06$

Check: _____

Check: _____

Check: _____

NAME

CLASS

ANSWERS

EXERCISE SET 3.3 (pp. 97–98)

Solve the following equations. Check.

1. $5x + 6 = 31$

2. $3x + 6 = 30$

3. $4x - 6 = 34$

Check: ————

Check: ————

Check: ————

4. $6x - 3 = 15$

5. $7x + 2 = -54$

6. $5x + 4 = -41$

7. $5x + 7x = 72$

8. $4x + 5x = 45$

9. $4y - 2y = 10$

10. $8y - 5y = 15$

11. $10.2y - 7.3y = -58$

12. $6.8y - 2.4y = -88$

1. ————

2. ————

3. ————

4. ————

5. ————

6. ————

7. ————

8. ————

9. ————

10. ————

11. ————

12. ————

ANSWERS

13.

_____

14.

_____

15.

_____

16.

_____

17.

_____

18.

_____

19.

_____

20.

_____

21.

_____

22.

_____

23.

_____

24.

_____

13. $x + \frac{1}{3}x = 8$

14. $x + \frac{1}{4}x = 10$

15. $-\frac{1}{3}x + x = 20$

16. $-\frac{2}{3}x + \frac{5}{3}x = 21$

17. $8y - 35 = 3y$

18. $4x - 6 = 6x$

19. $4x - 7 = 3x$

20. $y = 15 - 4y$

21. $x + 1 = 16 - 4x$

22. $6y + 8 - 4y = 18$

23. $2x - 1 = 4 + x$

24. $5y - 2 = 28 - y$

25. $5x + 2 = 3x + 6$ **26.** $6x + 3 = 2x + 11$ **27.** $5y + 3 = 2y + 15$

25.

26.

28. $5 - 2x = 3x - 7x + 25$ **29.** $10 - 3x = 2x - 8x + 40$

27.

28.

30. $4 + 3x - 6 = 3x + 2 - x$ **31.** $5 + 4x - 7 = 4x + 3 - x$

29.

30.

32. $4y - 4 + y = 6y + 20 - 4y$ **33.** $5y - 7 + y = 7y + 21 - 5y$

31.

32.

33.

ANSWERS

34. $\dfrac{5}{2}x + \dfrac{1}{2}x = 3x + \dfrac{3}{2} + \dfrac{5}{2}x$ **35.** $\dfrac{7}{8}x - \dfrac{1}{4} + \dfrac{3}{4}x = \dfrac{1}{16} + x$

34. _____

35. _____

36. $2.1x + 45.2 = 3.2 - 8.4x$ **37.** $7\dfrac{1}{2}y - \dfrac{1}{2}y = 4y + 39$

36. _____

37. _____

38. $\dfrac{1}{5}t - .4 + \dfrac{2}{5}t = .6 - \dfrac{1}{10}t$ **39.** $1.7t + 8 - 1.62t = .4t - .32 + 8$

38. _____

39. _____

NAME

CLASS

ANSWERS

EXERCISE SET 3.4 (p. 99)

Solve the following equations. Check.

1. $3(2y + 3) = 27$

2. $4(2y - 3) = 28$

3. $40 = 5(3x + 2)$

4. $9 = 3(5x - 2)$

5. $2(3 + 4m) - 9 = 45$

6. $2(4m + 3) - 9 = 85$

1.

2.

3.

4.

5.

6.

ANSWERS

7. $5r - (2r + 8) = 16$

8. $6b - (3b + 8) = 16$

7.

_____

8.

_____

9. $3g - 3 = 3(7 - g)$

10. $5(d + 4) = 3d + 10$

9.

_____

10.

_____

11. $6 - 2(3x - 1) = 2$

12. $10 - 3(2x - 1) = 1$

11.

_____

12.

_____

13. $5(d + 4) = 7(d - 2)$

14. $3(t - 2) = 9(t + 2)$

15. $3(x - 2) = 5(x + 2)$

16. $5(y + 4) = 3(y - 2)$

17. $8(2t + 1) = 4(7t + 7)$

18. $7(5x - 2) = 6(6x - 1)$

13.

14.

15.

16.

17.

18.

ANSWERS

19. $3(r - 6) + 2 = 4(r + 2) - 21$

20. $5(t + 3) + 9 = 3(t - 2) + 6$

19. _____

20. _____

21. $3 - (x + 2) = 5 + 3(x + 2)$

22. $13 - (2c + 2) = 2(c + 2) + 3c$

21. _____

22. _____

23. $\frac{1}{4}(8y + 4) - 17 = -\frac{1}{2}(4y - 8)$

24. $\frac{7}{4}\left(\frac{1}{2}x + \frac{4}{7}\right) = -3 - (2x + 7)$

23. _____

24. _____

EXERCISE SET 3.5 (p. 100)

Solve the equations.

1. $(x + 3)(x + 2) = 0$

2. $(x - 3)(x - 2) = 0$

3. $(x - 3)(x + 2) = 0$

4. $x(x - 5) = 0$

5. $0 = y(y + 10)$

6. $0 = \frac{1}{2}y\left(\frac{2}{3}y - 12\right)$

7. $(2x - 5)(x - 4) = 0$

8. $2x(3x - 2) = 0$

9. $(2x - 1)(x + 2) = 0$

10. $(5x + 1)(4x - 12) = 0$

11. $(7x - 14)(14x - 7) = 0$

12. $(4x - 9)(4x + 9) = 0$

13. $(3x - 9)(x + 3) = 0$

14. $0 = 5x(5x - 6)$

ANSWERS

1.

2.

3.

4.

5.

6.

7.

8.

9.

10.

11.

12.

13.

14.

ANSWERS

15.

_____

16.

_____

17.

_____

18.

_____

19.

_____

20.

_____

21.

_____

22.

_____

23.

_____

24.

_____

25.

_____

26.

_____

27.

_____

28.

_____

15. $\left(\dfrac{1}{3} - 3x\right)\left(\dfrac{1}{5} - 2x\right) = 0$

16. $2.5x(2.5x - 5) = 0$

17. $\left(\dfrac{1}{5}x - 2\right)(x - 3) = 0$

18. $(7.2x - 14.4)(8.1x - 24.3) = 0$

19. $8.3x(x - 5.3) = 0$

20. $\dfrac{7}{4}x\left(\dfrac{3}{4}x - \dfrac{1}{2}\right) = 0$

21. $0 = \left(\dfrac{1}{3}y + \dfrac{2}{3}\right)\left(\dfrac{1}{4}y - \dfrac{1}{2}\right)$

22. $0 = \left(\dfrac{7}{4}x - \dfrac{1}{12}\right)\left(\dfrac{2}{3}x - \dfrac{12}{11}\right)$

23. $(.03x - .01)(.05x - 1) = 0$

24. $\left(.01x + \dfrac{1}{10}\right)\left(.03x - \dfrac{1}{10}\right) = 0$

25. $5.1x(.001x - 1) = 0$

26. $(.07x - 14)\left(\dfrac{1}{3}x - 3\right) = 0$

27. $(50 - .01x)(5 - .1x) = 0$

28. $\left(\dfrac{1}{5}y - .2\right)\left(\dfrac{1}{4}y - .5\right) = 0$

1.

EXERCISE SET 3.6 (pp. 101–103)

Solve the following problems.

1. If 18 is subtracted from six times a certain number, the result is 96. What is the number?

2.

2. If twice a number is increased by 16, the result is $\frac{2}{5}$ of the original number. Find the original number.

3.

3. A 180 foot rope is cut into three pieces so that the second piece is twice as long as the first and the third piece is three times as long as the second. How long is each piece of rope?

4. If a number is increased by two-fifths of itself the result is 56. What is the number?

4.

5. One angle of a triangle is four times as large as another. The third angle is 60° larger than the smaller. What is the measure of each angle?

5.

6.

6. Consecutive odd integers are next to each other, such as 5 and 7. The larger is 2 plus the smaller. The sum of the two consecutive odd integers is 76. What are the integers?

7. The sum of three consecutive integers is 108. What are the integers?

7.

8. The perimeter of a rectangle is 310 meters. The length is 25 meters more than the width. Find the width and length of the rectangle.

8.

9. One angle of a triangle is four times as large as another. The third angle is 45° less than the sum of the other two angles. Find the measure of the smallest angle of the triangle.

9.

10. If 18 is added to a number and the result is doubled, the new number will be six times the original number. Find the original number.

10.

ANSWERS

EXERCISE SET 3.7 (p. 104)

1. Solve $E = IR$ for I.

2. Solve $P = 4s$ for s.

1. _____

2. _____

3. _____

3. Solve $d = rt$ for t.

4. Solve $d = rt$ for r.

4. _____

5. _____

5. Solve $p = 2l + 2w$ for w.

6. Solve $P = 2l + 2w$ for l.

6. _____

7. _____

7. Solve $I = prt$ for t.

8. Solve $I = prt$ for pr.

8. _____

9. _____

9. Solve $a = \dfrac{b}{c}x$ for x.

10. Solve $\dfrac{a}{d} = \dfrac{b}{c}x$ for x.

10. _____

11. _____

11. Solve $a = x + b$ for x.

12. Solve $a = x - b$ for x.

12. _____

ANSWERS

13. _____

14. _____

15. _____

16. _____

17. _____

18. _____

19. _____

20. _____

21. _____

22. _____

23. _____

24. _____

13. Solve $A = \frac{1}{2}bh$ for b.

14. Solve $v = \frac{3k}{t}$ for k.

15. Solve $v = \frac{3k}{t}$ for t.

(Hint: first multiply on both sides by t.)

16. Solve $b = \frac{2A}{h}$ for A.

17. Solve $b = \frac{2A}{h}$ for h.

(Hint: first multiply on both sides by h.)

18. Solve $r = \frac{d}{t}$ for t.

(Hint: first multiply on both sides by t.)

19. Solve $L = \frac{6}{E}$ for E.

(Hint: first multiply on both sides by E.)

20. Solve $A = 2\pi rh$ for h.

21. Solve $V = \frac{1}{3}bh$ for b.

22. Solve $V = \frac{1}{2}gt^2$ for t^2.

23. Solve $\frac{2S}{t} = v$ for S.

24. Solve $F = \frac{9}{5}C + 32$ for C.

NAME_____

CHAPTER 3 TEST   CLASS_____SCORE_____GRADE_____

Before taking the test *be sure* to allow yourself a day or so for review. Use the objectives listed
in the margins to guide your study. The test will evaluate your progress and aid your
preparation for a possible classroom test. Allow about an hour for the test. Remove the test
from the book. When you finish read the test analysis on the answer page at the end
of the book.

1.
_____

2.
_____

Solve the equations.

3.

1. $t - 9 = 17$ **2.** $x + 7 = 15$
_____

4.
_____

3. $a - 6 = -9$ **4.** $5x = 35$
5.
_____

5. $3x = -18$ **6.** $-7x = -28$ **6.**
_____

7.
_____

7. $3t + 7 = 2t - 5$ **8.** $3(x + 2) = 27$
8.
_____

9.

9. $\frac{1}{2}x - \frac{3}{5} = \frac{2}{5}$ **10.** $9x - 3 = 24x$
_____

10.
_____

11. $12t = 19t - 15 - 20$ **12.** $-3x + 6(x + 4) = 9$ **11.**
_____

12.
_____

13. $2(3n - 5) = -15$ **14.** $3(x - 1) - 2(x + 2) = 0$
13.
_____

14.
_____

15. $(x + 5)(x - 3) = 0$ **16.** $x(3x - 8) = 0$
15.
_____

16.
_____

17. _____

18. _____

19. _____

20. _____

21. _____

22. _____

23. _____

17. Solve $A = 2\pi rh$ for r

18. Solve $b = \dfrac{2A}{h}$ for A

19. Solve $A = \dfrac{3h}{t}$ for t

Solve the following problems.

20. If 12 is subtracted from twice a certain number, the result is 30. Find the number.

21. The sum of two consecutive odd integers is 68. Find the integers.

22. One angle of a triangle is twice as large as another. The third angle is three times as large as the first. How large is each angle?

23. The perimeter of a rectangle is 36 cm. The length is 4 cm greater than the width. Find the width and length.

4 POLYNOMIALS

4.1 WHAT ARE POLYNOMIALS?

Expressions like these are called polynomials.

$3x + 2, -7x + 8, 2x^2 + 7x - 2$

$a^4 - 3a^2 + \frac{1}{2}a - 7, b^5 + 2b^4 - 7b + 5b^6 + \frac{1}{4}$

They show additions and subtractions. Each part to be added or subtracted is a number or a number times a variable to some power.

A polynomial can also consist of just one of the parts. Thus the following are also polynomials.

$4x^2, -7a, \frac{1}{2}y^5, 6t^4$

$x, a, 5, -2, \frac{1}{4}$

Do exercise 1 at the right.

Each of the above expressions contains only one variable. Thus they are called *polynomials in one variable*. Later we shall study polynomials with more than one variable.

Polynomials can represent numbers. When we replace the variable by a number, the polynomial then represents a number.

Examples. What does each polynomial represent when x stands for 2?

a) $3x + 2$: $3 \cdot 2 + 2 = 8$
b) $2x^2 + 7x + 3$: $2 \cdot 2^2 + 7 \cdot 2 + 3 = 2 \cdot 4 + 14 + 3 = 25$

Do exercises 2 through 5 at the right.

TERMS OF A POLYNOMIAL

Recall that for any expression showing subtractions we can find an equivalent one showing only additions. (See page 70.)

Examples

a) $-5x^2 - x = -5x^2 + {}^-x$
b) $4x^5 - 2x^6 - 4x = 4x^5 + {}^-2x^6 + {}^-4x$

Do exercises 6 and 7 at the right.

You should be able to:

a) Give examples of polynomials.

b) For a given polynomial, find an equivalent one having only plus signs between its terms.

c) Given a polynomial, name its terms.

d) Given a polynomial, identify its terms.

e) Given a polynomial, collect like terms.

f) Given a polynomial, tell what it represents when the variable stands for a given number.

1. Write three polynomials.

What does each polynomial represent when x stands for 3?

2. $-4x - 7$

3. $-5x^3 + 7x + 10$

What does each polynomial represent when x stands for -2?

4. $5x + 7$

5. $2x^2 + 5x - 4$

For each expression, find an equivalent one having only plus signs between its terms.

6. $-9x^3 - 4x^5$

7. $-2x^3 + 3x^7 - 7x$

125

Identify the terms of each polynomial.

8. $3x^2 + 6x + \dfrac{1}{2}$

9. $-4y^5 + 7y^2 - 3y - 2$

10. $7x^4 - 5x^2 + 3x - 5$

Identify the like terms in each polynomial.

11. $4x^3 - 1x^3 + 2$

12. $2x^3 + 4x^2 - 5x^2$

13. $4x^4 - 9x^3 - 7x^4 + 10x^3$

Collect like terms.

14. $3x^2 + 5x^2$

15. $4x^3 - 2x^3 + 2$

16. $5x^4 + 4x^3 + 2x^4$

17. $\dfrac{1}{2}x^5 - \dfrac{3}{4}x^5 + 4x^2 - 2x^2$

When a polynomial shows only additions, the terms are the parts separated by plus signs. It is not actually necessary to write a polynomial without minus signs to determine its terms, however.

Examples. What are the terms of each polynomial?

a) $4x^3 + {}^-3x + 12 + {}^-8x^3 + 5x$

Terms: $4x^3$, $-3x$, 12, $-8x^3$, $5x$

b) $3x^4 - 2x^6 - 4x + 2$

Terms: $3x^4$, $-2x^6$, $-4x$, 2

Do exercises 8 through 10 at the left.

COLLECTING LIKE TERMS

Terms which have the same variable and the same exponent are *like* terms, or *similar* terms.

Example. Identify the like terms in this polynomial.

$4x^3 + 5x - 4x^2 + 2x^3 + 5x^2$

Like terms: $4x^3$ and $2x^3$; $-4x^2$ and $5x^2$

Do exercises 11 through 13 at the left.

We can often simplify polynomials by collecting like terms, or "combining similar terms". To do this we use the distributive laws.

Examples. Collect like terms.

a) $3x^2 + 6x + 5x^2 = (3 + 5)x^2 + 6x$ (using the distributive law of multiplication over addition)
$$= 8x^2 + 6x$$

b) $6x^3 - 2x^3 = (6 - 2)x^3$ (using the distributive law of multiplication over subtraction)
$$= 4x^3$$

c) $5x^2 + 4x^4 - 2x^2 - 2x^4 = (5 - 2)x^2 + (4 - 2)x^4$
$$= 3x^2 + 2x^4$$

Do exercises 14 through 17 at the left.

In collecting like terms we may get zero.

Example

$5x^3 - 5x^3 = (5 - 5)x^3 = 0x^3$ or 0

When this happens we need not write anything if there are other terms remaining.

Example

$3x^4 - 3x^4 + 2x^2 = 0x^4 + 2x^2 = 0 + 2x^2 = 2x^2$

Do exercises 18 through 21 at the right.

Recall that 1 is the multiplicative identity. Thus we can always multiply any term of a polynomial by 1 without changing the polynomial.

Examples

a) $5x^2 + x = 5x^2 + 1x$ (we multiplied the second term by 1)

b) $3x^4 - x^3 = 3x^4 - 1x^3$ (we multiplied the second term by 1)

In combining similar terms we sometimes need to supply a 1 so that we can use the distributive laws.

Examples. Combine similar terms.

a) $5x^2 + x^2 = 5x^2 + 1x^2$

$\qquad\qquad = (5 + 1)x^2$

$\qquad\qquad = 6x^2$

b) $5x^4 - 6x^3 - x^4 = 5x^4 - 6x^3 - 1x^4$

$\qquad\qquad\qquad = (5 - 1)x^4 - 6x^3$

$\qquad\qquad\qquad = 4x^4 - 6x^3$

Do exercises 22 through 26 at the right.

Do exercise set 4.1, p. 145.

Collect like terms.

18. $24 - 4x^3 - 24$

19. $6x^3 - 6x^3$

20. $5x^3 - 8x^5 + 8x^5$

21. $-2x^4 + 16 + 2x^4 + 9 - 3x^5$

Collect like terms.

22. $7x - x$

23. $5x^3 - x^3 + 4$

24. $3x^2 - x^2 + 7x^4$

25. $\frac{3}{4}x^3 + 4x^2 - x^3 + 7$

26. $8x^2 + 1 - x^2 + x^3 - 1 - 4x^2$

OBJECTIVES

You should be able to:

a) Arrange a polynomial in descending order.

b) Identify the degrees of the terms of a polynomial.

c) Identify the coefficients of the terms of a polynomial.

d) Tell whether a given polynomial is a monomial, binomial, trinomial, or none of these.

e) Add polynomials.

Arrange each polynomial in descending order.

27. $x + 3x^5 + 4x^3 + 5x^2 + 6x^7 - 2x^4$

28. $4x^2 - 3 + 7x^5 + 2x^3 - 5x^4$

29. $-14 + 7x^2 - 10x^5 + 14x^7$

Collect like terms and then arrange in descending order.

30. $3x^2 - 2x + 3 - 5x^2 - 1 - x$

31. $-x + \frac{1}{2} + 14x^4 - 7x - 1 - 4x^4$

32. What is the degree of each term of this polynomial?

$-6x^4 + 8x^2 - 2x + 9$

4.2 ASCENDING OR DESCENDING ORDER AND ADDITION

ASCENDING OR DESCENDING ORDER

The associative and commutative laws allow us to rearrange the terms of a polynomial.

Examples

a) $4x^5 + 4x^7 + x^2 + 2x = 4x^7 + 4x^5 + x^2 + 2x$

b) $3 + 4x^5 + 4x^2 + 5x + 3x^3 = 4x^5 + 3x^3 + 4x^2 + 5x + 3$

Note how the terms on the right have been rearranged. The term with the largest exponent is first. The term with the next largest exponent is second, and so on. Polynomials arranged like this are said to be in *descending order*.
Descending order is a standard way of writing polynomials. We will usually arrange them this way. When the opposite order is used, the order is called *ascending*. It is not incorrect to use ascending order. We just choose to consider descending order our standard way to write polynomials.

Do exercises 27 through 31 at the left.

DEGREES

A term of a polynomial may be a number times a variable to some power.

Examples

$5x^3, 8x^2, -5x^4, 7x$ (in $7x$, x is to the first power)

A term may also be just a number. For example, 5, −9, or 3 may be terms of a polynomial.

The *degree* of a term is its exponent. For example, in the polynomial $8x^4 + 3x + 7$

the degree of $8x^4$ is 4

the degree of $3x$ is 1   (Recall $x = x^1$)

the degree of 7 is 0   (We could name 7 as $7x^0$)

The degree of a polynomial is its largest exponent.

Do exercise 32 at the left.

COEFFICIENTS

In the polynomial

$$3x^4 - 4x^3 + 7x^2 + x - 8$$

the *coefficient* of the first term is 3

the *coefficient* of the second term is -4

the *coefficient* of the third term is 7

the *coefficient* of the fourth term is 1

the *coefficient* of the fifth term is -8

Do exercise 33 at the right.

Sometimes coefficients are zero. In this case we usually do not write the term. We may also say that a term with a zero coefficient is a *missing term*. For example, in the polynomial

$$8x^5 - 2x^3 + 5x + 7$$

There is no term with x^4. We can say "the x^4 term (or the *fourth degree* term) is missing", or "the x^4 term has a *zero coefficient*".
There is no term with x^2. We can say "the x^2 term (or the *second degree* term) is missing", or the "x^2 term has a *zero coefficient*."

We could write terms with zero coefficients. For example,

$$3x^2 + 9 = 3x^2 + 0 \cdot x + 9$$

It is just shorter not to write terms with zero coefficients.
Polynomials with just one term are called *monomials*. Polynomials with just two terms are called *binomials*. Those with just three terms are called trinomials. Some examples are below.

Examples

Monomials	*Binomials*	*Trinomials*
$4x^2$	$2x + 4$	$3x^3 + 4x + 7$
9	$3x^5 + 6x$	$6x^7 - 7x^2 + 4$
$-23x^{19}$	$-9x^7 - 6$	$4x^2 - 6x - \frac{1}{2}$

Do exercises 34 through 41 at the right.

33. What is the coefficient of each term of this polynomial?

$$5x^9 + 6x^3 + x^2 - x + 4$$

Use this list of polynomials to do exercises 34–41.

a) $2x^3 + 4x^2 - 2$

b) $-3x^4$

c) $x^3 + 1$

d) $x^4 - x^2 + 3x + \frac{1}{4}$

34. For a), b), c) and d), which terms are missing?

35. Rewrite a), c) and d), putting the "missing" terms with zero coefficients.

36. Which of the above are binomials?

37. Which of the above are trinomials?

38. Which of the above are monomials?

39. What is the coefficient of the third degree term in a)?

40. What is the coefficient of the fourth degree term in b)?

41. What is the coefficient of the second degree term in d)?

Add.

42. $3x^2 + 2x - 2$ and $-2x^2 + 5x + 5$.

43. $-4x^5 + 3x^3 + 4$ and $7x^4 + 2x^2$

44. $31x^4 + x^2 + 2x - 1$ and

$-7x^4 + 5x^3 - 2x + 2$

45. $17x^3 - x^2 + 3x + 4$ and

$-15x^3 + x^2 - 3x - \dfrac{2}{3}$

Simplify

46. $(4x^2 - 5x + 3) + (-2x^2 + 2x - 4)$

47. $(3x^3 - 4x^2 - 5x + 3) +$

$\left(5x^3 + 2x^2 - 3x - \dfrac{1}{2}\right)$

ADDITION OF POLYNOMIALS

Suppose we want to add the polynomials $x^3 + 4$ and $2x^3 + 9$. We could write the following for the sum.

$$(x^3 + 4) + (2x^3 + 9)$$

We would like to get a polynomial in standard form for the sum. A way to do this is to drop the parentheses and collect like terms.

$$x^3 + 4 + 2x^3 + 9 = 3x^3 + 13$$

The *sum* of two polynomials can be obtained by writing a plus sign between them and then collecting like terms.

Example 1. Add $-3x^3 + 2x - 4$ and $4x^3 + 3x^2 + 2$.

$$(-3x^3 + 2x - 4) + (4x^3 + 3x^2 + 2) = (-3 + 4)x^3 + 3x^2 + 2x$$
$$+ (-4 + 2)$$
$$= x^3 + 3x^2 + 2x - 2$$

Note that we still arrange the terms in descending order.

Example 2. Add $\frac{2}{3}x^4 + 3x^2 - 2x + \frac{1}{2}$ and $-\frac{1}{3}x^4 + 5x^3 - 3x^2 + 3x - \frac{1}{2}$

$(\frac{2}{3}x^4 + 3x^2 - 2x + \frac{1}{2}) + (-\frac{1}{3}x^4 + 5x^3 - 3x^2 + 3x - \frac{1}{2}) =$

$(\frac{2}{3} - \frac{1}{3})x^4 + 5x^3 + (3 - 3)x^2 + (-2 + 3)x + (\frac{1}{2} - \frac{1}{2}) =$

$\frac{1}{3}x^4 + 5x^3 + x$

Do exercises 42 through 45 at the left.

To simplify an expression like the one in the following example, we add the polynomials.

Example 3. Simplify.

$(3x^2 - 2x + 2) + (5x^3 - 2x^2 + 3x - 4) =$

$5x^3 + (3 - 2)x^2 + (-2 + 3)x + (2 - 4) =$

$5x^3 + x^2 + x - 2$

Do exercises 46 and 47 at the left.

Do exercise set 4.2, p. 147.

4.3 PROPERTIES OF ADDITION AND SUBTRACTION

Compare (a) and (b) below.

a) $(2x^2 + x + 4) + (3x - 9) = 2x^2 + (1 + 3)x + (4 - 9)$
$$= 2x^2 + 4x - 5$$

b) $(3x - 9) + (2x^2 + x + 4) = 2x^2 + (3 + 1)x + (-9 + 4)$
$$= 2x^2 + 4x - 5$$

Do exercises 48 and 49 at the right.

You have probably noticed that we can add polynomials in any order. Compare (c) and (d) below.

c) $(2x + 3) + [(4x^2 + 1) + (-x^2 + x)] = (2x + 3) + [(4 - 1)x^2 + x + 1]$
$$= (2x + 3) + [3x^2 + x + 1]$$
$$= 3x^2 + 3x + 4$$

d) $[(2x + 3) + (4x^2 + 1)] + (-x^2 + x) = [4x^2 + 2x + 4] + (-x^2 + x)$
$$= (4 - 1)x^2 + (2 + 1)x + 4$$
$$= 3x^2 + 3x + 4$$

Do exercises 50 and 51 at the right.

You have probably noticed that when adding polynomials we can group them in any manner.

Addition of polynomials is commutative and associative.

ADDITIVE INVERSES

Look for a pattern.

a) $2x + {}^-2x = 0$ b) $-6x^2 + 6x^2 = 0$

c) $(5x^3 + 2) + (-5x^3 - 2) = 0$

d) $(7x^3 - 6x^2 - x + 4) + (-7x^3 + 6x^2 + x - 4) = 0$

Do exercises 52 through 54 at the right.

You should be able to:

a) Find the additive inverse of a polynomial by replacing each coefficient by its inverse.

b) Find an equivalent expression for an additive inverse like $^-(3x + 2)$, by replacing each coefficient by its inverse.

c) Subtract polynomials.

Add.

48. $(x^3 + 3x^2) +$

$\quad (-3x^3 + 2x^2 - 2)$

49. $(-3x^3 + 2x^2 - 2) + (x^3 + 3x^2)$

Add.

50. $(x + 1) + \left[\left(x^2 + \dfrac{1}{4}\right) + (-4x^2 - 3)\right]$

51. $\left[(x + 1) + \left(x^2 + \dfrac{1}{4}\right)\right] + (-4x^2 - 3)$

Add.

52. $^-3x + 3x$

53. $^-9x^3 + 9x^3$

54. $(5x^4 + 2x^3) + (-5x^4 - 2x^3)$

Find the additive inverse of each polynomial.

55. $12x^4 - 3x^2 + 4x$

56. $-4x^4 + 3x^2 - 4x$

57. $-13x^6 + 2x^4 - 3x^2 + x - \dfrac{5}{13}$

58. $-7x^3 + 2x^2 - x + 3$

Rename each of the following.

59. $^-(4x^3 - 6x + 3)$

60. $^-(5x^4 + 3x^2 + 7x - 5)$

61. $^-\left(14x^{10} - \dfrac{1}{2}x^5 + 5x^3 - x^2 + 3x\right)$

Perhaps you noticed that polynomials have additive inverses. To find the additive inverse of a polynomial replace each coefficient by its additive inverse.

The additive inverse of any polynomial is found by replacing each coefficient by its additive inverse.

Example 1. Find the additive inverse of

$$4x^5 - 7x - 8 + \frac{5}{6}.$$

Inverse: $-4x^5 + 7x + 8 - \dfrac{5}{6}$

Do exercises 55 through 58 at the left.

SYMBOLS FOR ADDITIVE INVERSES

The additive inverse of $8x^2 - 4x + \frac{1}{4}$ is the polynomial obtained by replacing each coefficient by its additive inverse. Thus the inverse is

$$-8x^2 + 4x - \frac{1}{4}$$

We can also represent the additive inverse of $8x^2 + {}^-4x + \frac{1}{4}$ like this

$$^-\left(8x^2 - 4x + \frac{1}{4}\right)$$

Thus

$$^-\left(8x^2 - 4x + \frac{1}{4}\right) = -8x^2 + 4x - \frac{1}{4}$$

Example 2. Rename: $^-\left(-7x^4 - \dfrac{5}{9}x^3 + 8x^2 - x + 67\right)$

$$7x^4 + \frac{5}{9}x^3 - 8x^2 + x - 67$$

Do exercises 59 through 61 at the left.

SUBTRACTION OF POLYNOMIALS

Recall that we can subtract rational numbers by adding an additive inverse. This also works for polynomials. (See page 72.)

Example 3

$(9x^5 + x^3 - 2x^2 + 4) - (2x^5 + x^4 - 4x^3 - 3x^2) =$

$(9x^5 + x^3 - 2x^2 + 4) + {}^-(2x^5 + x^4 - 4x^3 - 3x^2) =$

$(9x^5 + x^3 - 2x^2 + 4) + (-2x^5 - x^4 + 4x^3 + 3x^2) =$

$7x^5 - x^4 + 5x^3 + x^2 + 4$

Do exercises 62 through 64 at the right.

Perhaps you noticed that we might skip some steps by mentally thinking through finding the inverse of the subtrahend. Consider this example:

$(9x^5 + x^3 - 2x) - (-2x^5 + 5x^3 + 6)$

We know we have to add the inverse of the subtrahend. This inverse can be found by taking the inverse of each term. We try to do this mentally.

Think: The inverse of $-2x^5$ is $2x^5$. This added to $9x^5$ gives $11x^5$.

Think: The inverse of $5x^3$ is $-5x^3$. This added to x^3 gives $-4x^3$.

Think: There is no x term in the subtrahend, so we have just $-2x$.

Think: The additive inverse of 6 is -6. There is no term to add this to so in the answer we have -6.

Thus we have for the difference

$11x^5 - 4x^3 - 2x - 6$

Such thinking can speed up our work.

Do exercises 65 through 69 at the right.

Do exercise set 4.3, p. 151.

Subtract.

62. $(5x^2 + 4) - (2x^2 + 1)$

63. $(7x^3 + 2x + 4) - (5x^3 - 4)$

64. $(-3x^2 + 5x - 4) -$
$(-4x^2 + 11x - 2)$

Subtract.

65. $(-6x^4 + 3x^2 + 6)$
$-(2x^4 + 5x^3 - 5x^2 + 7)$

66. $(-2x + 5) - (2x^2 + 3)$

67. $(3x^2 - 2) - (2x^6 - 5x^4 + 5x^2 - 5)$

68. $\left(\frac{3}{2}x^3 - \frac{1}{2}x^2 + .3\right)$
$-\left(\frac{1}{2}x^3 + \frac{1}{2}x^2 + \frac{4}{3}x + .2\right)$

69. $\left(-\frac{1}{3}x^2 - \frac{4}{5}x + 1.7\right)$
$-\left(-\frac{1}{3}x^2 + \frac{2}{5}x + .4\right)$

You should be able to add and subtract polynomials using columns.

70. Add

$$-2x^3 + 5x^2 - 2x + 4$$
$$x^4 \qquad\quad + 6x^2 + 7x - 10$$
$$-9x^4 + \quad 6x^3 + x^2 \qquad\quad - 2$$

71. Add $-3x^3 + 5x + 2$, and $x^3 + x^2 + 5$, and $x^3 - 2x - 4$

Subtract the second polynomial from the first. Use columns.

72. $4x^3 + 2x^2 - 2x - 3$,

$\quad 2x^3 - 3x^2 + 2$

73. $2x^3 + x^2 - 6x + 2$,

$\quad x^5 + 4x^3 - 2x^2 - 4x$

4.4 ADDING AND SUBTRACTING IN COLUMNS

When we add polynomials we can write a plus sign between them and then collect like terms. We do not always have to do it that way. It is sometimes easier to write the polynomials one under the other. When doing this, we write like terms in a column, leaving spaces for missing terms.

Example 1. Add $9x^5 - 2x^3 + 6x^2 + 3$, and $5x^4 - 7x^2 + 6$, and

$3x^6 - 5x^5 + x^2 + 5$

Arrange the polynomials with like terms in columns.

$$9x^5 \qquad\quad - 2x^3 + 6x^2 + 3$$
$$5x^4 \qquad\quad - 7x^2 + 6$$
$$3x^6 - 5x^5 \qquad\qquad\quad + x^2 + 5$$
$$\overline{3x^6 + 4x^5 + 5x^4 - 2x^3 \qquad\quad + 14}$$

Do exercises 70 and 71 at the left.

To subtract polynomials using columns we write the polynomial to be subtracted underneath. Recall that we are to add the inverse of the subtrahend. To find the inverse we replace each coefficient by its inverse. You can do this mentally or, if helpful, by writing plus or minus signs like the shaded ones in the following example.

Example 2. Subtract $9x^2 - 5x - 3$ from $5x^2 - 3x + 6$.

$$5x^2 \qquad -3x \quad + \quad 6$$
$$\boxed{-} \qquad\quad \boxed{+} \qquad\qquad \boxed{+}$$
$$9x^2 \qquad -5x \qquad -3$$
$$\overline{-4x^2 \quad + \quad 2x \quad + \quad 9}$$

Do exercises 72 and 73 at the left.

Do exercise set 4.4, p. 153.

4.5 MULTIPLICATION OF POLYNOMIALS

MONOMIALS

Consider the product of $3x$ and $4x$. We use parentheses to show the multiplication.

$(3x)(4x) = (3 \cdot 4)(x \cdot x)$ (by the commutative and associative laws)

$\qquad\quad = 12x^2$

Consider the product of $3x$ and $-x$

$(3x)(-x) = (3x)(-1x)$

$\qquad\quad = -3x^2$

Consider the product of $-7x^5$ and $4x^3$.

$(-7x^5)(4x^3) = (-7 \cdot 4)(x^5 \cdot x^3)$

$\qquad\qquad\quad = -28x^8$

Multiply.

74. $3x$ and -5

75. $-x$ and x

76. $-x$ and $-x$

Do exercises 74 through 76 at the right.

Perhaps you have discovered how we multiply monomials. We multiply their coefficients. Then we use properties of exponents to get just one variable symbol.

To multiply two monomials such as $2x^3$ and $5x^2$, we multiply the coefficients, and then use properties of exponents. We get $2 \cdot 5 x^{3+2}$, or $10x^5$.

Example. $(2x^4)(-5x^2) = (2)(-5)x^{4+2}$

$\qquad\qquad\qquad = -10x^6$

Do exercises 77 through 81 at the right.

Multiply.

77. $-x^2$ and x^3

78. $3x^5$ and $4x^2$

79. $4x^5$ and $-2x^6$

80. $-7x^4$ and $-x$

81. $7x^5$ and 0

Multiply.

82. $4x$ and $2x + 4$

83. $3x^2$ and $-5x + 2$

Multiply.

84. $x + 8$ and $x + 5$

85. $(x + 5)(x - 4)$

OTHER POLYNOMIALS

The distributive laws are the basis for multiplying polynomials other than monomials. Let's see how this works. We will use parentheses to show the multiplication.

Example 1. Multiply $2x$ and $5x + 3$.

$$(2x)(5x + 3) = (2x)(5x) + (2x)(3) \quad \text{(by the distributive law)}$$
$$= 10x^2 + 6x \quad\quad \text{(multiplying the monomials)}$$

Do exercises 82 and 83 at the left.

Sometimes we have to use the distributive law more than once.

Example 2. Multiply $x + 5$ and $x + 4$.

$$(x + 5)(x + 4) = \underset{a}{(x + 5)x} + \underset{b}{(x + 5)4} \quad \text{(distributive law)}$$

Now let's consider the two parts *a* and *b* separately.

$$a \quad (x + 5)x = xx + 5x \quad\quad \text{(distributive law)}$$
$$= x^2 + 5x \quad\quad \text{(multiplying the monomials)}$$
$$b \quad (x + 5)4 = (x)4 + (5)4 \quad \text{(distributive law)}$$
$$= 4x + 20 \quad\quad \text{(multiplying the monomials)}$$

Now we replace the parts *a* and *b* in the original expression with their answers.

$$(x + 5)(x + 4) = (x^2 + 5x) + (4x + 20)$$
$$= x^2 + 9x + 20 \quad \text{(collecting like terms and arranging in descending order)}$$

Do exercises 84 and 85 at the left.

Example 3. Multiply $4x + 3$ and $x - 2$.

$$(4x + 3)(x - 2) = \underset{a}{(4x + 3)x} + \underset{b}{(4x + 3)(^-2)} \quad \text{(distributive law)}$$

(Continued on next page)

Again let's consider the two parts *a* and *b* separately.

a $(4x + 3)x = (4x)x + (3)x$ (distributive law)

$\quad\quad\quad = 4x^2 + 3x$ (multiplying the monomials)

b $(4x + 3)(-2) = 4x(-2) + 3(-2)$ (distributive law)

$\quad\quad\quad\quad = -8x - 6$ (multiplying the monomials)

Now we replace the parts *a* and *b* in the original expression with their answers.

$(4x + 3)(x - 2) = (4x^2 + 3x) + (-8x - 6)$

$\quad\quad\quad = 4x^2 - 5x - 6$ (collecting like terms and arranging in descending order)

Do exercises 86 and 87 at the right.

Example 4. Find the product $(x^2 + 2x - 3)(x^2 + 4)$

$(x^2 + 2x - 3)(x^2 + 4) = (x^2 + 2x - 3)(x^2) + (x^2 + 2x - 3)(4)$

$\quad\quad\quad\quad\quad\quad\quad\quad\quad a \quad\quad\quad\quad\quad\quad\quad\quad b$

Consider the two parts *a* and *b* separately.

a $(x^2 + 2x - 3)x^2 = (x^2)(x^2) + 2x(x^2) - 3(x^2)$

$\quad\quad\quad\quad = x^4 + 2x^3 - 3x^2$

b $(x^2 + 2x - 3)4 = (x^2)4 + (2x)4 + (-3)4$

$\quad\quad\quad\quad = 4x^2 + 8x - 12$

Now we replace the parts *a* and *b* in the original expression with their answers.

$(x^2 + 2x - 3)(x^2 + 4) = (x^4 + 2x^3 - 3x^2) + (4x^2 + 8x - 12)$

$\quad\quad\quad\quad\quad\quad = x^4 + 2x^3 + x^2 + 8x - 12$

Do exercises 88 and 89 at the right.

Multiply.

86. $5x + 3$ and $x - 4$

87. $(2x - 3)(3x - 5)$

Find the products.

88. $(x^2 + 3x - 4)(x^2 + 5)$

89. $(2x^3 - 2x + 5)(3x^2 - 7)$

Note that parentheses are sometimes needed. To show a product we use parentheses around both polynomials like this

$(9x^5 + 3)(x - 4)$

Without the parentheses we would have

$9x^5 + 3x - 4$

and this does not have the same meaning. For a monomial and a binomial we need parentheses, for example

$4x^2(3x + 2)$

For two monomials we do not need parentheses. For example

$(3x^2)(7x)$ and $3x^27x$

mean the same thing. (We could, of course, simplify $3x^27x$ to $21x^3$).

We usually write parentheses only when we need them. It is not incorrect to use parentheses when they are not needed, but it is incorrect to omit them when they are needed.

You probably discovered in the previous exercises that multiplying two polynomials amounts to multiplying each term of one polynomial by every term of the other polynomial.

To multiply two polynomials (where at least one of them is not a monomial), multiply each term of one by every term of the other. Then add the results.

There are many ways to write symbols to find products, just as we had several ways to find sums. One way is to write one polynomial under the other. Then we multiply each term at the top by every term of the bottom one. Then add the results and collect like terms. To help in adding we can write like terms in columns.

Example 1. Multiply $4x^2 - 2x + 3$ by $x + 2$.

$4x^2 - 2x + 3$

$\underline{x + 2}$

$4x^3 - 2x^2 + 3x$ (multiplying the top row by x)

$\underline{8x^2 - 4x + 6}$ (multiplying the top row by 2)

$4x^3 + 6x^2 - x + 6$ (adding)

Multiply.

90. $3x^2 - 2x + 4$ and $x + 5$

Example 2. Multiply $5x^3 - 3x + 4$ by $-2x^2 - 3$.

$5x^3 - 3x + 4$

$\underline{-2x^2 - 3}$

$-10x^5 + 6x^3 - 8x^2$ (multiplying by $-2x^2$)

$\underline{- 15x^3 + 9x - 12}$ (multiplying by -3)

$-10x^5 - 9x^3 - 8x^2 + 9x - 12$ (adding)

Note how we left space for "missing terms". This helps in adding.

91. $(-5x^3 + 4x^2 + 2)(-4x^2 - 8)$

Do exercises 90 and 91 at the right.

In the exercises you will check some examples to see that multiplication of polynomials is commutative and associative. The polynomial 1 is the multiplicative identity. All of these properties are like the rational numbers. There is a distinguishing property. Most polynomials do not have polynomials for multiplicative inverses.

Do exercise set 4.5, p. 157.

OBJECTIVES

You should be able to:

a) Multiply a monomial and a polynomial mentally.

b) Find the product of two binomials mentally.

Multiply mentally. Just write the answer.

92. $4x(2x^2 - 3x + 4)$

93. $2x^3(5x^3 + 4x^2 - 5x)$

Multiply. Look for patterns.

94. $(x + 2)(x^3 + 4)$

95. $(x^3 + 2)(x + 1)$

4.6 SPECIAL PRODUCTS OF POLYNOMIALS

Now we want to speed up our work by learning some patterns. You have probably discovered a quick way to find the product of a monomial and a binomial, such as

$$5x(2x^2 - 3x + 4)$$

You can use the distributive law mentally and just write the answer.

$$10x^3 - 15x^2 + 20x$$

Do exercises 92 and 93 at the left.

PRODUCTS OF TWO BINOMIALS

We use the distributive law three times when we multiply two binomials. It turns out that we multiply each term of one binomial by every term of the other. As you do these exercises look for other patterns.

Do exercises 94 and 95 at the left.

Usually the product of two binomials will have four terms, although in some cases we can collect some terms. In any case we can always write four terms. Perhaps you also noticed that we can think mentally of such products as follows:

$$(a + b)(c + d) = a \cdot c + a \cdot d + b \cdot c + b \cdot d$$

We can multiply a by c and a by d. Then we multiply b by c and b by d. Sometimes like terms occur. If they do we collect them mentally.
We can also explain the multiplication this way: Multiply the first terms, the outside terms, the inside terms, and the last terms. This rule is sometimes abbreviated FOIL.

Example 1. $(x + 8)(x + 5) = x^2 + 5x + 8x + 40$

$$= x^2 + 13x + 40$$

Example 2. $(x + 6)(x - 6) = x^2 - 6x + 6x - 36$

$$= x^2 - 36$$

Example 3. $(x + 3)(x - 2) = x^2 - 2x + 3x - 6$

$$= x^2 + x - 6$$

Do exercises 96 through 102 at the right.

When two binomials have no terms higher than first degree and contain the same variable, we can always collect some like terms. Then the product will not have four terms when written in the standard way.

Example 4. $(x + 5)(x - 3) = x^2 - 3x + 5x - 15$

$$= x^2 + 2x - 15$$

Example 5. $(3x + 4)(2x + 7) = 6x^2 + 21x + 8x + 28$

$$= 6x^2 + 29x + 28$$

When we multiply the first terms we get the first term of the product. When we multiply the "outside" terms and the "inside" terms and then add, we get the second term of the product. These can always be "collected". When we multiply the second terms we get the last term of the product. This thinking can help speed up our work.

Do exercises 103 through 107 at the right.

Do exercise set 4.6, p. 159.

Multiply mentally. Just write the answer.

96. $(x + 3)(x + 4)$

97. $(x + 3)(x - 5)$

98. $(2x + 1)(x + 4)$

99. $(2x^2 - 3)(x - 2)$

100. $(6x^2 + 5)(2x^3 + 1)$

101. $(x^3 + 7)(x^3 - 7)$

102. $(2x^4 + x^2)(-x^3 + x)$

Multiply mentally. Just write the answer.

103. $(x + 3)(x - 2)$

104. $(3x + 4)(x - 2)$

105. $(x - 2)(x + 8)$

106. $(4x - 5)(2x - 1)$

107. $(4x + 3)(-x - 2)$

You should be able to:

a) Multiply the sum and difference of two terms mentally.
b) Square a binomial mentally.
c) Find special products such as those in a) and b) and those in the previous section, mentally.

Multiply.

108. $(x + 5)(x - 5)$

109. $(2x - 3)(2x + 3)$

Multiply mentally.

110. $(x + 2)(x - 2)$

111. $(x - 7)(x + 7)$

112. $(3x + 5)(3x - 5)$

113. $(2x^3 - 1)(2x^3 + 1)$

4.7 MORE SPECIAL PRODUCTS

MULTIPLYING THE SUM AND DIFFERENCE OF TWO TERMS

Look for a pattern in the following.

a) $(x + 2)(x - 2) = x^2 - 2x + 2x - 4$

$$= x^2 - 4$$

b) $(3x - 5)(3x + 5) = 9x^2 + 15x - 15x - 25$

$$= 9x^2 - 25$$

Do exercises 108 and 109 at the left.

Perhaps you discovered the following.

The product of the sum and difference of two terms is the difference of their squares.

$$(a + b)(a - b) = a^2 - b^2$$

Example 1. $(x + 4)(x - 4) = x^2 - 4^2 = x^2 - 16$

Example 2. $(2x + 5)(2x - 5) = (2x)^2 - 5^2 = 4x^2 - 25$

Example 3. $(3x^2 - 7)(3x^2 + 7) = (3x^2)^2 - 7^2 = 9x^4 - 49$

Example 4. $(-4x - 10)(-4x + 10) = (-4x)^2 - 10^2 = 16x^2 - 100$

Do exercises 110 through 113 at the left.

SQUARING BINOMIALS

In this special product we multiply a binomial by itself. This is also called "squaring a binomial". Look for a pattern in what follows.

a) $(x + 2)(x + 2) = x^2 + 2x + 2x + 4 = x^2 + 4x + 4$

b) $(3x + 5)(3x + 5) = 9x^2 + 15x + 15x + 25 = 9x^2 + 30x + 25$

c) $(x - 2)(x - 2) = x^2 - 2x - 2x + 4 = x^2 - 4x + 4$

d) $(3x - 5)(3x - 5) = 9x^2 - 15x - 15x + 25 = 9x^2 - 30x + 25$

Do exercises 114 through 118 at the right.

Perhaps you discovered a quick way to square a binomial.

The square of a binomial is the sum of the square of the first term, twice the product of the terms, and the square of the last term.

$(a + b)^2 = a^2 + 2 \cdot a \cdot b + b^2$

$(a - b)^2 = a^2 - 2 \cdot a \cdot b + b^2$

Example 1. $(x + 3)^2 = x^2 + 2 \cdot x \cdot 3 + 3^2 = x^2 + 6x + 9$

Example 2. $(x - 5)^2 = x^2 - 2 \cdot x \cdot 5 + 5^2 = x^2 - 10x + 25$

Example 3. $(2x + 7)^2 = (2x)^2 + 2 \cdot 2x \cdot 7 + 7^2 = 4x^2 + 28x + 49$

Example 4. $(3x^2 - 5x)^2 = (3x^2)^2 - 2 \cdot 3x^2 \cdot 5x + (5x)^2$
$$= 9x^4 - 30x^3 + 25x^2$$

Do exercises 119 through 124 at the right.

Multiply.

114. $(x + 3)(x + 3)$

115. $(2x - 4)(2x - 4)$

116. $(x - 5)(x - 5)$

117. $(4x + 2)(4x + 2)$

118. $(3x - 4)(3x - 4)$

Multiply mentally.

119. $(x + 2)^2$

120. $(x - 4)^2$

121. $(2x + 5)^2$

122. $(4x^2 - 3x)^2$

123. $(x + 9)(x + 9)$

124. $(3x^2 - 5)(3x^2 - 5)$

Multiply mentally.

125. $(x + 5)(x + 6)$

126. $(x - 4)(x + 4)$

127. $4x^2(-2x^3 + 5x^2 + 10)$

128. $(9x^2 + 1)^2$

129. $(2x - 5)(2x + 8)$

130. $\left(2x - \dfrac{1}{2}\right)^2$

Now that you have learned how to multiply quickly certain kinds of polynomials, try doing it when several kinds are mixed together. You should first look to see what kind of problem you have. Then you should use the best method of multiplying.

Example 5.

$(x + 3)(x - 3) = x^2 - 9$ — (this is the product of a sum and a difference of the same two terms.)

Example 6.

$(x + 7)(x - 5) = x^2 + 2x - 35$ — (multiply each term of one by every term of the other and then combine.)

Example 7.

$(x + 7)(x + 7) = x^2 + 14x + 49$ — (the square of a binomial)

Example 8.

$2x^3(9x^2 + x - 7) = 18x^5 + 2x^4 - 14x^3$ — (the product of a monomial by a trinomial. Multiply each term of the trinomial by the monomial.)

Example 9.

$(3x^2 - 7x)^2 = 9x^4 - 42x^3 + 49x^2$ — (the square of a binomial)

Do exercises 125 through 130 at the left.

Do exercise set 4.7, p. 161.

NAME

CLASS

ANSWERS

EXERCISE SET 4.1 (pp. 125–127)

What does each polynomial represent when x stands for 4?

1. $-5x + 2$ **2.** $x^2 - 3x + 1$ **3.** $x^3 + 5x^2 + x$

4. $-5 - 6x + x^2$ **5.** $2x^2 - 5x + 7$ **6.** $7 + x + 3x^2$

What does each polynomial represent when x stands for -1?

7. $3x + 5$ **8.** $6 - 2x$ **9.** $x^2 - 2x + 1$

10. $5x - 6 + x^2$ **11.** $3x^3 - 2x^2 - 1$ **12.** $5x^4 - 3x^2 + 6x - 1$

For each expression, find an equivalent one having only plus signs between its terms.

13. $7x - 1$ **14.** $6x - {}^-3$

15. $-7x - \dfrac{2}{3}$ **16.** $-3x^3 + 7x^2 - 3x - 2$

17. $2 - 3x - {}^-x^2$ **18.** $2x^3 - {}^-3x - 2$

1.

2.

3.

4.

5.

6.

7.

8.

9.

10.

11.

12.

13.

14.

15.

16.

17.

18.

ANSWERS

19. _____

_____

20. _____

_____

21. _____

_____

22. _____

_____

23. _____

_____

24. _____

_____

25. _____

_____

26. _____

_____

27. _____

28. _____

29. _____

30. _____

31. _____

32. _____

33. _____

34. _____

35. _____

36. _____

37. _____

Identify the terms in each expression.

19. $5x^2 - 6x - 3$ **20.** $-4x^4 + 6x - 1$ **21.** $-2 - 3x + 5x^4$

Identify the like terms in each polynomial.

22. $5x^3 + 6x^2 - 3x^3$ **23.** $-3 + x - 2 - 5x$

24. $x^4 - x^4 + 5x^4 - x^3$ **25.** $x^5 - 6x^3 + 4x^5 - 3x^4 + 7x^3$

26. $1 - 5x^2 - 3 + 7x^2 + 8$

Collect like terms.

27. $5x^3 + 6x^3 + 4$ **28.** $6x^4 - 3x^4 + 7$

29. $-3x^4 - 6x^5 + 7x^4 - 4$ **30.** $5x - 7x$

31. $6x^4 - 2x^4 + 5$ **32.** $5 + 6x - 4 - 7x$

33. $-4x^3 + 2x^2 - 3x^2 + 5$ **34.** $8x + 2x^2 - 6x + 3x^2$

35. $\frac{1}{4}x^5 - 5 + \frac{1}{2}x^5 - 2x$ **36.** $1 - 2x^3 + 8x - 3 + 7x^3 + 7x$

37. $\frac{1}{5}x^4 + \frac{1}{5} - 2x^2 + \frac{1}{10} - \frac{3}{15}x^4 + 2x^2$

NAME

CLASS

ANSWERS

EXERCISE SET 4.2 (pp. 128–130)

Arrange each polynomial in descending order.

1. $x^5 + x + 6x^3 + 1 + 2x^2$ **2.** $3 + 2x^4 - 5x^6 - 2x^2 + 3x$

3. $5x^3 + 15x^9 + x - x^2 + 7x^8$ **4.** $9x - 5 + 6x^3 - 5x^4 + x^5$

Collect like terms and then arrange in descending order.

5. $3x^4 - 5x^6 - 2x^4 + 6x^6$ **6.** $-1 + 5x^3 - 3 - 7x^3 + x^4 + 5$

7. $-x + \dfrac{3}{4} + 15x^4 - x - \dfrac{1}{2} - 3x^4$ **8.** $-5x^7 - 3x^5 + x^4 + 3 - 6x^5 - x^4$

What is the degree of each term of each polynomial?

9. $-7x^3 + 6x^2 + 3x + 7$ **10.** $8x - 3x^2 + 9 - 8x^3$

What is the coefficient of each term of each polynomial.

11. $6x^3 + 7x^2 - 8x - 2$ **12.** $-2 + 8x - 3x^2 + 6x^3 - 5x^4$

ANSWERS

1.

2.

3.

4.

5.

6.

7.

8.

9.

10.

11.

12.

13. _____

14. _____

15. _____

16. _____

17. _____

18. _____

19. _____

20. _____

21. _____

22. _____

Use this list of polynomials to do exercises 13–22.

a) $5x^3 - 6x^2 - 3$

d) $x^2 + 1$

b) $x^5 + x$

e) $7x^4 - 7x^3 + 6x^2 - 5x + 2$

c) $-6x^3$

f) $5x^2 + 6x + 1$

13. For a), b), c) and d) which terms are missing?

14. Rewrite a), b), c) and d) putting in the missing terms with zero coefficients.

15. Which of the above are monomials?

16. Which of the above are binomials?

17. Which of the above are trinomials?

18. What is the coefficient of the second term in a)?

19. What is the coefficient of the fourth term in e)?

20. What is the coefficient of the third term in f)?

21. What is the degree of the polynomial in c)?

22. What is the degree of the polynomial in e)?

NAME _____

CLASS _____

ANSWERS

Add:

23. $3x + 2$ and $-4x + 3$

24. $5x^2 + 6x + 1$ and $-7x + 2$

25. $-6x + 2$ and $x^2 + x - 3$

26. $6x^4 + 3x^3 - 1$ and

$4x^2 - 3x + 3$

27. $3x^5 + 6x^2 - 1$ and

$7x^2 + 6x - 2$

28. $7x^3 + 3x^2 + 6x$ and

$-3x^2 - 6$

29. $-4x^4 + 6x^2 - 3x - 5$ and

$6x^3 + 5x + 9$

30. $5x^3 + 6x^2 - 3x + 1$ and

$5x^4 - 6x^3 + 2x - 5$

31. $(7x^3 + 6x^2 + 4x + 1) +$

$(-7x^3 + 6x^2 - 4x + 5)$

32. $(3x^4 - 5x^2 - 6x + 5) +$

$(-4x^3 + 6x^2 + 7x - 1)$

23. _____

_____

24. _____

_____

25. _____

_____

26. _____

_____

27. _____

_____

28. _____

_____

29. _____

_____

30. _____

_____

31. _____

_____

32. _____

_____

Add.

33. _____

34. _____

35. _____

36. _____

37. _____

38. _____

39. _____

40. _____

33. $5x^4 - 6x^3 - 7x^2 + x - 1$

and $4x^3 - 6x + 1$

34. $8x^5 - 6x^3 + 6x + 5$ and

$-4x^4 + 3x^3 - 7x$

35. $9x^8 - 7x^4 + 2x^2 + 5$ and

$8x^7 + 4x^4 - 2x$

36. $4x^5 - 6x^3 - 9x + 1$ and

$6x^3 + 9x^2 + 9x$

37. $\frac{1}{4}x^4 + \frac{2}{3}x^3 + \frac{5}{8}x^2 + 7$ and

$-\frac{3}{4}x^4 + \frac{3}{8}x^2 - 7$

38. $.02x^5 - .2x^3 + x + .08$ and

$-.01x^5 + x^4 - .8x - .02$

39. $\left(\frac{1}{3}x^9 + \frac{1}{5}x^5 - \frac{1}{2}x^2 + 7\right) +$

$\left(-\frac{1}{5}x^9 + \frac{1}{4}x^4 - \frac{3}{5}x^5 + \frac{3}{4}x^2 + \frac{1}{2}\right)$

40. $(.03x^6 + .05x^3 + .22x + .05) +$

$\left(\frac{7}{100}x^6 - \frac{3}{100}x^3 + .5\right)$

NAME

CLASS

ANSWERS

EXERCISE SET 4.3 (pp. 131–133)

Add.

1. $(x^4 + 5x^2) + (-5x^4 + 2x^2 - 2)$ **2.** $(-5x^2 + 3x - 4) + (2x^2 - 5)$

3. $(-x^5 + 4x^3 - 5) + (6x^5 + 2x^2 - 6)$

4. $(6x^4 + 2x) + (-3x^4 - 2x^2 - 2x + 1)$

5. $[(x + 5) + (x^2 - 2)] + (-x + 3)$

6. $(x + 3) + [(x^3 - 2x) + (x - 2)]$

7. $[(x^2 - 2) + (3x^3 - 5)] + [(2x^3 + x) + (x^3 + 7)]$

8. $(x^5 - x^3 - 3x) + [-2x^5 + (-2x^3 + 2x)]$

9. $5x - 5x$ **10.** $6x^4 - 6x^4$ **11.** $(6x^3 + 2x) + (-6x^3 - 2x)$

Find the additive inverse of each of the following.

12. $12x^4 - 3x^3 + 3$ **13.** $4x^3 - 6x^2 - 8x + 1$ **14.** $5x^5 - \frac{1}{3}x^3 + 2x - \frac{5}{4}$

ANSWERS
1.
2.
3.
4.
5.
6.
7.
8.
9.
10.
11.
12.
13.
14.

ANSWERS

Rename each of the following.

15. _____

15. $^-(4x^2 - 3x + 2)$

16. $^-(-6x^4 + 2x^3 + 9x - 1)$

16. _____

17. $^-\left(-4x^3 - 6x^2 + \dfrac{3}{4}x - 8\right)$

17. _____

Subtract.

18. _____

18. $(5x^2 + 6) - (3x^2 - 8)$

19. _____

19. $(7x^3 - 2x^2 + 6) - (7x^2 + 2x - 4)$

20. _____

20. $(6x^5 - 3x^4 + x + 1) - (8x^5 + 3x^4 - 1)$

21. _____

21. $\left(\dfrac{1}{2}x^2 - \dfrac{3}{2}x + 2\right) - \left(\dfrac{3}{2}x^2 + \dfrac{1}{2}x - 2\right)$

22. _____

22. $(6x^2 + 2x) - (-3x^2 - 7x + 8)$

23. _____

23. $7x^3 - (-3x^2 - 2x + 1)$

24. _____

24. $\left(\dfrac{5}{8}x^3 - \dfrac{1}{4}x - \dfrac{1}{3}\right) - \left(-\dfrac{1}{8}x^3 + \dfrac{1}{4}x - \dfrac{1}{3}\right)$

25. _____

25. $\left(\dfrac{1}{5}x^3 + 2x^2 - .1\right) - \left(-\dfrac{2}{5}x^3 + 2x^2 + .01\right)$

26. _____

26. $(.08x^3 - .02x^2 + .01x) - (.02x^3 + .03x^2 - 1)$

27. _____

27. $(.8x^4 + .2x - 1) - \left(\dfrac{7}{10}x^4 + \dfrac{1}{5}x - .1\right)$

EXERCISE SET 4.4 (p. 134)

Add.

1. $-3x^4 + 6x^2 + 2x - 1$
   $ - 3x^2 + 2x + 1$
   _____

2. $-4x^3 + 8x^2 + 3x - 2$
   $ - 4x^2 + 3x + 2$
   _____

3. $3x^5 - 6x^3 + 3x$
   $ - 3x^4 + 3x^3 + x^2$
   _____

4. $4x^5 - 5x^3 + 2x$
   $ - 4x^4 + 2x^3 + 2x^2$
   _____

5. $ - 3x^2 + x$
   $5x^3 - 6x^2 + 1$
   $ 3x - 8$
   _____

6. $ - 4x^2 + 2x$
   $3x^3 - 5x^2 + 3$
   $ 5x - 5$
   _____

7. $-\frac{1}{2}x^4 - \frac{3}{4}x^3 + 6x$
   $\phantom{-\frac{1}{2}x^4} \frac{1}{2}x^3 + x^2 + \frac{1}{4}x$
   $\frac{3}{4}x^4 + \frac{1}{2}x^2 + \frac{1}{2}x + \frac{1}{4}$
   _____

8. $-\frac{1}{4}x^4 - \frac{1}{2}x^3 + 2x$
   $\phantom{-\frac{1}{4}x^4} \frac{3}{4}x^3 - x^2 + \frac{1}{2}x$
   $\frac{1}{2}x^4 + \frac{1}{2}x^2 + \frac{1}{2}x + \frac{1}{2}$
   _____

9. $ - 4x^2$
   $4x^4 - 3x^3 + 6x^2 + 5x$
   $ 6x^3 - 8x^2 + 1$
   $-5x^4$
   $ 6x^2 - 3x$
   _____

10. $ 3x^2$
    $5x^4 - 2x^3 + 4x^2 + 5x$
    $ 5x^3 - 5x^2 + 2$
    $-7x^4$
    $ 3x^2 - 2x$
    _____

11. $3x^4 - 6x^2 + 7x$
    $ 3x^2 - 3x + 1$
    $-2x^4 + 7x^2 + 3x$
    $ 5x - 2$
    _____

12. $5x^4 - 8x^2 + 4x$
    $ 5x^2 - 2x + 3$
    $-3x^4 + 3x^2 + 5x$
    $ 3x - 5$
    _____

NAME

CLASS

ANSWERS

1. _____

2. _____

3. _____

4. _____

5. _____

6. _____

7. _____

8. _____

9. _____

10. _____

11. _____

12. _____

ANSWERS

13.

_____

14.

_____

15.

_____

16.

_____

17.

_____

18.

_____

19.

_____

20.

_____

13.

$$
\begin{array}{l}
3x^5 - 6x^4 + 3x^3 \qquad\quad\ - 1 \\
\qquad\ 6x^4 - 4x^3 + 6x^2 \\
3x^5 \qquad\quad\ + 2x^3 \\
\qquad\ - 6x^4 \qquad\quad - 7x^2 \\
-5x^5 \qquad\quad + 3x^3 \qquad\quad + 2 \\
\hline
\end{array}
$$

14.

$$
\begin{array}{l}
4x^5 - 3x^4 + 2x^3 \qquad\quad\ - 2 \\
\qquad\ 6x^4 + 5x^3 + 3x^2 \\
5x^5 \qquad\quad\ + 4x^3 \\
\qquad\ - 6x^4 \qquad\quad - 5x^2 \\
-3x^5 \qquad\quad + 2x^3 \qquad\quad + 5 \\
\hline
\end{array}
$$

15.

$$
\begin{array}{l}
\quad\ - x^3 + 6x^2 + 3x + 5 \\
x^4 \qquad\ - 3x^2 \qquad + 2 \\
\qquad\qquad\qquad - 5x + 3 \\
6x^4 \qquad\ + 4x^2 \qquad - 1 \\
\quad\ - x^3 \qquad\quad + 6x \\
\hline
\end{array}
$$

16.

$$
\begin{array}{l}
\quad\ - 2x^3 + 3x^2 + 5x + 3 \\
x^4 \qquad\ - 5x^2 \qquad + 1 \\
\qquad\qquad\qquad - 7x + 4 \\
4x^4 \qquad\ + 6x^2 \qquad - 2 \\
\quad\ - x^3 \qquad\quad + 5x \\
\hline
\end{array}
$$

17.

$$
\begin{array}{l}
\quad\ - 3x^4 + 6x^3 - 6x^2 + 5x + 1 \\
5x^5 \qquad\quad - 3x^3 \qquad\quad - 5x \\
\qquad\ 4x^4 + 7x^3 \qquad\quad + 3x + 1 \\
-2x^5 \qquad\qquad\qquad + 7x^2 \qquad - 8 \\
\hline
\end{array}
$$

18.

$$
\begin{array}{l}
\quad\ - 5x^4 + 4x^3 - 7x^2 + 3x + 2 \\
3x^5 \qquad\quad - 7x^3 \qquad\quad - 6x \\
\qquad\ 3x^4 + 5x^3 \qquad\quad + 5x - 3 \\
-5x^5 \qquad\qquad\qquad + 10x^2 \qquad + 4 \\
\hline
\end{array}
$$

19.

$$
\begin{array}{l}
.15x^4 + .10x^3 - .9x^2 \\
\qquad\ - .01x^3 + .01x^2 + x \\
1.25x^4 \qquad\qquad + .11x^2 \qquad + .01 \\
\qquad\quad .27x^3 \qquad\qquad\qquad + .99 \\
-.35x^4 \qquad\qquad + 15x^2 \qquad - .03 \\
\hline
\end{array}
$$

20.

$$
\begin{array}{l}
.05x^4 + .12x^3 - .5x^2 \\
\qquad\ - .02x^3 + .02x^2 + 2x \\
1.5x^4 \qquad\qquad + .01x^2 \qquad + .15 \\
\qquad\quad .25x^3 \qquad\qquad\qquad + .85 \\
-.25x^4 \qquad\qquad + 10x^2 \qquad - .04 \\
\hline
\end{array}
$$

21. Add $6x^4 - 3x^2 + 9x - 2$ and $7x^4 + 2x^2 - 3$ and $-3x^3 + 6x + 1$

21.

_____

22. Add $5x^4 - 2x^2 + 3x - 3$ and $8x^4 + 4x^2 - 5$ and $-5x^3 + 7x + 2$

22.

_____

23. Add $3x^3 - 6x^2 + 2x - 1$ and $6x^2 - 3x + 1$ and $-2x + 1$ and $4x^4 - 7x^2 + 5$.

23.

_____

24.

24. Add $5x^3 - 3x^2 + 4x - 2$ and $8x^2 - 4x + 1$ and $-3x + 4$ and $6x^4 - 3x^2 + 8$

_____

25.

25. Add $-4x^5 + 6x^3 - 2x^2 + x$ and $3x^5 - 6x^4 - 2x^2 + \frac{1}{2}$ and

$2x^2 - \frac{3}{2}x - \frac{3}{4}$

_____

ANSWERS

26. _____

Subtract the second polynomial from the first.

26. $3x^2 - 6x + 1$ $\qquad\qquad$ $6x^2 + 8x - 3$

_____

27. _____

27. $5x^4 + 6x^3 - 9x^2 + 1$ $\qquad\qquad$ $-6x^3 + 8x + 9$

_____

28. _____

28. $5x^5 + 6x^2 - 3x + 6$ $\qquad\qquad$ $6x^3 + 7x^2 - 8x - 9$

_____

29. _____

29. $3x^4 + 6x^2 + 8x - 1$ $\qquad\qquad$ $4x^5 - 6x^4 - 8x - 7$

_____

30. _____

30. $6x^5 + 3x^2 - 7x + 2$ $\qquad\qquad$ $10x^5 + 6x^3 - 5x^2 - 2x + 4$

EXERCISE SET 4.5 (pp. 135–139)

Multiply.

1. $3x$ and -4

2. $4x$ and 5

3. $6x^2$ and 7

4. $-4x$ and -3

5. $-5x$ and -6

6. $3x^2$ and -7

7. x^2 and $-2x$

8. $-x^3$ and $-x$

9. x^4 and x^2

10. $-x^5$ and x^3

11. $3x^4$ and $2x^2$

12. $-\dfrac{1}{5}x^3$ and $-\dfrac{1}{3}x$

13. $-4x^4$ and 0

14. $7x^5$ and x^5

15. $-.1x^6$ and $.2x^4$

Multiply.

16. $3x$ and $-x + 5$

17. $2x$ and $4x - 6$

18. $4x^2$ and $3x + 6$

19. $-6x^2$ and $x^2 + x$

20. $3x^2$ and $6x^4 + 8x^3$

21. $4x^4$ and $x^2 - 6x$

Multiply.

22. $(x + 6)(x + 3)$

23. $(x + 6)(-x + 2)$

24. $(x + 3)(x - 3)$

ANSWERS
1.
2.
3.
4.
5.
6.
7.
8.
9.
10.
11.
12.
13.
14.
15.
16.
17.
18.
19.
20.
21.
22.
23.
24.

ANSWERS

25. $(x - 5)(2x - 5)$

26. $(2x + 5)(2x + 5)$

25. _____

26. _____

27. $(3x - 5)(3x + 5)$

28. $(3x + 1)(3x + 1)$

27. _____

28. _____

29. $\left(2x - \dfrac{1}{2}\right)\left(x + \dfrac{3}{2}\right)$

30. $(2x + .1)(3x - .1)$

29. _____

30. _____

31. $(x^2 + 6x + 1)(x + 1)$

32. $(2x^3 + 6x + 1)(2x + 1)$

31. _____

32. _____

33. $(-5x^2 - 7x + 3)(2x + 1)$

34. $(2x^3 - 5x + 6)(2x + 2)$

33. _____

34. _____

35. $(3x^2 - 6x + 2)(x^2 - 3)$

36. $(x^2 + 6x - 1)(-3x^2 + 2)$

35. _____

36. _____

NAME

CLASS

ANSWERS

EXERCISE SET 4.6 (pp. 140–141)

Multiply. Just write the answer.

1. $4x(x + 1)$ **2.** $3x(x + 2)$ **3.** $-3x(x - 1)$

4. $-5x(-x - 1)$ **5.** $x^2(x^3 + 1)$ **6.** $-2x^3(x^2 - 1)$

7. $3x(2x^2 - 6x + 1)$ **8.** $-4x(2x^3 - 6x^2 - 5x + 1)$

Multiply.

9. $(x + 1)(x^2 + 3)$ **10.** $(x^2 - 3)(x - 1)$

11. $(x^3 + 2)(x + 1)$ **12.** $(x^4 + 2)(x + 12)$

Multiply mentally. Just write the answer.

13. $(x + 2)(x - 3)$ **14.** $(x + 2)(x + 2)$

15. $(3x + 2)(3x + 3)$ **16.** $(4x + 1)(2x + 2)$

17. $(5x - 6)(x + 2)$ **18.** $(x - 8)(x + 8)$

19. $(3x - 1)(3x + 1)$ **20.** $(2x + 3)(2x + 3)$

21. $(4x - 2)(x - 1)$ **22.** $(2x - 1)(3x + 1)$

23. $\left(x - \dfrac{1}{4}\right)\left(x + \dfrac{1}{4}\right)$ **24.** $\left(x + \dfrac{3}{4}\right)\left(x + \dfrac{3}{4}\right)$

1.
2.
3.
4.
5.
6.
7.
8.
9.
10.
11.
12.
13.
14.
15.
16.
17.
18.
19.
20.
21.
22.
23.
24.

ANSWERS

25. _____

26. _____

27. _____

28. _____

29. _____

30. _____

31. _____

32. _____

33. _____

34. _____

35. _____

36. _____

37. _____

38. _____

39. _____

40. _____

41. _____

42. _____

43. _____

44. _____

45. _____

46. _____

47. _____

48. _____

25. $(x - .1)(x + .1)$

26. $(3x^2 + 1)(x + 1)$

27. $(2x^2 + 6)(x + 1)$

28. $(2x^2 + 3)(2x - 1)$

29. $(-2x + 1)(x + 6)$

30. $(3x + 4)(2x - 4)$

31. $(x + 7)(x + 7)$

32. $(2x + 5)(2x + 5)$

33. $(1 + 2x)(1 - 3x)$

34. $(-3x - 2)(x + 1)$

35. $(x^2 + 3)(x^3 - 1)$

36. $(x^4 - 3)(2x + 1)$

37. $(x^2 - 2)(x - 1)$

38. $(x^3 + 2)(x - 3)$

39. $(3x^2 - 2)(x^4 - 2)$

40. $(x^{10} + 3)(x^{10} - 3)$

41. $(3x^5 + 2)(2x^2 + 6)$

42. $(1 - 2x)(1 + 3x^2)$

43. $(8x^3 + 1)(x^3 + 8)$

44. $(4 - 2x)(5 - 2x^2)$

45. $(4x^2 + 3)(x - 3)$

46. $(7x - 2)(2x - 7)$

47. $(4x^4 + x^2)(x^2 + x)$

48. $(5x^6 + 3x^3)(2x^6 + 2x^3)$

NAME

CLASS

ANSWERS

EXERCISE SET 4.7 (pp. 142–144)

Multiply mentally.

1. $(x + 4)(x - 4)$ **2.** $(x + 1)(x - 1)$

3. $(2x + 1)(2x - 1)$ **4.** $(x^2 + 1)(x^2 - 1)$

5. $(5x - 2)(5x + 2)$ **6.** $(3x^4 + 2)(3x^4 - 2)$

7. $(2x^2 + 3)(2x^2 - 3)$ **8.** $(6x^5 - 5)(6x^5 + 5)$

9. $(3x^4 - 4)(3x^4 + 4)$ **10.** $(x^2 - .2)(x^2 + .2)$

11. $(x^6 - 1x^2)(x^6 + 1x^2)$ **12.** $(2x^3 - .3)(2x^3 + .3)$

13. $(x^4 + 3x)(x^4 - 3x)$ **14.** $\left(\dfrac{3}{4} + 2x^3\right)\left(\dfrac{3}{4} - 2x^3\right)$

15. $(x^{12} - 3)(x^{12} + 3)$ **16.** $(12 - 3x^2)(12 + 3x^2)$

17. $(2x^8 + 3)(2x^8 - 3)$ **18.** $\left(x - \dfrac{2}{3}\right)\left(x + \dfrac{2}{3}\right)$

19. $(x + 2)^2$ **20.** $(2x - 1)^2$

21. $(3x^2 + 1)^2$ **22.** $\left(3x + \dfrac{3}{4}\right)^2$

23. $(x - \frac{1}{2})^2$ **24.** $\left(2x - \dfrac{1}{5}\right)^2$

1.	
2.	
3.	
4.	
5.	
6.	
7.	
8.	
9.	
10.	
11.	
12.	
13.	
14.	
15.	
16.	
17.	
18.	
19.	
20.	
21.	
22.	
23.	
24.	

ANSWERS

25. _____

26. _____

27. _____

28. _____

29. _____

30. _____

31. _____

32. _____

33. _____

34. _____

35. _____

36. _____

37. _____

38. _____

39. _____

40. _____

41. _____

42. _____

43. _____

44. _____

45. _____

46. _____

47. _____

48. _____

25. $(3 + x)^2$

26. $(x^3 - 1)^2$

27. $(x^2 + 1)^2$

28. $(8x - x^2)^2$

29. $(2 - 3x^4)^2$

30. $(6x^3 - 2)^2$

31. $(5 + 6x^2)^2$

32. $(3x^2 - x)^2$

33. $(3 - 2x^3)^2$

34. $(x - 4x^3)^2$

35. $4x(x^2 + 6x - 3)$

36. $8x(-x^5 + 6x^2 + 9)$

37. $\left(2x^2 - \frac{1}{2}\right)\left(2x^2 - \frac{1}{2}\right)$

38. $(-x^2 + 1)^2$

39. $\left(\frac{3}{4}x + 1\right)\left(\frac{3}{4}x + 2\right)$

40. $(-3x + 2)(3x + 2)$

41. $(6x^3 - 1)(2x^2 + 1)$

42. $-6x^2(x^3 + 8x - 9)$

43. $(6x^4 + 4)^2$

44. $(3x - 5)(x + 2)$

45. $(3x + 2)(4x^2 + 5)$

46. $(2x^2 - 7)(3x^2 + 9)$

47. $(8 - 6x^4)^2$

48. $\left(\frac{1}{5}x^2 + 9\right)\left(\frac{3}{5}x^2 - 7\right)$

CHAPTER 4 TEST

NAME_____

CLASS_____SCORE_____GRADE_____

ANSWERS

Before taking the test, *be sure* to allow yourself a day or so for review. Use the objectives listed in the margins to guide your study. The test will evaluate your progress and aid your preparation for a possible classroom test. Allow about an hour for the test. Remove the test from the book. When you finish, read the test analysis on the answer page at the end of the book.

In the given expression identify a) the degree of the term, and b) the coefficient of the term.

1. $-7x$ **2.** x^5 **3.** 4 **4.** x

Collect like terms and then arrange in descending order.

5. $3x^2 - 5x + 7x - 8x^2 + 1$ **6.** $\frac{2}{5}x + 7 - 12x - \frac{3}{5} + \frac{1}{5}(x^3 - 32)$

Add.

7. $3x^4 - x^3 \qquad + x - 4$ **8.** $3x^5 - 4x^4 + x^3 \qquad - 3$
    $\quad x^5 \qquad + 7x^3 - 3x^2 \qquad - 5$ $3x^4 - 5x^3 + 3x^2$
    $\quad - 5x^4 \qquad + 6x^2 - x$ $4x^5 \qquad + 4x^3$
    $2x^5 + 2x^4 - 5x^3 - 8x^2 + x - 7$ $-5x^5 \qquad - 5x^2$
    $\overline{\qquad\qquad\qquad\qquad\qquad}$ $\quad - 5x^4 + 2x^3 \qquad + 5$

Subtract.

9. $4x^3 - 3x^2 + 6x - 5$ **10.** $3x^5 - 4x^4 \qquad + 2x^2 + 3$
    $5x^4 - x^3 + 5x^2 \qquad + 7$ $2x^5 - 4x^4 + 3x^3 + 4x^2 - 5$

Simplify.

11. $2(x^3 + 1) - 5(x - 2) + x(x^2 + x + 2)$

12. $3x^2 - 7(x + 6) + \frac{2}{3}(x^2 + 3x - 6)$

13. $\frac{1}{2}\left(x^2 - \frac{1}{2}\right) + 2\left(\frac{1}{4}x^2 - \frac{1}{2}x\right)$

14. $\frac{3}{4}x^2 - \frac{1}{4}(2x + 12) + \frac{1}{4}(x^2 + 8x - 4)$

1. a)

   b)

2. a)

   b)

3. a)

   b)

4. a)

   b)

5.

6.

7.

8.

9.

10.

11.

12.

13.

14.

ANSWERS

15. _____

16. _____

17. _____

18. _____

19. _____

20. _____

21. _____

22. _____

23. _____

24. _____

25. _____

26. _____

27. _____

28. _____

29. _____

30. _____

Simplify.

15. $.01(x^2 - .5) + .3(.1x^2 - x)$

16. $.2(x^3 - .3x^2 + .5) - .3(.2x^2 + 3x - 2)$

17. Add $3x^2 - 4x + 2$ and $-2x^3 - 3x^2 + 4x + 5$ and $x^3 + x^2 + x - 3$

18. Add $-x^5 + 7x^3 + 2x - 5$ and $2x^4 - x^3 + 3x^2 + 2$ and $5x^3 - 2x^2 + 5x - 3$

Subtract the second polynomial from the first.

19. $3x^2 + 7,$ $5x^2 - 4x + 1$

20. $3x^4 - 2x^2 - 3,$ $x^4 + 3x^3 + 2x^2 + x - 3$

Multiply.

21. $-6x^2(3x^2 - 5x - 2)$ **22.** $(3x - 2)(4x^2 - 5x + 1)$

23. $(2x + 3)(x - 7)$ **24.** $(7x - 1)^2$

25. $(2x^2 + 3)(3x^3 - 5)$ **26.** $(2x^3 - 7)(2x^3 + 7)$

27. $(2x + 3)(3x^2 - 2x - 1)$ **28.** $(3x + 2)^2$

29. $(5x - 7)(3x + 2)$ **30.** $5x^4(3x^3 - 8x^2 + 10x + 2)$

5 POLYNOMIALS AND FACTORING

5.1 FACTORING POLYNOMIALS

Factoring is the reverse of multiplication. So to learn to factor quickly, we study the quick methods of multiplication. Factoring is important for doing many things in algebra, for example solving equations.

FACTORING MONOMIALS

To factor a monomial we find two monomials whose product is that monomial.

Compare

Multiply	*Factor*
a) $(4x)(5x) = 20x^2$	$20x^2 = (4x)(5x)$
b) $(2x)(10x) = 20x^2$	$20x^2 = (2x)(10x)$
c) $(-4x)(-5x) = 20x^2$	$20x^2 = (-4x)(-5x)$
d) $(x)(20x) = 20x^2$	$20x^2 = (x)(20x)$

The monomial $20x^2$ thus has many factorizations. There are still other ways to factor $20x^2$.

Do exercises 1 and 2 at the right.

A way to start factoring a monomial is to factor the coefficient and then shift some of the letters to one factor and some to the other.

Example. Find three factorizations of $15x^3$.

a) $15x^3 = (3 \cdot 5)x^3 = (3x)(5x^2)$

b) $15x^3 = (3 \cdot 5)x^3 = (3x^2)(5x)$

c) $15x^3 = (1 \cdot 15)x^3 = (x)(15x^2)$

Do exercises 3 through 5 at the right.

FACTORING WHEN TERMS HAVE A COMMON FACTOR

Recall that to multiply a monomial and a polynomial which has more than one term, we use the distributive law. We multiply each term by the monomial (see page 70). We factor by doing the reverse.

You should be able to:

a) Find several factorizations of a given monomial.

b) Factor polynomials when the terms have a common factor.

c) Factor by grouping, expressions with four terms, like $x^2 + 3x + 4x + 12$.

1. a) Multiply $(3x)(4x)$

   b) Factor. $12x^2$

2. a) Multiply. $(2x)(8x^2)$

   b) Factor. $16x^3$

Find three factorizations of each monomial.

3. $8x^4$

4. $15x^2$

5. $6x^5$

6. a) Multiply $3(x + 2)$

b) Factor. $3x + 6$

7. a) Multiply.

$2x(x^2 + 5x + 4)$

b) Factor.

$2x^3 + 10x^2 + 8x$

Factor.

8. $x^2 + 3x$

9. $3x^2 + 6x$

10. $3x^6 - 5x^3 + 2x^2$

11. $20x^3 + 12x^2 - 16$

12. $9x^4 - 15x^3 + 3x^2$

13. $22x^3 + 11x^2 - 33x - 44$

14. $20x^4 + 30x^3 - 20x^2 + 50x$

15. $\frac{3}{4}x^3 + \frac{5}{4}x^2 + \frac{7}{4}x + \frac{1}{4}$

16. $35x^7 - 49x^6 + 14x^5 - 63x^3$

Compare.

Multiply

a) $5(x + 3) = 5x + 15$

b) $3x(x^2 + 2x - 4) = 3x^3 + 6x^2 - 12x$

Factor

$5x + 15 = 5 \cdot x + 5 \cdot 3$
$\qquad\qquad = 5(x + 3)$

$3x^3 + 6x^2 - 12x =$
$3x \cdot x^2 + 3x \cdot 2x - 3x \cdot 4 =$
$3x(x^2 + 2x - 4)$

Do exercises 6 and 7 at the left.

The trick in doing this kind of factoring is to find a factor common to all the terms. There may not always be one other than 1. In some cases there is more than one common factor. Then we use the one with the largest possible coefficient and the largest exponent. In this way we "factor completely".

Example 1. Factor.

$3x^2 + 6 = 3 \cdot x^2 + 3 \cdot 2 = 3(x^2 + 2)$

Example 2. Factor.

$5x^4 + 20x^3 = 5x^3 \cdot x + 5x^3 \cdot 4 = 5x^3(x + 4)$

Example 3. Factor.

$16x^3 + 20x^2 = (4x^2)(4x) + 4x^2 \cdot 5 = 4x^2(4x + 5)$

Example 4. Factor.

$15x^5 - 12x^4 + 27x^3 - 3x^2 =$

$(3x^2)(5x^3) - (3x^2)(4x^2) + (3x^2)(9x) - (3x^2)(1) =$

$3x^2(5x^3 - 4x^2 + 9x - 1)$

If you can spot the common factor without factoring each term, you can just write the answer.

Do exercises 8 through 16 at the left.

FACTORING BY GROUPING

Consider this expression.

$x^2 + 3x + 4x + 12$

There is no common factor other than 1. But, let's look at the parts of the expression $x^2 + 3x$ and $4x + 12$. We can factor $x^2 + 3x$.

$x^2 + 3x = x(x + 3)$

We can also factor $4x + 12$.

$4x + 12 = 4(x + 3)$

Then

$x^2 + 3x + 4x + 12 = x(x + 3) + 4(x + 3)$

Now we can use the distributive law, like this

$x(x + 3) + 4(x + 3) = (x + 4)(x + 3)$

Thus we factored the expression $x^2 + 3x + 4x + 12$ by using the distributive law with the parts. Then we used the distributive law again with the common binomial factor. We can factor many expressions this way. We must use trial and error to see which terms are to be grouped together.

Example 1. Factor.

$$\begin{aligned} x^2 + 7x + 2x + 14 &= (x^2 + 7x) + (2x + 14) \\ &= x(x + 7) + 2(x + 7) \\ &= (x + 2)(x + 7) \end{aligned}$$

Example 2. Factor.

$$\begin{aligned} x^2 + 3x - 4x - 12 &= (x^2 + 3x) + (-4x - 12) \\ &= x(x + 3) - 4(x + 3) \\ &= (x - 4)(x + 3) \end{aligned}$$

Example 3. Factor.

$$\begin{aligned} 5x^2 - 10x + 2x - 4 &= (5x^2 - 10x) + (2x - 4) \\ &= 5x(x - 2) + 2(x - 2) \\ &= (5x + 2)(x - 2) \end{aligned}$$

Example 4. Factor.

$$\begin{aligned} 2x^2 - 3x + 12x - 18 &= 2x^2 + 12x - 3x - 18 \\ &= (2x^2 + 12x) + (-3x - 18) \\ &= 2x(x + 6) - 3(x + 6) \\ &= (2x - 3)(x + 6) \end{aligned}$$

Do exercises 17 through 22 at the right.

Do exercise set 5.1, p. 179.

Factor.

17. $x^2 + 5x + 2x + 10$

18. $x^2 - 4x + 3x - 12$

19. $x^2 + 5x - 4x - 20$

20. $5x^2 - 10x + 2x - 4$

21. $2x^2 + 8x - 3x - 12$

22. $16x^2 + 20x - 12x - 15$

OBJECTIVES

You should be able to factor trinomials by examining the coefficients of the first and last terms.

Factor.

23. $x^2 + 7x + 12$

24. $x^2 + 12x + 35$

25. $x^2 - x - 2$

5.2 FACTORING TRINOMIALS

To see how to factor trinomials (polynomials with three terms), consider this product.

$$(x + 2)(x + 5) = x^2 + 2x + 5x + 10$$
$$= x^2 + 7x + 10$$

Note that the coefficient 7 is the sum of 2 and 5, and the 10 is the product of 2 and 5. In general,

$$(x + a)(x + b) = x^2 + (a + b)x + ab$$

When we are factoring we can use this equation in reverse. That is,

$$x^2 + (a + b)x + ab = (x + a)(x + b)$$

Now suppose we are trying to factor a polynomial such as

$$x^2 + 5x + 6$$

We think of pairs of integers whose product is 6 and whose sum is 5.

Pairs of factors of 6	Sum of factors
1, 6	7
$-1, -6$	-7
2, 3	5
$-2, -3$	-5

Thus the desired integers are 2 and 3. Then

$$x^2 + 5x + 6 = (x + 2)(x + 3)$$

You can always check by multiplying. You should try to perform the above process mentally to speed up your work.

Example 1. Factor $x^2 - 8x + 12$.

We look for two numbers whose product is 12 and whose sum is -8. They are -2 and -6.

$$x^2 - 8x + 12 = (x - 2)(x - 6)$$

Do exercises 23 through 25 at the left.

On the preceding page we were working with equations of the type

$x^2 + ax + b$

Now let's consider trinomials for which the coefficient of the first term is not 1. Note that

$(2x + 5)(3x + 4) = 6x^2 + 8x + 15x + 20$
$\qquad\qquad\quad = 6x^2 + 23x + 20$

The 6 is the product of 2 and 3 (the coefficients of the first terms).
The 20 is the product of 5 and 4 (the coefficients of the last terms).
The 23 is the product of the inside coefficients plus the product of the outside coefficients.
In general, to factor $ax^2 + bx + c$, we think of two binomials with blanks like this

$(\underline{\quad}x + \underline{\quad})(\underline{\quad}x + \underline{\quad})$

The numbers in the first blanks must have the product a.
The numbers in the last blanks must have the product c.
The product of the inside ones plus the product of the outside ones must be b.

When the coefficient of the first term is not 1, the first thing we do is look for a factor common to all three terms. If there is one, we remove that common factor before proceeding. This will be illustrated in Example 3, on the following page.

Example 2. Factor $3x^2 + 5x + 2$.

We first look for a factor common to all the terms. There is none, other than 1. Next we look for two numbers whose product is 3. These are

1,   3
$-1, -3$

We have these possibilities:

$(x + \quad)(3x + \quad)$ or $(-x + \quad)(-3x + \quad)$

Now we look for numbers whose product is 2. These are

1,   2
$-1, -2$ (both positive or both negative)

Then we have these as some of the possiblities of factorizations.

a) $(x + 1)(3x + 2)$ c) $(-x + 1)(-3x + 2)$

b) $(x - 1)(3x - 2)$ d) $(-x - 1)(-3x - 2)$

Factor.

26. $6x^2 + 7x + 2$

27. $6x^2 + 15x + 9$

28. $2x^2 + x - 6$

Factor.

29. $6x^2 + 17x + 7$

30. $4x^2 + 2x - 6$

31. $6x^2 - 5x + 1$

Now we multiply each of these. We must get $3x^2 + 5x + 2$. We see that either (a) or (d) gives the desired factorization. We prefer to have the first coefficients positive when that is possible. Thus the factorization we choose is $(x + 1)(3x + 2)$.

Example 3. Factor $8x^2 + 8x - 6$.

We first look for a factor common to all three terms. The number 2 is a common factor, so we begin by factoring it out.

$2(4x^2 + 4x - 3)$

We continue by factoring the trinomial in parentheses.

We look for pairs of numbers whose product is 4. These are

4, 1   and   2, 2   (both positive)

We then have these possibilities

$(4x + \ )(x + \ )$ and $(2x + \ )(2x + \ )$

Next we look for pairs of numbers whose product is -3. They are

3, -1 and -3, 1   (one positive and one negative)

Then we have these as some possibilities for factorizations.

$(4x + 3)(x - 1)$
$(4x - 1)(x + 3)$
$(4x + 1)(x - 3)$
$(4x - 3)(x + 1)$
$(2x + 3)(2x - 1)$

We multiply to see if we get $4x^2 + 4x - 3$, and find that the desired factorization is

$(2x + 3)(2x - 1)$.

But don't forget the common factor we removed at the beginning. We must supply it in order to have a factorization of the original polynomial.

$8x^2 + 8x - 6 = 2(2x + 3)(2x - 1)$

Do exercises 26 through 28 on the preceding page, 29 through 31 at the left.

Note that when we factor trinomials *we must use trial and error*. As you practice, you will find that you can make better and better guesses, however. Don't forget, when factoring any polynomials, *always look first for common factors!*

Do exercise set 5.2, p. 181.

5.3 TRINOMIAL SQUARES, DIFFERENCES OF TWO SQUARES

TRINOMIAL SQUARES

We can use the equation for squaring a binomial (see page 143) in reverse to factor trinomials which are squares.

$$a^2 + 2 \cdot a \cdot b + b^2 = (a + b)^2$$

$$a^2 - 2 \cdot a \cdot b + b^2 = (a - b)^2$$

We must first be able to recognize when a trinomial is a square (also called a *trinomial square*). Note, from the above equations, the following:

1. Two of the terms are squares (a^2 and b^2). Thus they must be terms such as 4 or x^2 or $25x^4$ or $16x^2$. There must be no minus signs preceding these terms.

2. If we multiply a and b (the square roots of these terms) and double the result, we get the third term, or the additive inverse of that term.

Examples

a) The trinomial $x^2 + 6x + 9$ is a square because:

Two of the terms are squares, x^2 and 9, and

If we multiply the square roots, x and 3, and then double the result, we get the third term, $6x$.

b) The trinomial $x^2 + 6x + 11$ is not a square because only one of the terms is a square.

c) The trinomial $16x^2 - 56x + 49$ is a square because:

Two of the terms are squares, $16x^2$ and 49, and

If we multiply the square roots, $4x$ and 7, and then double the result, we get the additive inverse of the third term.

Do exercise 32 at the right.

To factor a trinomial square we use the above equations. From these we see that we should use the square roots of the squared terms and the sign of the third term.

Example 1. Factor $x^2 + 6x + 9$.

$$x^2 + 6x + 9 = x^2 + 2 \cdot 3 \cdot x + 3^2 = (x + 3)^2$$

Example 2. Factor $x^2 - 14x + 49$.

$$x^2 - 14x + 49 = x^2 - 2 \cdot 7 \cdot x + 7^2 = (x - 7)^2$$

Example 3. Factor $16x^2 - 40x + 25$.

$$16x^2 - 40x + 25 = (4x)^2 - 2 \cdot (4x) \cdot 5 + 5^2 = (4x - 5)^2$$

Do exercises 33 through 37 at the right.

OBJECTIVES

You should be able to:

a) recognize trinomial squares and differences of two squares.

b) factor trinomial squares and differences of two squares.

32. Which of the following are trinomial squares?

a) $x^2 + 8x + 16$ b) $x^2 - 10x + 25$

c) $x^2 - 12x + 4$ d) $4x^2 + 20x + 25$

e) $5x^2 - 14x + 16$ f) $16x^2 + 40x + 25$

Factor.

33. $x^2 + 2x + 1$

34. $x^2 - 2x + 1$

35. $x^2 + 4x + 4$

36. $25x^2 - 70x + 49$

37. $16x^2 - 56x + 49$

Factor and check by multiplying. (Always factor completely, even though such directions are not given.)

38. $x^2 - 9$

39. $x^2 - 64$

40. $5 - 20x^6$

41. $81x^4 - 1$

42. $49x^4 - 25x^{10}$

DIFFERENCES OF TWO SQUARES

We can use the equation $(a - b)(a + b) = a^2 - b^2$ (see page 142) in reverse to factor polynomials which are differences of two squares. The reversed equation is

$$a^2 - b^2 = (a - b)(a + b)$$

To recognize a difference of two squares, note that both terms must be squares, such as $4x^2$ or 9 or $25x^2$. There *must be a minus sign between them.* When we factor a difference of two squares we get the product of a sum and difference of the same two terms (see the equation above).

Example 1. Factor $x^2 - 4$.

$$x^2 - 4 = x^2 - 2^2 = (x - 2)(x + 2)$$

Example 2. Factor $18x^2 - 50x^6$.

Remember, you should *always* look first for a factor common to all terms. This time there is one.

$$18x^2 - 50x^6 = 2x^2(9 - 25x^4) = 2x^2[3^2 - (5x^2)^2]$$
$$= 2x^2(3 - 5x^2)(3 + 5x^2)$$

Example 3. Factor $49x^4 - 9x^6$.

$$49x^4 - 9x^6 = x^4(49 - 9x^2) = x^4(7 - 3x)(7 + 3x)$$

FACTORING COMPLETELY

Whenever you obtain a factor that can still be factored, you should factor it. When no factor can be factored further, we say that we have *factored completely.*

Example 4. Factor $1 - 16x^{12}$.

$$1 - 16x^{12} = (1 - 4x^6)(1 + 4x^6) \qquad \text{(The first factor can be factored,}$$
$$\text{again as a difference of two squares.)}$$
$$= (1 - 2x^3)(1 + 2x^3)(1 + 4x^6)$$

Be careful *not* to try to factor a sum of two squares, because it is not true that $a^2 + b^2 = (a + b)(a + b)$.

A factorization can always be checked by multiplying the factors to see if the original polynomial is obtained. You should do this. In the exercises, remember the following:

1. Always look first for a common factor.

2. Always factor completely, even if the directions do not say so.

3. Never try to factor a sum of two squares.

Do exercises 38 through 42 at the left.

Do exercise set 5.3, p. 183.

5.4 COMPLETING THE SQUARE

We just factored expressions which were differences of two squares, where there were only two terms. There are also differences of two squares having more than two terms. This can happen if one of the squares is a trinomial.

Example 1. Factor $x^2 + 2x + 1 - 25$.

Since $x^2 + 2x + 1 = (x + 1)^2$, we have
$$x^2 + 2x + 1 - 25 = (x + 1)^2 - 25 = (x + 1 - 5)(x + 1 + 5)$$
$$= (x - 4)(x + 6)$$

Example 2. Factor $x^2 + 6x + 9 - 16$.

$$x^2 + 6x + 9 - 16 = (x + 3)^2 - 4^2$$
$$= (x + 3 - 4)(x + 3 + 4)$$
$$= (x - 1)(x + 7)$$

Do exercises 43 through 48 at the right.

Let us try to use the same method to factor $x^2 + 2x - 3$. This trinomial is not a square because the last term is not 1. Let's add zero, using $1 - 1$, like this:

$$x^2 + 2x - 3 = x^2 + 2x + (1 - 1) - 3$$
$$= (x^2 + 2x + 1) - 4$$
$$= (x + 1)^2 - 4$$
$$= (x + 1 - 2)(x + 1 + 2)$$
$$= (x - 1)(x + 3)$$

In this example we added zero, naming it $1 - 1$. Why? If we look at the first two terms of our polynomial, we see that the third term would have to be 1 for it to be a square. So we add $1 - 1$ to make a trinomial square. To use this method we need to figure out what the third term of a trinomial must be for the trinomial to be a square.

Example 3. What must be added to $x^2 + 8x$ to make it a trinomial square? To find the third term we take half the coefficient of x. Then square it. Half of 8 is 4, and $4^2 = 16$. Thus we must add 16, and $x^2 + 8x + 16$ is a square.

OBJECTIVES

You should be able to factor polynomials by completing the square.

Factor.

43. $x^2 + 2x + 1 - 16$

44. $x^2 + 2x + 1 - 49$

45. $x^2 + 4x + 4 - 25$

46. $x^2 - 2x + 1 - 64$

47. $x^2 + 8x + 16 - 9$

48. $x^2 + 16x + 64 - 100$

What must be added to make a trinomial square?

49. $x^2 + 12x$ **50.** $x^2 - 6x$

51. $x^2 + 14x$ **52.** $x^2 - 22x$

Factor by completing the square.

53. $x^2 + 2x - 3$

54. $x^2 - 6x + 8$

55. $x^2 + 6x + 8$

56. $x^2 + 10x + 21$

57. $x^2 - 8x - 9$

Example 4. What must be added to $x^2 - 10x$ to make it a trinomial square?

We take half of -10, obtaining -5. We square it to get 25. Thus 25 must be added to get the trinomial square $x^2 - 10x + 25$.

Do exercises 49 through 52 at the left.

Example 5. Factor $x^2 - 4x - 21$.

This polynomial is not a square. The third term would have to be 4. To see this we take half the coefficient of x and square it. Thus we shall add zero to the polynomial, naming it $4 - 4$.

$$
\begin{aligned}
x^2 - 4x - 21 &= x^2 - 4x + (4 - 4) - 21 \\
&= (x^2 - 4x + 4) - 4 - 21 \\
&= (x - 2)^2 - 5^2 \\
&= (x - 2 - 5)(x - 2 + 5), \quad \text{or} \quad (x - 7)(x + 3)
\end{aligned}
$$

The method of factoring in Example 5 is called *completing the square*.

Example 6. Factor $x^2 + 8x + 7$ by completing the square.

The third term of $x^2 + 8x$ will have to be 16 (half of 8 squared) to make a square. So we shall add zero, naming it $16 - 16$.

$$
\begin{aligned}
x^2 + 8x + 7 &= x^2 + 8x + (16 - 16) + 7 \\
&= x^2 + 8x + 16 - 9 \\
&= (x + 4)^2 - 3^2 \\
&= (x + 4 - 3)(x + 4 + 3), \quad \text{or} \quad (x + 1)(x + 7)
\end{aligned}
$$

Although you can factor most trinomials without completing the square, it is important that you learn to do it this way. Completing the square is an important technique, used in several different ways.

Do exercises 53 through 57 at the left.

Do exercise set 5.4, p. 185.

5.5 SOLVING EQUATIONS BY FACTORING

In a previous chapter we saw how to solve factored equations by using the *principle of zero products*.

The product of factors is zero if and only if at least one of the factors is zero.

Consider an equation like this:

$x^2 + 5x + 6 = 0$

Factoring the polynomial on the left and using the principle of zero products we have

$x^2 + 5x + 6 = 0$

$(x + 2)(x + 3) = 0$

These equations have the same solutions. Let's solve.

$(x + 2)(x + 3) = 0$

$x + 2 = 0 \quad \text{or} \quad x + 3 = 0 \quad \text{(principle of zero products)}$

$x = -2 \quad \text{or} \quad x = -3$

The solutions are -2 and -3.

Example 1. Solve $x^2 + 2x - 15 = 0$

$(x + 5)(x - 3) \qquad \text{(factoring)}$

$x + 5 = 0 \quad \text{or} \quad x - 3 = 0 \quad \text{(principle of zero products)}$

$x = -5 \quad \text{or} \quad x = 3$

The solutions are -5 and 3.

Example 2. Solve $x^2 - x = 6$

$x^2 - x - 6 = 0 \qquad \text{(adding } -6)$

$(x - 3)(x + 2) = 0 \qquad \text{(factoring)}$

$x - 3 = 0 \quad \text{or} \quad x + 2 = 0 \quad \text{(principle of zero products)}$

$x = 3 \quad \text{or} \quad x = -2$

The solutions are 3 and -2.

Do exercises 58 through 61 at the right.

Solve.

58. $x^2 + 5x + 6 = 0$

59. $x^2 - 4x = 0$

60. $x^2 - 3x = 28$

61. $25x^2 - 16 = 0$

Translate to equations. Then solve and check.

62. One more than a number times one less than the number is 24.

63. Seven less than a number times eight less than the number is 0.

64. A number times one less than the number is zero.

SOLVING PROBLEMS

Let's solve some applied problems. Recall that to do this we first translate the problem situation to an equation and then solve the equation.

Example 1. Solve this problem.

One more than a number times one less than that number is 8

$$(x + 1) \cdot (x - 1) = 8$$

Solve: $(x + 1)(x - 1) = 8$

$$x^2 - 1 = 8 \quad \text{(multiplying)}$$
$$x^2 - 1 - 8 = 0 \quad \text{(adding } -8 \text{ to get 0 on one side)}$$
$$x^2 - 9 = 0$$
$$(x - 3)(x + 3) = 0 \quad \text{(factoring)}$$
$$x - 3 = 0 \quad \text{or} \quad x + 3 = 0 \quad \text{(principle of zero products)}$$
$$x = 3 \quad \text{or} \quad x = -3$$

Check:

$(x + 1)(x - 1) = 8$		$(x + 1)(x - 1) = 8$	
$(3 + 1)(3 - 1)$	8	$(-3 + 1)(-3 - 1)$	8
$4 \cdot 2$		$-2 \cdot -4$	
8		8	

Thus there are two such numbers 3 and -3.

Do exercises 62 through 64 at the left.

Example 2. Solve this problem.

The square of a number minus twice the number is 48.

$$\underbrace{x^2} \quad \underbrace{-} \quad \underbrace{2x} \quad \underbrace{=} \; \underbrace{48}$$

Solve: $x^2 - 2x = 48$

$x^2 - 2x - 48 = 0$ (adding -48)

$(x - 8)(x + 6) = 0$ (factoring)

$x - 8 = 0$ or $x + 6 = 0$ (principle of zero products)

  $x = 8$ or $x = -6$

Thus there are two such numbers 8 or -6. They both check.

Do exercises 65 and 66 at the right.

Sometimes it is helpful to reword a problem before translating.

Example 3. The height of a triangle is 7 cm more than the base. The area of the triangle is 30 cm². Find the height and base.

Rewording: $\frac{1}{2}$ times the base times the base plus 7 is 30

Translate: $\frac{1}{2} \; \cdot \; b \; \cdot \; (b + 7) \; = \; 30$

Solve: $\frac{1}{2} \cdot b \cdot (b + 7) = 30$

$\frac{1}{2}(b^2 + 7b) = 30$ (multiplying)

$b^2 + 7b = 60$ (multiplying by 2)

$b^2 + 7b - 60 = 0$ (adding -60)

$(b + 12)(b - 5) = 0$ (factoring)

$b + 12 = 0$ or $b - 5 = 0$

  $b = -12$ or $b = 5$

The solutions are -12 and 5. Now think about the problem. Can the base of a triangle have a negative length? No. Thus the base is 5 cm. The height is 7 cm more than the base, so the height is 12 cm.

Do exercise 67 at the right.

Do exercise set 5.5, p. 187.

Translate to equations. Then solve and check.

65. The square of a number minus the number is 20.

66. Twice the square of a number plus 1 is 73.

67. The width of a rectangle is 2 cm less than the length. The area is 15 cm². Find the length and width.

NAME

CLASS

ANSWERS

EXERCISE SET 5.1 (pp. 165–167)

1. a) Multiply: $7 \cdot 3x$ **2.** a) Multiply: $3x \cdot 5x^2$ **3.** a) Multiply: $(2x)8x^3$

   b) Factor: $21x$ b) Factor: $15x^3$ b) Factor: $16x^4$

Find three factorizations of each monomial.

4. $6x^3$ **5.** $9x^4$ **6.** $-9x^5$

7. $15x^5$ **8.** $12x^4$ **9.** $24x^4$

10. a) Multiply: $4(x + 1)$ **11.** a) Multiply: $6x(x + 2)$

   b) Factor: $4x + 4$ b) Factor: $6x^2 + 12x$

12. a) Multiply: $3x(x^2 + 2x + 5)$ **13.** a) Multiply: $4x^2(2x^2 - 3x + 6)$

   b) Factor: $3x^3 + 6x^2 + 15x$ b) Factor: $8x^4 - 12x^3 + 24x^2$

14. a) Multiply: $-2x(x^3 + 6x - 1)$ **15.** a) Multiply: $2x^2(x^2 - 6x + 3)$

   b) Factor: $-2x^4 - 12x^2 + 2x$ b) Factor: $2x^4 - 12x^3 + 6x^2$

1.	
2.	
3.	
4.	
5.	
6.	
7.	
8.	
9.	
10.	
11.	
12.	
13.	
14.	
15.	

ANSWERS

16. _____

17. _____

18. _____

19. _____

20. _____

21. _____

22. _____

23. _____

24. _____

25. _____

26. _____

27. _____

28. _____

29. _____

30. _____

31. _____

32. _____

33. _____

Factor.

16. $x^2 + 6x$

17. $x^3 + 7x^2$

18. $2x^2 + 2x - 4$

19. $17x^5 + 34x^3 + 51x$

20. $\frac{5}{3}x^6 + \frac{4}{3}x^5 + \frac{1}{3}x^4 + \frac{1}{3}x^3$

21. $\frac{5}{7} + \frac{3}{7}x^2 - \frac{6}{7}x^3 - \frac{1}{7}$

Factor.

22. $y^2 + 4y + y + 4$

23. $x^2 + 5x + 2x + 10$

24. $x^2 - 4x - x + 4$

25. $a^2 + 5a - 2a - 10$

26. $6x^2 + 4x + 9x + 6$

27. $9x^2 - 6x + 9x - 6$

28. $3x^2 - 4x - 12x + 16$

29. $24 - 18y - 20y + 15y^2$

30. $35x^2 - 40x + 21x - 24$

31. $8x^2 - 6x - 28x + 21$

32. $4x^2 + 6x - 6x - 9$

33. $2x^4 - 6x^2 - 5x^2 + 15$

NAME

CLASS

ANSWERS

EXERCISE SET 5.2 (pp. 168–170)

Factor by examining the coefficients of the first and last terms.

1. $x^2 + 8x + 15$ **2.** $x^2 + 5x + 6$

3. $x^2 - 2x - 15$ **4.** $x^2 - 10x + 25$

5. $x^2 + 7x + 12$ **6.** $x^2 + x - 42$

7. $x^2 + 2x - 15$ **8.** $x^2 + 8x + 12$

9. $y^2 + 9y + 8$ **10.** $x^2 - 6x + 9$

11. $x^4 + 5x^2 + 6$ **12.** $y^2 + 16y + 64$

13. $x^2 + 3x - 28$ **14.** $x^2 - 11x + 10$

15. $y^2 - .2y - .08$ **16.** $35x^2 - 57x - 44$

17. $15 + 8x + x^2$ **18.** $6 + 5x + x^2$

19. $24x^2 + 47x - 2$ **20.** $x^2 + \frac{2}{3}x + \frac{1}{9}$

21. $20 + 6x - 2x^2$ **22.** $4x^2 + 4x - 15$

Answers:
1. _____
2. _____
3. _____
4. _____
5. _____
6. _____
7. _____
8. _____
9. _____
10. _____
11. _____
12. _____
13. _____
14. _____
15. _____
16. _____
17. _____
18. _____
19. _____
20. _____
21. _____
22. _____

Factor by examining the coefficients of the first and last terms.

23. _____

24. _____

23. $3x^2 + 4x + 1$ **24.** $6x^2 + 13x + 6$

25. _____

26. _____

25. $x^2 - 5x - 24$ **26.** $9x^2 + 6x - 8$

27. _____

28. _____

27. $2x^2 - x - 1$ **28.** $15x^2 - 19x + 6$

29. _____

30. _____

29. $9x^2 + 18x - 16$ **30.** $2x^2 + 5x + 2$

31. _____

32. _____

31. $3x^2 - 5x - 2$ **32.** $18x^2 - 3x - 10$

33. _____

34. _____

33. $12x^2 + 31x + 20$ **34.** $15x^2 + 19x - 10$

35. _____

36. _____

35. $14x^2 + 19x - 3$ **36.** $35x^2 + 34x + 8$

37. _____

38. _____

37. $9x^4 + 18x^2 + 8$ **38.** $6 - 13x + 6x^2$

39. _____

40. _____

39. $9x^2 - 42x + 49$ **40.** $15x^4 - 19x^2 + 6$

41. _____

42. _____

41. $a^2 - 13a + 12$ **42.** $16a^2 + 78a + 27$

43. _____

44. _____

43. $x^4 + 5x^2 + 6$ **44.** $9a^2 + 12a - 5$

NAME

CLASS

ANSWERS

EXERCISE SET 5.3 (pp. 171–172)

Factor completely. Remember to look first for a common factor.

1. $x^2 + 6x + 9$ **2.** $y^2 - 6y + 9$

3. $2x^2 - 4x + 2$ **4.** $2y^2 - 40y + 200$

5. $x^3 - 18x^2 + 81x$ **6.** $144x + 24x^2 + x^3$

7. $y^2 - 12y + 36$ **8.** $81 + 18x + x^2$

9. $64 - 16y + y^2$ **10.** $a^4 + 14a^2 + 49$

11. $12x^2 + 36x + 27$ **12.** $2x^2 - 44x + 242$

13. $x^6 - 16x^3 + 64$ **14.** $y^6 + 26y^3 + 169$

15. $49 - 42x + 9x^2$ **16.** $20x^2 + 100x + 125$

17. $5y^4 + 10y^2 + 5$ **18.** $16x^8 - 8x^4 + 1$

Factor.

19. $x^2 - 4$ **20.** $x^2 - 36$

21. $8x^2 - 98$ **22.** $98x^4 - 2$

1.
2.
3.
4.
5.
6.
7.
8.
9.
10.
11.
12.
13.
14.
15.
16.
17.
18.
19.
20.
21.
22.

ANSWERS

23.

_____

24.

_____

25.

_____

26.

_____

27.

_____

28.

_____

29.

_____

30.

_____

31.

_____

32.

_____

33.

_____

34.

_____

35.

_____

36.

_____

37.

_____

38.

_____

39.

_____

40.

_____

41.

_____

42.

_____

43.

_____

44.

_____

45.

_____

46.

_____

23. $4x^2 - 25$

24. $100a^2 - 9$

25. $16x - 81x^3$

26. $36b - 49b^3$

27. $16x^2 - 25x^4$

28. $x^{16} - 9x^2$

29. $49x^4 - 81$

30. $25a^2 - 9$

31. $a^{12} - 4a^2$

32. $81y^6 - 25$

33. $121a^8 - 100$

34. $x^6 - \dfrac{1}{9}$

Factor.

35. $\dfrac{1}{16} - y^2$

36. $\dfrac{1}{25} - x^2$

37. $25 - \dfrac{1}{49}x^2$

38. $4 - \dfrac{1}{9}y^2$

39. $36x^2 - \dfrac{1}{25}y^2$

40. $4x + 4x^2 + 1$

41. $1 - 2a^3 + a^6$

42. $25y^6 + 10y^3 + 1$

43. $\dfrac{1}{25}y^2 - \dfrac{1}{16}$

44. $36a^2 - 15a + \dfrac{25}{16}$

45. $\dfrac{1}{81}x^6 - \dfrac{8}{27}x^3 + \dfrac{16}{9}$

46. $\dfrac{1}{9}a^2 + \dfrac{1}{3}a + \dfrac{1}{4}$

NAME

CLASS

ANSWERS

EXERCISE SET 5.4 (pp. 173–174)

Factor.

1. $x^2 + 12x + 36 - 49$ **2.** $x^2 + 2x + 1 - 9$

3. $x^2 + 14x + 49 - 16$ **4.** $x^2 - 16x + 64 - 100$

5. $x^2 + 18x + 81 - 1$ **6.** $x^2 + 14x + 49 - 121$

7. $x^4 + 20x^2 + 100 - 4$ **8.** $x^2 - 16x + 64 - 36$

9. $2x^6 + 16x^3 + 32 - 50$ **10.** $2x^2 + 4x + 2 - \dfrac{2}{9}$

11. $a^4 - 6a^2 + 9 - 1$ **12.** $4x^4 + 12x^2 + 9 - 144$

1. _____

2. _____

3. _____

4. _____

5. _____

6. _____

7. _____

8. _____

9. _____

10. _____

11. _____

12. _____

ANSWERS

Factor by completing the square.

13. _____

14. _____

15. _____

16. _____

17. _____

18. _____

19. _____

20. _____

21. _____

22. _____

23. _____

24. _____

13. $x^2 + 6x + 8$ **14.** $y^2 + 8y + 15$

15. $x^2 - 8x + 12$ **16.** $y^2 + 10y + 16$

17. $2x^2 - 16x - 256$ **18.** $2x^2 - 28x + 26$

19. $a^2 - 24a - 25$ **20.** $c^2 - 8c + 15$

21. $10b^2 - 50b - 840$ **22.** $10t^2 - 200t + 640$

23. $x^2 - 14x - 72$ **24.** $r^2 - 22r + 21$

NAME

CLASS

ANSWERS

EXERCISE SET 5.5 (pp. 175–177)

Solve each equation and check.

1. $x^2 + 6x + 5 = 0$

Check

2. $x^2 + 7x + 6 = 0$

3. $x^2 - 5x = 0$

4. $4x^2 - 9 = 0$

5. $x^2 + 6x + 9 = 0$

6. $3t^2 + t = 2$

7. $6y^2 - 4y - 10 = 0$

ANSWERS

1.

2.

3.

4.

5.

6.

7.

ANSWERS

8. $5x^2 = 6x$

Check

8. _____

_____

9. $3x^2 = 7x + 20$

9. _____

_____

10. $0 = 2y^2 + 12y + 10$

10. _____

_____

11. $-5x = -12x^2 + 2$

11. _____

_____

12. $14 = x(x - 5)$

12. _____

_____

13. $0 = -3x + 5x^2$

13. _____

_____

14. $x^2 - 5x = 18 + 2x$

14. _____

_____

Translate to equations. Then solve the equations.

15. If you subtract a number from four times its square, the result is 3. Find the number.

15. _____

16. If the square of a number is increased by 7, the result is 32. Find the number.

16. _____

17. Eight more than the square of a number is six times the number. Find the number.

17. _____

18. The product of two positive consecutive integers is 182. Find the numbers.

18. _____

19. The length of a rectangle is 4 cm greater than the width. The area of the rectangle is 96 cm². Find the length of the rectangle.

19. _____

20. If the sides of a square are lengthened by 3 m, the area becomes 81 m². Find the length of a side of the original square.

20. _____

21. Find the number whose square is 28 more than 12 times the number.

21. _____

22. Adding four to the square of a number gives the same result as subtracting three from eight times the number. Find the number.

22. _____

23. The product of two consecutive even integers is 168. Find the numbers.

23. _____

24. The number of square inches in a square is five more than the number of inches in its perimeter. Find the length of a side of the square.

24. _____

25. The sum of the squares of two consecutive odd positive integers is 74. Find the numbers.

25. _____

NAME_____

CHAPTER 5 TEST

CLASS_____SCORE_____GRADE_____

Before taking the test *be sure* to allow yourself a day or so for review. Use the objectives listed in the margins to guide your study. The test will evaluate your progress and aid your preparation for a possible classroom test. Allow about an hour for the test. Remove the test from the book. When you finish, read the test analysis on the answer page at the end of the book.

ANSWERS

1. _____

2. _____

3. _____

4. _____

5. _____

Factor completely.

1. $3x + 3$

2. $10x - 20$

6. _____

7. _____

3. $9x^2 - 4$

4. $2x^2 - 50$

8. _____

5. $x^2 - 8x + 15$

6. $x^3 - 7x^2 + 12x$

9. _____

7. $3x^2 + 18x - 48$

8. $x^3 - 8x^2 + 15x$

10. _____

11. _____

9. $10x^2 - 101x + 10$

10. $x^2 - 6x + 9$

12. _____

11. $12x^2 + 60x + 75$

12. $9x^2 - 30x + 25$

13. _____

14. _____

13. $16x^4 - 1$

14. $3(x + 7) - 7(x + 7)$

15. _____

15. $x^2 + 2x + 1 - 25$

16. $x^2 - 6x + 9 - 16$

16. _____

ANSWERS

17. _____

18. _____

19. _____

20. _____

21. _____

22. _____

23. _____

24. _____

25. _____

17. What must be added to $x^2 - 20x$ to make it a trinomial square?

18. What must be added to $x^2 + 30x$ to make it a trinomial square?

Solve for x.

19. $x^2 + 2x - 35 = 0$

20. $2x^2 + 5x = 12$

21. $16 = x(x - 6)$

22. $x^2 - 6x = 20 + 2x$

23. The square of a number is 6 more than the number. Find the number.

24. The product of two consecutive even integers is 288. Find the integers.

25. Adding 5 to the square of a number gives the same result as subtracting 7 from 13 times the number. Find the number.

6 SYSTEMS OF EQUATIONS AND GRAPHS

6.1 GRAPHS AND EQUATIONS

On a number line we associate numbers with points. There is one and only one point for each number. On a flat surface (plane) we can associate pairs of numbers with points, using two number lines at right angles to each number. In this drawing there are two number lines, called *axes*. The point where they cross is called the *origin* of the graph and is labeled 0. The arrows on the axes show the positive directions. The positive directions can be chosen as we wish, but they are usually chosen as shown.

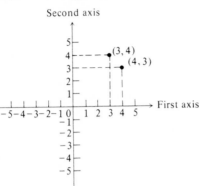

PLOTTING POINTS

To locate a point we need a pair of numbers, such as (4, 3). The first number is a distance in the direction of the first axis. The second number is the distance in the direction of the second axis. Notice that (4, 3) and (3, 4) give different points. Hence we speak of these as *ordered pairs* of numbers, meaning that it makes a difference which number of the pair is first.

The numbers that locate a point are called the *coordinates* of the point. If the coordinates of a point are (5, −2) we say that the first coordinate is 5 and the second coordinate is −2.*

The following drawing shows some points and their coordinates. Notice that in region I (called the first *quadrant*) both coordinates of a point are positive. In region II (the second quadrant) the first coordinate is negative and the second one positive.

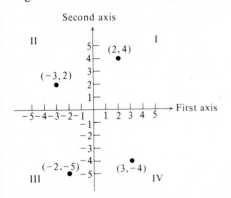

Do exercises 1 through 3 at the right.

---

* The first coordinate of a point is sometimes called its *abscissa*, and the second coordinate its *ordinate*.

You should be able to:

a) Given ordered pairs of numbers (coordinates) plot points.

b) Given points on a graph, find their coordinates.

c) Draw graphs of simple equations.

1. What can you say about the coordinates of a point in the third quadrant?

2. What can you say about the coordinates of a point in the fourth quadrant?

3. Use graph paper. Draw and label a first and second axis. Then plot these points.

a) (4, 6)   b) (6, 4)

c) (−2, 5)   d) (−3, −4)

e) (5, −3)

193

4. Find the coordinates of points

B, C, D, E, F and *G*.

FINDING COORDINATES

We can find, at least roughly, the coordinates of a point by finding how far to the right or left it is located and also how far up or down. In this drawing, for example, point A is about 4 units to the right and about 3 units up, so its coordinates are (4, 3).

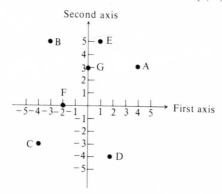

Do exercise 4 at the left.

GRAPHS OF EQUATIONS

If an equation has two variables, its solutions must be pairs of numbers. We usually take the variables in alphabetical order, and hence get ordered pairs of numbers for solutions.

Example

To show that (3, 7) is a solution of $y = 2x + 1$ we replace x by 3 and y by 7.

$$y = 2x + 1$$

$$\begin{array}{c|c} 7 & 2 \cdot 3 + 1 \\ \hline & 7 \end{array}$$

Actually, the solution set of an equation like $y = 2x + 1$ is infinite (unending).

To graph an equation, **we plot enough points to see what the pattern is and then draw a line or curve.**

Example 1.

Graph the equation $y - x = 1$.

We will use alphabetical order for the variables. Thus the first axis will be the *x*-axis and the second axis the *y*-axis.

It will be helpful to solve the equation for the second variable, *y*:

$$y = x + 1.$$

Next we find some solutions (ordered pairs), keeping the results in a table. To find an ordered pair we choose any number for x, substitute it in the equation, and then find y.

Suppose we choose 0 for x. Then

$$y = x + 1 = 0 + 1 = 1.$$

We chose 0 for x and substituted it in the equation. We found that when $x = 0$, it must follow that $y = 1$. This gives us the ordered pair $(0, 1)$. We enter these results in a table.

x	0				
y	1				

Now we continue to find ordered pairs. Suppose we choose -1 for x. Then

$$y = x + 1 = -1 + 1 = 0.$$

When $x = -1$, it must follow that $y = 0$. This gives us the ordered pair $(-1, 0)$. If we continue choosing numbers for x and computing y from the equation we can find many ordered pairs. We can fill in a table like this.

We choose any number we please for x. Then we find the y that goes with it.

x	0	-1	-5	1	3
y	1	0	-4	2	4

Now we plot these points.

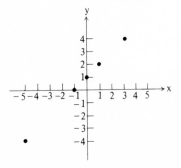

5. Use graph paper. Draw and label an x-axis and a y-axis. Then graph the equation $y - x = 3$.

In this case the points seem to lie on a straight line. In fact they do. If we take all of the solutions of the equation we get the entire line. So the graph of this equation is a straight line through the points we plotted. We can draw the line with a ruler. Here is the graph of the equation.

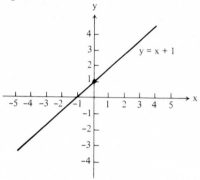

Do exercise 5 at the left.

Some equations have graphs that are not straight lines.

Example 2

Graph the equation $y = x^2$.
We select numbers for x and find the corresponding values of y.

x	0	-1	1	-2	2	-3	3
y	0	1	1	4	4	9	9

Next we plot these points and then connect them with a smooth curve.

6. Use graph paper. Draw a graph of $y = x^2 - 1$.

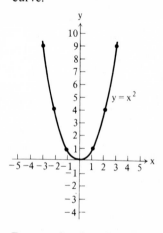

Do exercise 6 at the left.

Do exercise set 6.1, p. 221.

6.2 LINEAR EQUATIONS

The equations in which we are primarily interested in beginning algebra are those which have straight lines for their graphs. Such equations are called *linear equations*. There is an easy way to tell whether an equation is linear. The variables must occur to the first power only. There must be no products of variables or variables in denominators. The following equations are linear.* (Graphs are straight lines.)

$$3x + 5 = 3y, \; 5x = -4, \; y = 9x - 15y$$

These equations are not linear. (Graphs are not straight lines.)

$$xy = 2, \; 5x^3 = 17y + 4, \; 6x^2 + 7x = 4y + 3, \; x = \frac{3}{y}$$

Since in linear equations the variables occur to the first power only they are also called *first degree equations*.

GRAPHING LINEAR EQUATIONS

When we graph a first degree equation we know that the graph will be a straight line. Thus plotting two points is sufficient. We usually use at least one more point, as a check.

Example 1

Graph the equation $2y - 4 = 4x$.

We can solve for y: $y = 2x + 2$.

If $x = 1$, $y = 4$.

If $x = -2$, $y = -2$.

We plot these points and draw the line.

If $x = 3$, $y = 8$.

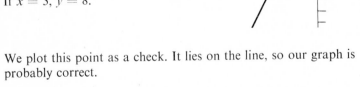

We plot this point as a check. It lies on the line, so our graph is probably correct.

Do exercises 7 and 8 at the right.

---

* Note that one variable can be missing.

OBJECTIVES

You should be able to:

a) Recognize linear equations.

b) Graph linear equations using intercepts.

c) Graph linear equations with a missing variable.

d) Tell whether graphs of linear equations are parallel.

7. Which of these equations are linear (first degree)?

a) $3y + 4x = 7$ b) $5y = 10$

c) $4x^2 + y = 5$ d) $3x^2y = 4$

e) $x = 3 + \dfrac{1}{y}$

8. Use graph paper.

Graph $6x - 2y = -2$

9. Use graph paper.

Graph $2y = 3x - 6$

GRAPHING USING INTERCEPTS

The points where a line crosses the axes are called the *intercepts*. We can usually find these points easily. Thus we often use them in graphing.

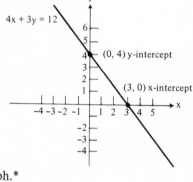

Example 2

Graph $4x + 3y = 12$.

If we choose 0 for x we find that $y = 4$. Thus (0, 4) is the y-intercept.

If we choose 0 for y we find that $x = 3$. Thus (3, 0) is the x-intercept.

We plot these points and draw the graph.*
A third point should be used as a check.

Do exercise 9 at the left.

EQUATIONS WITH A MISSING VARIABLE

Consider the equation $y = 3$. We can think of it as $y = 0 \cdot x + 3$. No matter what number we choose for x, we find that y is 3. Thus $(x, 3)$ is a solution no matter what x is. Hence if x is missing in an equation the graph is a line parallel to the x-axis. If y is missing the graph is a line parallel to the y-axis.

10. Graph these equations.

a) $x = 5$

b) $y = -2$

c) $x = 0$

Example 3

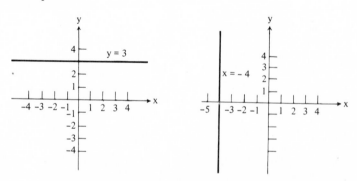

Do exercise 10 at the left.

---

* If a line goes through the origin we must use another point, not an intercept.

PARALLEL LINES

Parallel lines are lines on the same plane that do not meet, no matter how far they may be extended. When we draw graphs of two linear equations we may find that they meet, or intersect. We may find that they do not meet, hence are parallel.

Example 4

The graphs of $3x - y = -1$ and $2y = -3x - 2$ intersect.

Let us solve each equation for y.

$y = 3x + 1$

$y = -\dfrac{3}{2}x - 1$

Notice that the coefficients of the x-terms are different. The lines are not parallel.

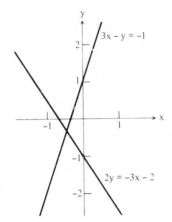

Example 5

The graphs of $y - 3x = 5$ and $3x - y = 2$ are parallel.

Let us solve each equation for y.

$y = 3x + 5$

$y = 3x - 2$

Notice that the coefficients of the x-terms are the same, but the constant terms are different. The lines are parallel.

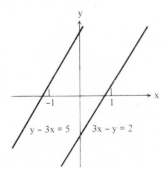

To tell whether lines are parallel, solve the equations for y. Parallel lines have the same x-terms but different constant terms.

Do exercise 11 at the right.

Do exercise set 6.2, p. 223.

11. Without graphing, tell whether these lines are parallel.

a) $y + 4 = x$

  $x - y = -2$

b) $y + 1 = 2x$

  $3x - y = -5$

OBJECTIVES

You should be able to:

Translate applied problems to systems of equations.

12. Translate to a system of equations. Do not attempt to solve.

The sum of two numbers is 115. The difference of the numbers is 21. Find the numbers.

6.3 TRANSLATING PROBLEMS TO EQUATIONS

The first, and often the hardest part of solving a problem is translating it to mathematical language. This problem of translating to equations becomes easier in many cases if we translate to more than one equation. When we do this we use more than one variable.

Example 1

The sum of two numbers is 15. One number is four times the other. Find the numbers.

There are two statements in the problem. We translate the first one:

The sum of two numbers is 15

$$x + y \qquad = 15$$

Note that we have used x and y for the two numbers. Now we translate the second statement, remembering to use x and y.

One number is four times the other

$$y \quad = \quad 4 \quad \cdot \quad x$$

For the second statement we could also have translated to $x = 4 \cdot y$. This would also have been correct.

The problem has been translated to a *pair* or *system* of equations. We list what the letters represent and then list the equations.

Let $x =$ one number and $y =$ the other number.

$$x + y = 15$$
$$y = 4x$$

Do exercise 12 at the left.

Example 2. Howie Doon is 21 years older than Izzi Retyrd. In six years Howie will be twice as old as Izzi. How old are they now?
We translate the first statement. Let us reword it first.

Howie's age is 21 more than Izzi's age

$$x \qquad = 21 \qquad + y$$

Of course x represents Howie's age and y represents Izzi's age.

We translate the second statement, but again it helps to reword it first.

Howie's age in six years will be twice Izzi's age in six years

$$x + 6 \qquad = \qquad 2 \qquad \cdot \qquad (y + 6)$$

We have now translated to the system of equations

Let x = Howie's age and y = Izzi's age.

$$x = 21 + y$$
$$x + 6 = 2(y + 6)$$

Making a drawing is often helpful. Whenever it makes sense to do so you should make a drawing before trying to translate.

Example 3. The perimeter of a rectangle is 90 cm. The width is 20 cm. greater than the length. Find the length and width.
First we make a drawing. We use l for length and w for width.

We translate the first statement.

The perimeter is 90 cm.

$$2l + 2w \quad = \quad 90$$

We translate the second statement.

The width is 20 cm greater than the length

$$w \quad = \quad 20 + l$$

We have translated to the system of equations

Let w = width and l = length.

$$2w + 2l = 90$$
$$w = 20 + l$$

Do exercises 13 and 14 at the right.

Do exercise set 6.3, p. 225. Save these for a later lesson!

Translate each problem to a system of equations. Do not try to solve. Draw a picture where helpful.

13. Chuck Rowst is half as old as Sammy Tary. Five years from now Sammy will be 56. How old is Chuck?

14. The perimeter of a rectangle is 76 cm. The length is 17 more than the width. Find the length and width.

You should be able to:

a) Decide whether an ordered pair is a solution of a system of equations.

b) Solve systems of equations using graphs.

6.4 SOLUTIONS OF SYSTEMS OF EQUATIONS, GRAPHS

Translating a problem to mathematical language is often rather easy when we translate to a system of equations. Thus it is important to consider solutions of systems of equations.

Recall from the preceding chapter that the set of solutions of a linear equation with two variables is infinite (unending). Each solution is an ordered pair of numbers.

By a solution of a system of two equations we mean an ordered pair that makes both equations true. Consider the system of equations

$$x + y = 8$$
$$2x - y = 1$$

Let us graph both equations of this system.

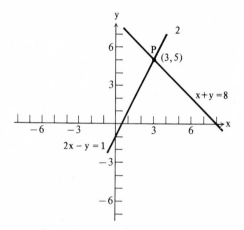

Every point on line 1 has coordinates which satisfy the equation $x + y = 8$.

For example, the ordered pairs (1, 7) and (−1, 9) satisfy this equation. We check:

$$x + y = 8$$

$1 + 7$	8
8	

$$x + y = 8$$

$-1 + 9$	8
8	

Similarly, every point on line 2 has coordinates which satisfy the equation $2x - y = 1$. For example (1, 1), satisfies it. We check:

$$2x - y = 1$$

$2 \cdot 1 - 1$	1
$2 - 1$	
1	

Which points (ordered pairs) satisfy *both* equations? From the graph we see that there is only one. It is the point P where the graphs cross. This point looks as if its coordinates are (3, 5). We check:

$x + y = 8$

$3 + 5$	8
8	

$2x - y = 1$

$2 \cdot 3 - 5$	1
$6 - 5$	
1	

Thus there is just one solution of the system of equations. It is (3, 5). In other words, $x = 3$ and $y = 5$.

Do exercises 15 and 16 at the right.

To solve a system of equations graphically we graph both equations and find the coordinates of the point(s) of intersection. If the lines are parallel there is no solution.

Example

Solve graphically $x + 2y = 7$

$\qquad\qquad\qquad x = y + 4$

We graph the equations

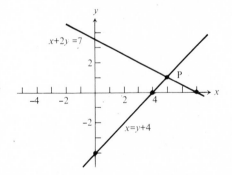

Point *P* looks as it if has coordinates (5, 1). This checks.

Do exercises 17 and 18 at the right.

Do exercise set 6.4, p. 229.

15. Decide if (2, −3) is a solution of the system

$x = 2y + 8$

$2x + y = 1$

16. Decide if (20, 40) is a solution of the system

$a = \dfrac{1}{2}b$ (Remember to use alphabetical order of variables)

$b + 5 = 45$

17. Solve graphically

$2x + y = 1$

$\quad x = 2y + 8$

18. $3x - 2y = -4$

$\quad 2y - 3x = -2$

You should be able to:
Solve a system of two equations by the substitution method.

a) When one of them has a variable alone on one side.

b) When neither has a variable alone on one side.

6.5 SOLVING BY THE SUBSTITUTION METHOD

Solving systems of equations graphically takes more time than other methods.
Also, solutions obtained graphically are often only approximate. Thus we shall develop other methods of solving systems of equations without graphing. The first method is called the *substitution method*. We illustrate with an example. Suppose we wish to solve the system

$x + y = 6$

$x = y + 2$

Look at the second equation $x = y + 2$. Remember, it says that x and $y + 2$ represent the same number. In other words, x and $y + 2$ are two names for the same thing. Thus wherever we see the name x we can use the name $y + 2$ if we wish. Let us go to the first equation and put in $y + 2$ instead of x:

$x + y = 6$

$(y + 2) + y = 6$ (substituting $y + 2$ for x)

Now we solve the last equation. Notice that it has only one variable.

$(y + 2) + y = 6$

$2y + 2 = 6$

$2y = 4$

$y = 2$

Now we return to the original pair of equations. We can now substitute 2 for y in either of them. Let's use the first equation.

$x + y = 6$

$x + 2 = 6$ (substituting 2 for y)

$x = 4$

Now we have a possible solution to the system, $x = 4$ and $y = 2$. In other words, the ordered pair (4, 2) may be a solution. We check:

$x + y = 6$		$x = y + 2$	
$4 + 2$	6	4	$2 + 2$
6			4

Since the answer checks, we have the solution. Here is another example.

Example 1

Solve: $s = 13 - 3t$

$\qquad s + t = 5$

We substitute $13 - 3t$ for s in the second equation.

$$s + t = 5$$
$$(13 - 3t) + t = 5$$

Now we solve for t:

$$13 - 2t = 5$$
$$13 - 5 = 2t$$
$$8 = 2t$$
$$4 = t$$

Next we substitute 4 for t in the second equation of the original pair:

$$s + t = 5$$
$$s + 4 = 5$$
$$s = 1$$

The ordered pair (1, 4) may be a solution. We check:

$s = 13 - 3t$			$s + t = 5$	
1	$13 - 3 \cdot 4$		$1 + 4$	5
	$13 - 12$		5	
	1			

Do exercise 19 at the right.

In both of the preceding examples one equation of the system had a variable alone on one side. Suppose neither equation of a pair has a variable alone. Then we solve one of the equations for one of the variables and proceed as before.

Example 2

Solve:

$$x + 2y = 7$$
$$x - y = 4$$

Let us solve the second equation for x:

$$x - y = 4$$
$$x = y + 4$$

Solve by the substitution method. Do not graph.

19. $x + y = 5$

$\qquad x = y + 1$

Solve using substitution method. First get *y* alone on one side of one equation.

20. $2x + y = 9$

   $2x - y = 7$

Now we substitute in the first equation of the original pair.

$$x + 2y = 7$$
$$(y + 4) + 2y = 7$$

Now we solve for *y*:

$$y + 4 + 2y = 7$$
$$3y + 4 = 7$$
$$3y = 3$$
$$y = 1$$

Next, we go back to either of the original equations and substitute 1 for *y*. This time, let's use the second one:

$$x - y = 4$$
$$x - 1 = 4$$
$$x = 5$$

The ordered pair (5, 1) may be a solution. We check:

$$
\begin{array}{c|c}
x + 2y = 7 & \\
\hline
5 + 2\cdot 1 & 7 \\
5 + 2 & \\
7 & \\
\end{array}
\qquad\qquad
\begin{array}{c|c}
x - y = 4 & \\
\hline
5 - 1 & 4 \\
4 & \\
\end{array}
$$

The solution is (5, 1); that is, $x = 5$, $y = 1$.

Do exercise 20 at the left.

In using the substitution method, as in exercise 9, we substituted for *y*. We could also have substituted for *x* but that would have been more work, because the coefficient of *x* is not 1 in either equation.

Do exercise set 6.5, p. 231.

6.6 THE ADDITION METHOD

Another method of solving systems of equations is called the *addition method*. The reason for the name will become clear. We will illustrate this method with an example. Suppose we wish to solve the system

$x + y = 5$

$x - y = 1$

We shall use the addition principle for equations. We shall add 1 on both sides of the first equation. But, according to the second equation, $x - y$ and 1 are the same thing. Thus we can add $x - y$ to the left side and 1 to the right side.

$$\begin{array}{r} x + y = 5 \\ x - y = 1 \\ \hline 2x + 0y = 6 \end{array}$$

The addition on the left is the addition of two expressions with variables. We do this addition by combining similar terms. Notice that after the addition one variable has disappeared. We have an equation with just one variable.

$2x = 6$

We solve for x:

$x = 3$

Now we can proceed as we did with the substitution method. We substitute 3 for x in either of the original equations.

$x + y = 5$

$3 + y = 5$

$y = 2$

The ordered pair (3, 2) may be a solution. We check:

$$\begin{array}{c|c} x + y = 5 \\ \hline 3 + 2 & 5 \\ 5 & \end{array} \qquad \begin{array}{c|c} x - y = 1 \\ \hline 3 - 2 & 1 \\ 1 & \end{array}$$

Do exercises 21 and 22 at the right.

Solve using the addition method.

21. $x + y = 5$

$2x - y = 4$

22. $3x - 3y = 6$

$3x + 3y = 0$

Solve. Multiply one equation by -1 first.

23. $5x + 3y = 17$

$\quad\ 5x - 2y = -3$

24. $-x - y = -5$

$\quad\ 2x - y = 4$

The addition method allows us to eliminate a variable. Sometimes we need to multiply by -1 in order that this will happen.

Example 1

Solve:

$2x + 3y = 8$

$\ x + 3y = 7$

This time if we add we do not eliminate a variable. However, if the $3y$ were $-3y$ in one equation we could. So we shall multiply in the second equation by -1 and then add:

$\quad 2x + 3y = 8$

$\quad \underline{-x - 3y = -7}$

$\quad\ \ x \qquad\ = 1$

Now we substitute in one of the original equations:

$x + 3y = 7$

$1 + 3y = 7$

$\quad\ 3y = 6$

$\quad\ \ y = 2$

Check:

$2x + 3y = 8$		$x + 3y = 7$	
$2 \cdot 1 + 3 \cdot 2$	8	$1 + 3 \cdot 2$	7
$2 + 6$		$1 + 6$	
8		7	

Do exercises 23 and 24 at the left.

In the preceding example we used the multiplication principle, multiplying by -1. Sometimes we need to multiply by something other than -1.

Example 2

Solve:

$3x + 6y = -6$

$5x - 2y = 14$

This time we multiply by 3 in the second equation.

$$3x + 6y = -6$$
$$15x - 6y = 42 \quad \text{(multiplying by 3)}$$
$$18x = 36$$
$$x = 2$$

We go back to the first equation and substitute 2 for x:

$$3 \cdot 2 + 6y = -6$$
$$6 + 6y = -6$$
$$6y = -12$$
$$y = -2$$

Check:

$$
\begin{array}{c|c}
3x + 6y = -6 & 5x - 2y = 14 \\
\hline
3 \cdot 2 + 6 \cdot -2 \;\big|\; -6 & 5 \cdot 2 - 2 \cdot -2 \;\big|\; 14 \\
6 + {}^{-}12 & 10 - {}^{-}4 \\
-6 & 14
\end{array}
$$

Do exercises 25 and 26 at the right.

In using the addition method for solving systems of equations we sometimes need to use the multiplication principle more than once.

Example 3

Solve:

$$3x + 5y = 30$$
$$5x + 3y = 34$$

This time we use the multiplication principle with both equations. In the first equation we shall multiply by 5. In the second equation we shall multiply by -3.

$$15x + 25y = 150$$
$$-15x - 9y = -102$$
$$ 16y = 48$$
$$y = 3$$

Solve.

25. $x + 2y = 3$
  $2x + 6y = 8$

26. $2s + 3t = 5$
  $4s + 7t = 11$

Solve.

27. $5x + 3y = 2$

$3x + 5y = -2$

Now we substitute in one of the original equations:

$3x + 5y = 30$

$3x + 5\cdot 3 = 30$

$3x + 15 = 30$

$3x = 15$

$x = 5$

Check:

$3x + 5y = 30$			$5x + 3y = 34$	
$3\cdot 5 + 5\cdot 3$	30		$5\cdot 5 + 3\cdot 3$	34
$15 + 15$			$25 + 9$	
30			34	

28. $6x + 2y = 4$

$10x + 7y = -8$

Do exercises 27 and 28 at the left.

Example 4

To solve

$5s + 7t = -9$

$-3s + 5t = -13$

We could multiply the first equation by 3 and the second by 5. We could also multiply the first one by 5 and the second one by -7.

INCONSISTENT AND DEPENDENT EQUATIONS

When we graph a system of two linear equations we obtain two straight lines. When two different lines intersect, they have just one common point, so in such a case a system of two equations has *just one* solution. Two straight lines may also be parallel. This means that they have no point in common. If the graph of a system of equations is two parallel lines, then the system of equations has no solution. Such a system is called *inconsistent*. What happens if we try to use the substitution or the addition method for solving inconsistent equations? The following example illustrates.

Example 5. Try to solve this inconsistent system.

$y = 3x + 2$
$y = 3x + 1$

If we try to use the substitution method or the addition method, we obtain

$1 = 2$

This is clearly false, so there is no solution. The graphs of the equations are parallel lines.

Two equations may have the same graph. In such a case any solution of one is a solution of the other. A system of equations like this is called *dependent*. If we graph a system of equations and find that we get the same line both times, we know that we have a dependent system. What happens if we try to use the addition method or the substitution method to solve a dependent system? The following example illustrates.

Example 6. Try to solve this dependent system.

$x + 2y = 1$
$2y = 1 - x$

Let us use the substitution method. We solve the first equation for x.

$x = 1 - 2y$

Now we substitute in the second equation. We obtain

$2y = 1 - (1 - 2y)$
$2y = 2y$

Any number for y will make this true. If we had used the addition method the same thing would have happened.

Since any number for y will work, we have an infinite set of solutions.

Inconsistent equations have no solution. Their graphs are parallel lines. If we try to solve the system, we get a false equation. Dependent equations have the same solutions. Their graphs are the same line. If we try to solve the system, we get an equation that is true for any real number.

Do exercise set 6.6, p. 233.

Do exercise set 6.6A, p. 235.

OBJECTIVES

You should be able to:

a) Translate applied problems to equations or systems of equations.

b) Solve applied problems.

c) Solve motion problems, using the relationship $r = \dfrac{d}{t}$.

29. One number is 18 greater than another. Twice the smaller number plus three times the larger is 74. What are the numbers?

6.7 APPLIED PROBLEMS

Translating an applied problem to mathematical language is often easier if one translates to a system of equations. Now that you know how to solve systems of equations, solving applied problems should be easier.

Example 1

A man took his son camping, to a spot 27 kilometers from town. They drove 11 km more than they walked in getting there. How far did they walk? We translate, noting that some rewording may help.

Distance walked plus distance driven is 27 km.

$$x \quad + \quad y \quad = 27$$

Distance driven is 11 km more than distance walked.

$$y \quad = \quad 11 \quad + \quad x$$

We have a system of equations.

Let $x =$ distance walked, $y =$ distance driven.

$$x + y = 27 \qquad x + y = 27$$
$$y = 11 + x \quad \text{or} \quad -x + y = 11$$

The solution of this system is $x = 8$, $y = 19$.

We check in the original problem:

Distance walked is 8 km. This plus distance driven, 19 km, is 27 km. Also, 19 is 11 more than 8, so the answer checks.

The answer is

Distance walked is 8 kilometers.

Do exercise 29 at the left.

In exercise set 6.3, p. 225 you translated to equations but did not solve. Now you should finish solving those problems.

Return to exercise set 6.3, p. 225. Finish solving the problems.

MOTION PROBLEMS

Many applied problems deal with distance, time and speed. To solve such problems one must recall and use the definition of speed.

$$\text{Speed} = \frac{\text{distance}}{\text{time}}$$

We most often use r for speed, d for distance and t for time. Thus the definition of speed is

$$r = \frac{d}{t}$$

Do exercises 30 and 31 at the right.

We can solve $r = \dfrac{d}{t}$ for d or for t easily, whenever that is needed. Thus one needs only to remember the definition of speed. Whenever a motion problem is to be solved, one should remember to use $r = \dfrac{d}{t}$ or one of its other forms $d = rt$ or $t = \dfrac{d}{r}$.

Example 1

Two cars leave town at the same time, going opposite directions. One of them travels 60 mph and the other 30 mph. In how many hours will they be 150 miles apart?

We first make a drawing: Note that d_1 is the distance for the first car, d_2 the distance for the second car.

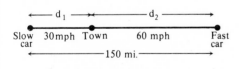

From the drawing we see that $d_1 + d_2 = 150$. Remember that $d = rt$ and substitute rt for d:

$$r_1 t_1 + r_2 t_2 = 150$$

Note that we use r_1 and r_2 for the two speeds and t_1 and t_2 for the two times. Since the cars travel the same amount of time, t_1 and t_2 are the same, and we can just use t for time.

30. Solve $r = \dfrac{d}{t}$ for d.

31. Solve your answer to exercise 30 for t.

32. Two cars leave town at the same time in the same direction. One travels 35 mph and the other 40 mph. In how many hours will they be 15 miles apart? (Hint: The times are the same. Be SURE to make a drawing.)

$$r_1 t + r_2 t = 150$$

$$30t + 60t = 150$$

Now we can solve this equation for t. We get

$t = \dfrac{15}{9}$, or $1\dfrac{2}{3}$ hours. This checks

Do exercises 32 and 33 at the left.

Example 2

A train leaves Podunk traveling east at 35 kilometers per hour (km/hr). An hour later another train leaves Podunk on a parallel track at 40 km/hr. How far from Podunk will the trains meet?

We first make a drawing:

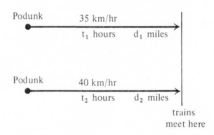

33. Two cars leave town at the same time in opposite directions. One travels 48 mph and the other 60 mph. How far apart will they be 3 hours later? (Hint: The times are the same. Be SURE to make a drawing.)

From the drawing we see that the distances are the same, $d_1 = d_2$, and from this we get $r_1 t_1 = r_2 t_2$ or

$$35t_1 = 40t_2$$

Now the slow train travels 1 hour longer than the other, so

$$t_1 = t_2 + 1$$

We now have a system of equations

$$35t_1 = 40t_2$$

$$t_1 = t_2 + 1$$

The solution is $t_1 = 8$ hours, $t_2 = 7$ hours.
The problem asks for distance, however. We can find it, remembering that $d = rt$. $d = 35 \cdot 8$ or $40 \cdot 7$. The distance is 280 kilometers. This checks in the problem.

34. An airplane flew for 5 hours with a 25 km/hr tail wind. The return flight against the same wind took 6 hours. Find the speed of the airplane in still air. (Hint: The distance is the same both ways. The speeds are $r + 25$ and $r - 25$, where r is the speed in still air.)

Do exercise 34 at the left.

Do exercise set 6.7, p. 237.

6.8 COIN AND MIXTURE PROBLEMS

In this lesson we shall see how to solve two more types of problems.

Example 1.

A man has some nickels and dimes. The value of the coins is $1.65. There are 12 more nickels than dimes. How many of each kind of coin has he?

We shall use d to represent the number of dimes and n to represent the number of nickels. We have one equation at once.

$d + 12 = n$

The value of the nickels, in cents, is $5n$, since each is worth 5¢. The value of the dimes, in cents, is $10d$, since each is worth 10¢. The total value is given as $1.65. Since we have the values of the nickels and dimes *in cents*, we must use cents for the total value. This is 165. Now we have another equation.

$10d + 5n = 165$

Thus we have a system of equations to solve.

$d + 12 = n$
$10d + 5n = 165$

The solution of this system is $d = 7$ and $n = 19$. This checks, so the man has 7 dimes and 19 nickels.

Do exercise 35 at the right.

Certain other problems are very much like coin problems. Although they are not about coins, they are solved the same way.

Example 2.

There were 411 people at a play. Admission for adults was $1.00 and for children $0.75. The receipts were $395.75. How many adults and how many children attended?

We shall use a for the number of adults and c for the number of children. Since the total number is 411, we have this equation.

$a + c = 411$

The amount paid in by the adults is $1 \cdot a$, since each paid $1.00. This amount is *in dollars*. In cents the amount is $100a$.

The amount paid in by the children is $0.75c$, in dollars, or $75 \cdot c$ in cents. If we use cents instead of dollars we can avoid decimal points. Then we have this equation.

$100a + 75c = 39575,$

since the total amount taken in is 39575, *in cents*.

We solve the system of equations, and find that $a = 350$ and $c = 61$, so there are 350 adults and 61 children.

Do exercise 36 at the right.

You should be able to solve coin and mixture problems.

35. On a table are 20 coins, quarters and dimes. Their value is $3.05. How many of each are there?

36. There were 166 paid admissions to a game. The price was $2.10 for adults and $0.75 for children. The amount taken in was $293.25. How many adults and how many children attended?

37. One solution is 50% alcohol and a second is 70% alcohol. How much of each should be used to make 30 kilograms of a solution that is 55% alcohol?

38. Grass seed A is worth $1.00 per pound and seed B is worth $1.35 per pound. How much of each would you use to make 50 lb of a mixture worth $1.14 per pound?

Example 3.

A chemist has a solution that is 80% acid and a solution that is 30% acid. He needs 100 liters of a solution that is 62% acid. He will prepare it by mixing the two solutions on hand. How much of each should he use?

Let us suppose that he uses x liters of the first solution and y liters of the second. Since the total is to be 100 liters we have

$$x + y = 100.$$

The amount of acid in the new mixture is to be 62% of 100 liters, or 62 liters. This is to be made up of the acid in the two solutions to be mixed. These amounts of acid are 80% x and 30% y. Thus we have another equation.

$$80\% \; x + 30\% \; y = 62, \quad \text{or} \quad .8x + .3y = 62.$$

We can eliminate fractions by multiplying on both sides by 10.

$$10(.8x + .3y) = 10 \times 62$$
$$8x + 3y = 620$$

We next solve this system of equations.

$$\begin{aligned} x + y &= 100 \\ 8x + 3y &= 620 \end{aligned}$$

The solution of this pair of equations is $x = 64$ and $y = 36$.

We check: The sum of 64 and 36 is 100. Now 80% of 64 is 51.2 and 30% of 36 is 10.8. These add up to 62. Hence the answer is 64 liters of the stronger acid and 36 liters of the other.

In example 3, note that we got one equation by considering the acid in the solutions. The acid in one plus the acid in the other must be the acid in the new mixture. This line of reasoning works for many mixture problems.

Do exercise 37 at the left.

Example 4.

A grocer wishes to mix some candy worth 45¢ per pound and some worth 80¢ per pound to make 350 pounds of a mixture worth 65¢ per pound. How much of each should he use?

We shall use x and y for the amounts. Then

$$x + y = 350.$$

Our second equation will come from the values. The value of the first candy, in cents, is $45x$ (x pounds at 45¢ per pound). The value of the second is $80y$, and the value of the mixture is 350×65. Thus we have

$$45x + 80y = 350 \times 65$$

Solving the system of equations, we get $x = 150$ lb and $y = 200$ lb.

Do exercise 38 at the left.

Do exercise set 6.8, p. 239.

6.9 LEVER OR TORQUE PROBLEMS (OPTIONAL)

Certain kinds of problems that arise in physics and engineering are solved using a basic scientific fact, sometimes called the "law of the lever." Actually the law applies to any object on which twists, or torques* are applied. Thus it applies to bridges, machinery, etc. The law is easily illustrated using levers or the idea of a see-saw.

A NUMBER LINE FOR REFERENCE

In applying the law of the lever it is helpful first to place a number line for reference. On this lever, for example, on which several forces are acting, a natural placement for the number line is as shown, with 0 at the hinge (fulcrum).

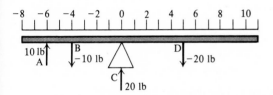

Once the line is placed, each force has a coordinate. The coordinate of force A is −6. The coordinate of force C is 0 and so on.
We shall call upward forces positive and downward forces negative. Thus forces A and C are positive and forces B and D are negative.

TORQUES

The torque due to each force is its twisting effect. This is calculated by multiplying the force by its coordinate. The torques in the above example are as follows:

A: $10 \times {}^-6$ or -60

B: $-10 \times {}^-4$ or 40

C: $20 \times$ 0 or   0

D: $-20 \times$ 5 or -100

The number line could have been placed in a different position. In that case the torques would be different.

---

* Pronounced *torks*.

OBJECTIVES

You should be able to:

a) Compute a torque when a force and a distance are given.

b) State the law of the lever.

c) Translate lever problems to equations and solve.

39. Two boys are on opposite sides of a see-saw, each 4 ft from the fulcrum. One weighs 120 lb and the other 100 lb. A 16-pound bag of sand is placed 5 ft from the fulcrum on the side with the lighter boy. Does this make a balance?

THE LAW OF THE LEVER

The law of the lever is a scientific law. As such it involves something besides mathematics. Physical experiments are required to verify it. One interesting thing about this law is that it does not matter where the number line is placed.

The Law of the Lever: **When forces act on an object, the twisting effects are balanced when the sum of the torques is zero.**

Example 1

Two boys balance each other on a see-saw. One weighs 80 lb and sits 6 ft from the hinge (fulcrum). The other sits 8 ft from the fulcrum. What is his weight?

We first make a drawing and place a number line. It is usually best to put the origin at the fulcrum if there is one.

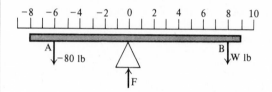

Now we compute the torques:
For force A the torque is -80 lb $\times$ $^-6$ ft or 480 ft-lb.*
For force B the torque is $W \times 8$ or $8W$ ft-lb.
The torque due to force F is 0 since its coordinate is 0.
For a balance we know the sum of the torques must be 0.

$$480 + 8W = 0$$

We solve for W and find that

$$W = -60 \text{ lb}$$

This means the second boy weighs 60 lb. We check:
The torque of force A is 480 ft-lb.
The torque of force B is -60 lb $\times$ 8 ft or -480 ft-lb.
The sum of these torques is 0, so the answer checks.*

In this example we have called the upward force at the fulcrum *F*. Since this force holds things up, it balances *A* and *B*. Therefore $F = 140$ lb. (This assumes that the lever has no weight.)

Do exercise 39 at the left.

---

* When we multiply lb by ft we get a new unit called foot-pounds (ft-lb).

Example 2

Two boys weighing 160 lb and 120 lb balance a see-saw by sitting 7 ft apart. How far is each from the fulcrum?

We first make a drawing. We shall put the origin at the fulcrum.

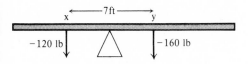

The torques are $-120x$ and $-160y$. Their sum must be 0.

$$-120x + {}^-160y = 0$$

The heavier boy is a distance of $|y|$ from the fulcrum. The other boy is a distance of $|x|$ from the fulcrum. We use absolute value here because we know that one of these numbers is negative, while we are considering *distance* which is not negative. Since the boys are 7 ft apart, we know that

$$|x| + |y| = 7.$$

Now we look at the drawing. If we choose the positive direction to the right, then y is positive and x is negative. Therefore

$$|y| = y \quad \text{and} \quad |x| = {}^-x.$$

Thus we have

$${}^-x + y = 7.$$

We now have a system of equations.

$$-120x + {}^-160y = 0$$
$${}^-x + y = 7$$

We solve and find that $x = -4$ and $y = 3$. This checks, so the answer is that the 160 lb boy is 3 ft from the fulcrum and the other boy is 4 ft from the fulcrum.

We could find the upward force at the fulcrum if we wish. Since it must balance the weights, it is 280 lb, assuming no weight for the lever itself. We do not need to know this force here because its distance from the origin is 0, hence its torque is 0.

Do exercise 40 at the right.

Do exercise set 6.9, p. 241.

40. Two boys weigh 100 lb and 150 lb respectively. They balance a see-saw by sitting 10 ft apart. How far is each from the fulcrum?

NAME

CLASS

ANSWERS

EXERCISE SET 6.1 (pp. 193–196)

1. In which quadrant is each point located?

a) (−5, 3)   b) (7, 3)   c) $\left(\frac{1}{3}, -5\right)$

d) (−12, 1)   e) (100, −1)   f) (−6, −29)

2. In Quadrant III the first coordinate is always a) _____ and the second coordinate is always b) _____.

3. In Quadrant II the a) _____ coordinate is always positive and the b) _____ coordinate is always negative.

4. Use graph paper. Draw a first axis and a second axis. Then plot these points.

a) (−1, 3)   b) (0, 4)   c) (−2, −4)

d) (5, 5)   e) (3, 0)   f) (3, −2)

5. Name the coordinates of points *A, B, C, D* and *E*.

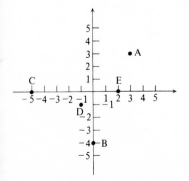

1.

a)_____   b)_____

c)_____   d)_____

e)_____   f)_____

_____

2.

a)_____

b)_____

_____

3.

a)_____

b)_____

_____

5.

A _____

B _____

C _____

D _____

E _____

_____

ANSWERS

Solve the following equations for y.

6. $x + y = 7$ **7.** $2x - y = 3$ **8.** $4 - 2y = x$

6.

9. Complete the table below for $y = 2x - 3$.

x	0	$\frac{3}{2}$	-1	1	-2	2
y						

7.

10. Plot the points of exercise 9 on graph paper. Draw the graph determined by these points.

Draw graphs of the following equations.

11. $y = -3x + 2$ **12.** $y = \frac{1}{2}x + 1$ **13.** $3y + 9x = 7$

8.

14. Complete the table for $y = x^2 + 1$.

x	0	-1	1	-2	2	-3	3
y							

15. Draw the graph determined by the table of exercise 14.

NAME

CLASS

ANSWERS

EXERCISE SET 6.2 (pp. 197–199)

Which of the following equations are linear? If an equation is not linear, give the reason.

1. $3x^2 = 2x + 1$ **2.** $y = 4x - 6$ **3.** $x = 7$

4. $2x - 3 = y^{-1}$ **5.** $3 + xy = x$ **6.** $2x - 5y = 4$

Find the x- and y-intercepts of the following equations:

7. $2x - 1 = y$ **8.** $3x + 4y = 12$ **9.** $5x - 7y = 3$

10. $7x = 2y + 6$ **11.** $y = -4$ **12.** $x = 5$

13–18. Graph the equations of exercises 7 through 12. Use graph paper and different axes for each equation. Label the intercepts and a third point used as a check.

19. What are the x- and y-intercepts of the line $x = 0$?

1. _____

2. _____

3. _____

4. _____

5. _____

6. _____

7. _____

8. _____

9. _____

10. _____

11. _____

12. _____

19. _____

ANSWERS

Without graphing, determine which of the following pairs of lines are parallel.

20. **20.** $x - y = 6$

$x - 7 = y$

21. $2x - 3y = 4$

$7 - 6y = -4x$

22. $2x - y = 8$

$2x + y = 3$

21.

22.

23. $3x + 6 = 2y$

$12 - 4y = -6x$

24. $x = 3$

$y = 4$

23.

24.

25. Write an equation whose graph will be parallel to:

a) $y + 3x = 5$

b) $x = -6$

25.

NAME

CLASS

ANSWERS

EXERCISE SET 6.3 (pp. 200–201)

Translate to a system of equations. Do not attempt to solve. Save these exercises for a later lesson.

1. The sum of two numbers is 27. One number is 3 more than the other. Find the numbers.

     1.

2. Find two numbers whose sum is 58 and whose difference is 16.

     2.

3. The difference between two numbers is 16. Three times the larger number is seven times the smaller number. What are the numbers?

     3.

4. Jerry Mander is 4 times as old as his daughter. Sixteen years from now he will be only twice as old as his daughter. How old are they now?

     4.

5. The perimeter of a rectangle is 400 meters. The length is 40 meters more than the width. Find the width.

5. _____

6. Three times one number is $\frac{1}{4}$ another number. The sum of 4 times the first number and 12 is three times the second number. What are the numbers?

6. _____

7. Two angles are supplementary. One is 8° more than 3 times the other. Find the angles. (Supplementary angles are angles whose sum is 180°).

7. _____

8. A 27-meter rope is cut into two pieces. One piece is 3 meters longer than the other. How long is the shorter piece?

8. _____

9. Al Batross is twice as old as his daughter. In four years Al's age will be three times what the daughter's age was six years ago. How old is each at present?

9.

_____

10. A rectangular lot has a perimeter of 700 feet. The lot is 50 feet longer than it is wide. What are the dimensions of the lot?

10.

11. The reserve football team scored 2 points more than twice the number of points the freshman team scored. A total of 26 points were scored. What were the scores?

_____

11.

12. Three hamburgers and four hot dogs cost a total of $2.15. The hamburgers cost $.25 more than the hot dogs. What is the cost of each hamburger and each hot dog?

_____

12.

_____

13. Ann has 150 coins all nickels and dimes. There are twice as many nickels as dimes. Find the number of each kind of coin.

13.

14. There are 37 students in an algebra class. There are 9 more boys than girls. How many boys and how many girls are in the class?

14.

15. Mary weighs three pounds more than Jane, and together they weigh 16 pounds less than the center on the football team who weighs 216 pounds. How much does each girl weigh?

15.

EXERCISE SET 6.4 (pp. 202–203)

Decide whether the given ordered pair is a solution of the system of equations following it. Remember to use alphabetical order of variables. Write yes or no.

1. (3, 2); $2r + 3s = 12$
$\qquad\qquad r - 4s = -5$

2. $\left(1, -\dfrac{1}{2}\right)$; $3a - 4b = 5$
$\qquad\qquad\qquad a + 7b = 10$

3. (1, 5); $5x - 2y = -5$
$\qquad\qquad 3x - 7y = -32$

4. (15, 20); $3x - 2y = 5$
$\qquad\qquad\quad 6x - 5y = -10$

5. (3, 2); $3t - 2s = 0$
$\qquad\qquad t + 2s = 15$

6. (12, 9) ; $\dfrac{n}{d} = \dfrac{3}{4}$
$\qquad\qquad\quad \dfrac{n + 7}{d} = \dfrac{4}{3}$

1. _____

2. _____

3. _____

4. _____

5. _____

6. _____

ANSWERS

Use graph paper. Solve graphically each system of equations. Check your "solution" on this paper.

7.

7. $x + 2y = 10$

$3x + 4y = 8$

Check: $x + 2y = 10$ $3x + 4y = 8$

8.

8. $u = v$

$4u = 2v - 6$

Check: $u = v$ $4u = 2v - 6$

9.

9. $8x - y = 29$

$2x + y = 11$

Check: $8x - y = 29$ $2x + y = 11$

10.

10. $4x - y = 10$

$3x + 5y = 19$

Check: $4x - y = 10$ $3x + 5y = 19$

11.

11. $a = 3b$

$3b - 6 = 2a$

Check: $a = 3b$ $3b - 6 = 2a$

12.

12. $x - 2y = 6$

$2x - 3y = 5$

Check: $x - 2y = 6$ $2x - 3y = 5$

NAME

CLASS

ANSWERS

EXERCISE SET 6.5 (pp. 204–206)

Solve by the substitution method. Remember, each solution is an ordered pair.

1. $x + y = 4$

$y = 2x + 1$

2. $x + y = 10$

$y = x + 8$

3. $y = x + 1$

$2x + y = 4$

4. $x - y = 6$

$x + y = -2$

5. $3x - 7y = 5$

$2x + y = 9$

6. $3x + 7y = 17$

$y = 2x$

7. $3x - 2y = -4$

$x = 12 - 2y$

8. $x - y = -3$

$2x + 3y = -6$

1.

2.

3.

4.

5.

6.

7.

8.

ANSWERS

9.

_____

10.

_____

11.

_____

12.

_____

13.

_____

14.

_____

15.

_____

16.

_____

9. $x + y = -2$

$y = 2x + 1$

10. $2x + y = 8$

$x - 2y = 1$

11. $x + 2y = 7$

$x - y = 5$

12. $2x + 3y = -2$

$2x - y = -9$

13. $x - 2y = 6$

$3x + 2y = 4$

14. $x + 2y = 10$

$3x + 4y = 8$

15. $y - 2x = -6$

$2y - x = 5$

16. $2a + 3b = 2$

$a - 2b = 8$

NAME

CLASS

ANSWERS

EXERCISE SET 6.6 (pp. 207–211)

Solve using the addition method.

1. $x + y = 10$

$x - y = 8$

2. $8x - y = 29$

$2x + y = 11$

3. $3a + 4b = 7$

$a - 4b = 5$

4. $x + 3y = 19$

$x - y = -1$

5. $x + y = 5$

$5x - 3y = 17$

6. $2a + b = -2$

$6a - 5b = 18$

7. $46 = 4x + 3z$

$14 = 2x - 3z$

8. $2r + 3s = 12$

$r - 4s = -5$

9. $5a - 2b = 0$

$2a - 3b = -11$

10. $2w - 3z = -1$

$3w + 4z = 24$

1.

2.

3.

4.

5.

6.

7.

8.

9.

10.

ANSWERS

11. $2a + 3b = -1$

$\quad\; 3a + 5b = -2$

12. $7p + 5q = 2$

$\quad\; 8p - 9q = 17$

11. _____

12. _____

13. $3x - 4y = 16$

$\quad\; 5x + 6y = 14$

14. $10t + 15u = 3$

$\quad\; 6t + 9u = 4$

13. _____

14. _____

15. $10x - 7 = y$

$\quad\; 3x + y = 6$

16. $0.3x + 0.2y = 0$

$\quad\; x + 0.5y + 0.5 = 0$

15. _____

16. _____

17. _____

17. $2x - 3z = 5$

$\quad\; 4x - 6z = 10$

18. $\dfrac{3}{4}x + \dfrac{1}{3}y = 8$

$\quad\; \dfrac{1}{2}x - \dfrac{5}{6}y = -1$

18. _____

19. _____

19. $3x - 8y = 11$

$\quad\; x + 6y - 8 = 0$

20. $4x - 5y = 2$

$\quad\; x = \dfrac{y + 7}{3}$

20. _____

NAME

CLASS

ANSWERS

EXERCISE SET 6.6A

Solve by any method.

1. $3x - y = 8$
   $x + 2y = 5$

2. $x - 3y = 0$
   $5x - y = -14$

3. $5x - 3y = 21$
   $x + 2y = -1$

4. $4x + y = 1$
   $6x + 2y = 3$

5. $6x - 2y = 11$
   $2x + y = 2$

6. $y = 1 - 2x$
   $y = x + 7$

7. $m - n = 32$
   $3m - 8n - 6 = 0$

8. $2z + 3t = 7$
   $z - 3t = 8$

9. $2p + 5q = 9$
   $3p - 2q = 4$

10. $3x - 2y = 10$
   $5x + 3y = 4$

1. _____

2. _____

3. _____

4. _____

5. _____

6. _____

7. _____

8. _____

9. _____

10. _____

ANSWERS

11. $5t - 3s = -4$

$\quad 4t + 7s = 25$

12. $x + 3y = 1$

$\quad 3x + y = 11$

11. _____

13. $a - 3b = 7$

$\quad 2a + 5b = -19$

14. $4x - y = 4$

$\quad y - x = 2$

12. _____

13. _____

15. $\dfrac{x}{3} + \dfrac{y}{2} = 3$

$\quad \dfrac{x}{4} - \dfrac{y}{3} = -6\dfrac{1}{4}$

16. $x + 2y = 1$

$\quad 6x - 5y = 3\dfrac{1}{6}$

14. _____

15. _____

17. $4(x - 2) = 2(y + 1)$

$\quad y + \dfrac{1}{3}(x + 1) = y + x - 3$

18. $6y = 5 + 3x$

$\quad y = 3x$

16. _____

19. $\dfrac{x + 3}{2} - \dfrac{4y - 11}{3} = 3x - 5y$

$\quad \dfrac{x + 1}{6} - \dfrac{2y + 2}{3} = \dfrac{2x - 6y}{2}$

17. _____

18. _____

20. $\dfrac{1}{2}(x + y - 7) - \dfrac{2}{3}(2x + y) = -8$

$\quad 8y - \dfrac{1}{3}(4y - x) = \dfrac{1}{4}(x - 5y + 65)$

19. _____

20. _____

EXERCISE SET 6.7 (pp. 212–214)

In the following exercises, translate to equations, then solve. Write the equations and the solution in the spaces provided.

1. The first of two numbers is three times the second. If 15 is added to each, the first result is twice the second result. What are the two numbers?

2. The sum of three numbers is 62. The second is half of the first and the third is 6 more than the second. What are the numbers?

3. Dick Jacobs has 27 coins in quarters and dimes. There are 7 more quarters than dimes. How many of each kind of coin does he have?

4. The perimeter of a rectangle is 100 centimeters. If the width is doubled and the length is decreased by 10 centimeters, the perimeter remains the same. What are the width and length of the rectangle?

5. Dicky and Micky start from home on their bikes and travel in opposite directions. Micky's speed is twice Dicky's. In 4 hours they are 72 kilometers apart. Find the speed at which each travels.

1.

2.

3.

4.

5.

ANSWERS

6.

_____

_____

_____

7.

_____

_____

_____

8.

_____

_____

_____

9.

_____

_____

_____

6. A train leaves a station and travels at 45 miles an hour. Three hours later a second train leaves and travels at 75 miles an hour. When will it overtake the first train?

7. It takes a passenger train 2 hours less time than it takes a freight train to make the trip from Central City to Clear Creek. The passenger train averages 60 miles per hour on the trip while the freight train averages 40 miles per hour. How far is it from Central City to Clear Creek?

8. An airplane is 75 miles directly behind a ship sailing a straight course. The speed of the plane is 120 mph and that of the ship is 20 mph. How long will it take the plane to overtake the ship?

9. Horace Cope paddles a canoe at a speed of 5 miles per hour in still water. He makes a trip down the river and then back in 10 hours when the river is flowing at 2 mph. How far down the river does Horace go?

EXERCISE SET 6.8 (pp. 215–216)

Translate the following problems to equations, then solve. Write the equations and the solution in the spaces provided.

1. A collection of dimes and quarters is worth $15.25. There are 103 coins in all. How many of each are there?

2. A collection of quarters and nickels is worth $1.25. There are 13 coins in all. How many of each are there?

3. A collection of quarters and half dollars is worth $20. There are 51 coins in all. How many of each are there?

4. A collection of nickels and dimes is worth $25. There are three times as many nickels as dimes. How many of each are there?

5. A collection of nickels and dimes is worth $2.90. There are 19 more nickels than dimes. How many of each are there?

1.

2.

3.

4.

5.

ANSWERS

6.

_____

_____

_____

7.

_____

_____

_____

8.

_____

_____

_____

9.

_____

_____

_____

10.

_____

_____

6. Solution A is 50% acid and solution B is 80% acid. How much of each should be used to make 100 grams of a solution that is 68% acid?

7. Solution A is 30% alcohol and solution B is 75% alcohol. How much of each should be used to make 100 liters of a solution that is 50% alcohol?

8. Farmer Jones has 100 liters of milk that is 4.6% butterfat. How much skim milk (no butterfat) should he mix with it to make milk that is 3.2% butterfat?

9. A candy mix sells for $1.10 per pound. It contains chocolates worth 90¢ per pound and other candy worth $1.50 per pound. How much of each are in 30 lb of the mixture?

10. There were 429 people at a play. Admission was $1 for adults and 75¢ for children. The receipts were $372.50. How many adults and how many children attended?

NAME

CLASS

ANSWERS

EXERCISE SET 6.9 (pp. 217–219)

Translate the following problems to equations, then solve. Write the equations and the solution in the spaces provided.

1. A bar 6 feet long is to be used as a lever. Where would a man weighing 175 pounds have to place the fulcrum in order to lift an object weighing 1400 pounds?

1. _____

_____

_____

2. V. Gates weighs 135 kilograms. What is the heaviest rock he can lift by standing on a lever if the fulcrum is 60 centimeters from the rock and the lever is 120 centimeters long?

2. _____

_____

_____

3. Kay Pubble and her younger sister sit on opposite sides of a balanced seesaw. Kay weighs 10 kilograms more than her sister. Kay sits 120 centimeters from the fulcrum and her sister sits 160 centimeters from the fulcrum. Find the weight of each girl.

3. _____

_____

_____

4. If a 240 kilogram weight is to rest on a beam 3 meters from the fulcrum, what weight must be placed on the other side of the fulcrum and 2 meters from it so that the beam will balance?

4. _____

_____

ANSWERS

5. Bud and Bob balance Neal on a see-saw. Bob weighs 76 pounds and sits 1 foot farther from the fulcrum than Bud, who weighs 88 pounds. Neal weighs 212 pounds and balances Bud and Bob by sitting 5 feet from the fulcrum. How far from the fulcrum do Bud and Bob sit?

5.

_____

6. A boy weighing 75 pounds sits on a see-saw 6 feet from the fulcrum, and another boy weighing 105 sits 8 feet from the fulcrum, but on the other side. To balance the board, where should a third boy sit who weighs 60 pounds?

_____

6.

_____

7. George Norge wants to move a 150-kilogram rock. He inserts the end of a board under the rock and then places a log under the board 30 centimeters from the rock. How much force (weight) must be used on the other end of the board to move the rock if the distance from the log to the other end is 1 meter?

7.

_____

8. A lever balances when weights of 12 kilograms and 16 kilograms are at the ends of the lever. If 4 kilograms are added to the heavier weight, the fulcrum must be moved 10 centimeters closer to it in order to balance the lever. How long is the lever?

8.

_____

NAME_____

CLASS_____SCORE_____GRADE_____

CHAPTER 6 TEST

Before taking the test *be sure* to allow yourself a day or so for review. Use the objectives listed in the margins to guide your study. The test will evaluate your progress and aid your preparation for a possible classroom test. Allow about an hour for the test. Remove the test from the book. When you finish read the test analysis on the answer page at the end of the book.

Solve the following equations for y:

1. $3x + y = 5$ **2.** $3y - 8 = 4x$

3. Complete the following table for $y = x^2 - 3$.

x	-2	-1	0	1	2
y					

4. Use graph paper. Graph $3x + 2y = 6$.

5. What are the x- and y-intercepts of $2x + 5y = 6$?

6. Which of the following pairs of lines are parallel?

a) $3x - y = 4$ b) $x = 4$ c) $2x - 3y = 6$

   $2x + 5y = 6$ $y = -1$ $10x - 6 = 15y$

ANSWERS

1.

_____

2.

_____

3.

See table.
_____

4.

See graph.
_____

5.

_____

6.

_____

ANSWERS

7.

_____

8.

_____

9.

_____

10.

_____

11.

_____

Use the table to approximate the following.

Translate the following to systems of equations. Do not solve.

7. The sum of two numbers is 22. One number is 8 more than the other. Find the numbers.

8. A newsboy notices that he has 4 times as many dimes as quarters. His dimes and quarters total $1.95. How many of each has he?

Determine whether the given ordered pair is a solution of the given system of equations. Write yes or no.

9. $(-1, 3)$; $x + 2y - 5 = 0$
 $2x + y = 1$

10. $(-2, -2)$; $2x - 3y = 2$
 $x + 2y = 8$

11. $(4, -2)$; $2x - y = 5$
 $3x + 2y = -4$

Solve by any method.

ANSWERS

12. $x + y = 14$
$\quad\;\; x - y = 22$

13. $3x - 2y = -7$
$\quad\;\; 2x - 5y = 10$

14. $2x - 3y - 22 = 0$
$\quad\;\;\;\; y = 6 - x$

15. $\quad x + 3y = 3.5$
$\quad\;\; 2x + 4y = 3$

16. $4x - 3y = 8$
$\quad\;\; 2x + y + 7 = 0$

17. $4c - 3d = 3$
$\quad\;\; 3c - 2d = 4$

18. $2x + 3y = 10$
$\quad\;\; 6y - 20 = -4x$

19. $\quad 4x + 5y = 5$
$\quad\;\; 6x + 7y = 7$

20. $2x + 3y = 13$
$\quad\;\; 3x - 5y = 9$

21. $2x + 4y = 10$
$\quad\;\; x + \;\; y = 2$

22. $3x + 2y = 12$
$\quad\;\; 2x + y = 5$

23. $\quad 2x - y = 5$
$\quad\;\; 3x + 2y = -4$

12.
_____

13.
_____

14.
_____

15.
_____

16.
_____

17.
_____

18.
_____

19.
_____

20.
_____

21.
_____

22.
_____

23.
_____

ANSWERS

Solve the problems.

24. A motorist traveling 55 mph is being pursued by a highway patrol car traveling 65 mph. The patrol car is 4 miles behind the motorist. How long will it take the patrol car to over-take the motorist?

24. _____

25. Weights of 60 and 90 kilograms are placed at the ends of a lever 9 meters long. Where should the fulcrum be placed to make the lever balance?

25. _____

26. The difference between two numbers is 16. Three times the larger number is seven times the smaller number. What are the numbers?

26. _____

7 INEQUALITIES

7.1 THE ADDITION PRINCIPLE

For solving equations there is an addition principle. For solving inequalities there is a similar one. Consider the inequality

$3 < 7$

It is true. Let us add 2 to both numbers. We get another true inequality.

$5 < 9$

Similarly if we add -3 to both numbers we get another true inequality

$0 < 4$

The Addition Principle for Inequalities: **If any number is added to both members of a true inequality another true inequality is obtained.**

The addition principle holds whether the inequality contains $<$ or $>$. The symbol $\leqslant$ means *less than or equal to*. The symbol $\geqslant$ means *greater than or equal to*. The addition principle also holds for inequalities containing $\leqslant$ or $\geqslant$.
Let's see how we use the addition principle to solve inequalities.

Example 1

Solve $x + 2 > 4$

We use the addition principle, adding -2:

$x + 2 + \boxed{-2} > 4 + \boxed{-2}$

$x > 2$

Any number greater than 2 makes the last sentence true, hence is a solution of that sentence. Any such number is also a solution of the original sentence. Hence we have it solved.
We cannot check the solutions of an inequality, by substitution, as we can check solutions of equations. There are too many of them. However, we don't really need to check. Let us see why. Consider the first and last inequalities

$x + 2 > 4 \quad \text{and} \quad x > 2$

You should be able to:

a) State the addition principle for inequalities.

b) Use the addition principle to solve simple inequalities, such as

$x + 8 < 9$ and $4y - 3 > 5y + 9$

Solve:

1. $x + 3 < 5$

2. $x - \dfrac{3}{4} > \dfrac{3}{8}$

3. $5x - 1 > 4x - 2$

Any number that makes the first one true must make the last one true. We know this by the addition principle. Now the question is, will any number that makes the last one true also be a solution of the first one? Let us use the addition principle again, adding 2.

$x > 2$

$x + 2 > 4$

Now we know that any number that makes $x > 2$ true also makes $x + 2 > 4$ true. Therefore the sentences $x > 2$ and $x + 2 > 4$ have the same solutions. Any time we use the addition principle a similar thing happens. Thus whenever we use the principle with inequalities the first and last sentences will have the same solutions.

Example 2

Solve $3x + 1 < 2x - 3$

We use the principle, adding $-2x$ and also -1:

$3x + 1 + \boxed{-2x - 1} < 2x - 3 + \boxed{-2x - 1}$

We collect like terms and simplify:

$x < -4$

Now we know that any number less than -4 is a solution.

4. $5y + 2 < -1 + 4y$

Example 3

Solve $x + \dfrac{1}{3} \geqslant \dfrac{3}{4}$

$x + \dfrac{1}{3} - \dfrac{1}{3} \geqslant \dfrac{3}{4} - \dfrac{1}{3}$ $\left(\text{adding } -\dfrac{1}{3}\right)$

$x \geqslant \dfrac{3}{4} \cdot \dfrac{3}{3} - \dfrac{1}{3} \cdot \dfrac{4}{4}$

$x \geqslant \dfrac{9}{12} - \dfrac{4}{12}$

$x \geqslant \dfrac{5}{12}$

Now we know that any number greater than or equal to $\dfrac{5}{12}$ is a solution.

Do exercises 1 and 2 on the preceding page and 3, 4 at the left.

Do exercise set 7.1, p. 255.

7.2 THE MULTIPLICATION PRINCIPLE

The multiplication principle for inequalities is somewhat different from the one for equations. Consider the inequality

$3 < 7$

It is true. Let us multiply both members by 2. We get another true inequality

$6 < 14$

Let us multiply both members by -3.

$-9 < -21$

This time the new inequality is false. However, if we reverse the inequality symbol (use $>$ instead of $<$) we will get a true inequality

$-9 > -21$

The Multiplication Principle for Inequalities: **If both members of a true inequality are multiplied by a positive number, another true inequality is obtained.**

If both members are multiplied by a negative number and the inequality sign is reversed, another true inequality is obtained.

The multiplication principle also holds for inequalities containing $\geq$ or $\leq$.

Example 1

Solve $-2y < \dfrac{5}{6}$

We use the multiplication principle, multiplying by $-\dfrac{1}{2}$.

$$-\frac{1}{2} \cdot (-2y) > -\frac{1}{2} \cdot \frac{5}{6}$$

$$y > -\frac{5}{12}$$

Any number greater than $-\dfrac{5}{12}$ is a solution.

Do exercises 5 and 6 at the right.

OBJECTIVES

You should be able to:

a) State the multiplication principle for inequalities.
b) Solve simple inequalities using the addition and multiplication principles.

Solve using the multiplication principle.

5. $-3x > \dfrac{3}{2}$

6. $5y < \dfrac{2}{3}$

Solve.

7. $3x + 2x > 11$

8. $4x + 2 < 2x + 3$

9. $-7x + 5 \geqslant -5x + 3$

We use the addition and multiplication principles together in solving inequalities in much the same way as for equations.

Example 2

Solve $6 - 5y > 7$

$-6 + 6 - 5y > -6 + 7$ (adding -6)

$-5y > 1$ (simplifying)

$-\dfrac{1}{5} \cdot (-5y) < -\dfrac{1}{5} \cdot 1$ $\left(\text{multiplying by } -\dfrac{1}{5}\right)$

$y < -\dfrac{1}{5}$ (simplifying)

Example 3

Solve $5x + 4x \leqslant 3$

$9x \leqslant 3$ (collecting like terms)

$\dfrac{1}{9} \cdot 9x \leqslant \dfrac{1}{9} \cdot 3$ $\left(\text{multiplying by } \dfrac{1}{9}\right)$

$x \leqslant \dfrac{1}{3}$ (simplifying)

Example 4

Solve $17 - 5y < 8y - 5$

$-17 + 17 - 5y < 8y - 5 - 17$ (adding -17)

$-5y < 8y - 22$ (simplifying)

$-8y - 5y < -8y + 8y - 22$ (adding $-8y$)

$-13y < -22$ (simplifying)

$-\dfrac{1}{13} \cdot (-13y) > -\dfrac{1}{13} \cdot (-22)$ $\left(\text{multiplying by } -\dfrac{1}{13}\right)$

$y > \dfrac{22}{13}$ (simplifying)

Do exercises 7 through 9 at the left.

Do exercise set 7.2, p. 257.

7.3 GRAPHS OF INEQUALITIES

We shall consider graphs of inequalities in one variable first. We graph them on a number line.

Example 1

Graph $x < 2$

The solutions of $x < 2$ are those numbers less than 2, and are shown on the graph by shading all points to the left of 2 on the line.

Example 2

Graph $x \geqslant -3$

The solutions of $x \geqslant -3$ are -3 and all numbers greater than -3. These are shown by shading the point for -3 and all points to the right of -3.

Example 3

Graph $3x + 2 < 5x - 1$

We first solve for x:

$2 < 2x - 1$ (adding $-3x$)

$3 < 2x$ (adding 1)

$\dfrac{3}{2} < x$ $\left(\text{multiplying by } \dfrac{1}{2}\right)$

Do exercises 10 and 11 at the right.

You should be able to:

a) Graph simple inequalities **in** one variable on a number line.

b) Graph linear inequalities in two variables on a plane.

Graph on a number line.

10. $x + 2 > 5$

11. $4 \leqslant 3x + 2$

Graph on a number line.

12. $|x| > 3$

13. $|x| \leqslant 4$

Example 4

Graph $|x| < 3$

The absolute value of a number is its distance from 0 on a number line. Thus for the absolute value of a number to be less than 3 the number must be between 3 and -3. Hence we shade the points between these two numbers.

Example 5

Graph $|x| \geqslant 2$

For the absolute value of a number to be greater than or equal to 2 its distance from 0 must be 2 or more. Thus the number must be 2 or greater; or it must be less than or equal to -2. Hence we shade the point for 2 and all points to its right. We also shade the point for -2 and all points to its left.

Do exercises 12 and 13 at the left.

INEQUALITIES IN TWO VARIABLES

We shall consider linear (first degree) inequalities in two variables only. We graph such inequalities on a plane. To see what these graphs look like, let us consider an example. We wish to graph the inequality $y > x + 2$. We first graph the equation $y = x + 2$ for comparison.

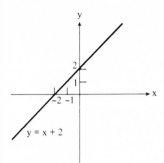

Now we compare $y = x + 2$ and $y > x + 2$. For all the points on the line, y is equal to $x + 2$. For all points above the line y will be greater than $x + 2$. Hence our graph consists of all points above the line $y = x + 2$ but not on it. To show this we draw a dashed line where $y = x + 2$ belongs and then shade the portion (a half-plane) above it.

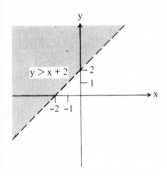

Example 6

Graph $y \leqslant x - 1$

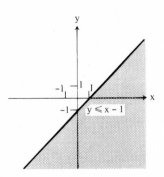

Use graph paper.

Graph these inequalities.

14. $y \geqslant 2x + 1$

We first sketch the line $y = x - 1$. All points on this line are in the graph, so we draw a solid line. Also, all points on the half-plane below the line are in the graph, so we shade the lower half-plane.

Any linear inequality in two variables has a graph which is either a half-plane or a half-plane and the line along its edge. So to graph such inequalities we first sketch the line obtained by using an equals sign instead of the inequality sign. Then determine which half-plane is to be shaded. This can be determined by trying just one point off the line.

Example 7

Graph $2x + 3y < 6$

We first graph the line $2x + 3y = 6$.

The intercepts are $(0,2)$ and $(3,0)$.

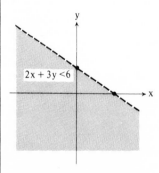

15. $2x - 5y < 10$

We try one point off the line. The origin is usually an easy one to use. $2 \cdot 0 + 3 \cdot 0 < 6$ is true, so the origin is in the graph. This means we shade the lower half-plane.

If the line goes through the origin, then we must test some other point not on the line. The point $(1, 1)$ is often a good one to try.

Do exercises 14 and 15 at the left.

Do exercise set 7.3, p. 259.

EXERCISE SET 7.1 (pp. 247–248)

1. Fill in the blanks to make the statement true. If any —————— is added to both members of a true inequality another true ——————— is obtained.

Which of the following are true? Write T or F.

2. If $x < 5$, then $x < 8$.

3. If $x - 3 < 7$, then $x < 7$.

4. If $x > -5$, then $x > -2$.

5. If $2x - 5 \geqslant 8$, then $2x - 3 \geqslant 9$.

6. If $x - 6 \leqslant -4$, then $x \geqslant 2$.

7. If $x - 7 > -12$, then $x > -10$.

Solve using the addition principle.

8. $x + 7 > 2$

9. $5x - 6 < 4x - 2$

10. $\dfrac{2}{3} - \dfrac{3}{4}x \leqslant \dfrac{1}{4}x - \dfrac{5}{12}$

11. $3x + 4 \geqslant 2x + 7$

12. $3 - 5y > -6 - 4y$

13. $\dfrac{2x - 3}{2} > \dfrac{x - 5}{2}$

14. $5 - x < \frac{2}{3} - 2x$

15. $.2x - 3.01 < .8x - 6.3$

ANSWERS

1.

2.

3.

4.

5.

6.

7.

8.

9.

10.

11.

12.

13.

14.

15.

ANSWERS

16. _____

17. _____

18. _____

19. _____

20. _____

21. _____

22. _____

23. _____

24. _____

25. _____

26. _____

27. _____

28. _____

29. _____

30. _____

31. _____

32. _____

33. _____

16. $6z + 7 < 2(3z + 2)$

17. $\dfrac{4}{5} - 3x > -\dfrac{1}{3} - 2x$

18. $.05 + 3(y - .2) > 4y - \dfrac{3}{5}$

19. $\dfrac{3}{5}x - 6 < -1 + .6x$

20. $-7 + c > 7$

21. $-5 \leqslant t - 1 + \dfrac{2}{3}$

22. $-14x + 21 > 21 - 15x$

23. $\dfrac{2x - 6}{5} \geqslant \dfrac{x - 1}{5} + 3$

24. $r + 9 > -2$

25. $2 + \dfrac{z}{3} < -2$

26. $\dfrac{2}{3}x + \dfrac{1}{4}x - 5 < -7 - \dfrac{1}{12}x$

27. $-5(.2 - x) > 5x - 2$

28. $5y + \dfrac{2}{9} < 4y - \dfrac{1}{6}$

29. $\dfrac{3}{7}(.1 + x) \leqslant \dfrac{2}{35} - \dfrac{4}{7}x$

30. $x + 2\left(\dfrac{3}{4} - x\right) < -2x$

31. $8 - 2k \geqslant -3 - 3k$

32. $4 - 3y < 2(3 - 2y)$

33. $2(r + 4) < 2r + 4$

EXERCISE SET 7.2 (pp. 249–250)

Fill in the blanks with inequality symbols to make true statements.

1. If $a < b$ and c is positive, then ac_____bc.

2. If c is negative and $a \leqslant b$, then ac_____bc.

Which of the following are true? Write T or F.

3. If $x < 8$, then $5x < 40$

4. If $x > 3$, then $-2x > -6$

5. If $-3 < 5$, then $-3x < 5x$

Solve the following inequalities.

6. $2x + 5 < 3$

7. $5b + 7 \geqslant b - 1$

8. $6 - 4y > 4 - 3y$

9. $m + 3 - 4m > 2m + 23$

10. $-2x + 5 < -3$

11. $-6(2 + s) \leqslant 2(3s - 6)$

12. $\dfrac{2x - 3}{4} > \dfrac{x - 1}{3}$

13. $3 - 6c > 15$

ANSWERS

1. _____

2. _____

3. _____

4. _____

5. _____

6. _____

7. _____

8. _____

9. _____

10. _____

11. _____

12. _____

13. _____

14. $5x - 3 < 2x + 3$

15. $\dfrac{x - 3}{2} + x > 4 - \dfrac{x - 5}{2}$

16. $10 - y > 8 - (y - 2)$

17. $3(3 - k) - 10 > 2$

18. $2(x - 5) - 6(x + 3) > 0$

19. $\dfrac{3 + x}{8} < \dfrac{15}{24}$

20. $5(12 - 3t) \geqslant 15(t + 4)$

21. $15x - 21 \geqslant 8x + 7$

22. $21 - 15x < -8x - 7$

23. $6(z - 5) < 15 + 5(7 - 2z)$

24. $3(2r + 3) - (3r + 2) > 12$

25. $3[1 - (a - 2)] \leqslant 3 - 4(1 - a)$

26. $-2(t - 5) - 1 \leqslant 5t + 7(1 - t)$

27. $4(y + 1) - 2(y + 3) > 2(y - 1)$

EXERCISE SET 7.3 (pp. 251–254)

Graph on a number line.

1. $7 < x + 2$

2. $4(x - 7) > x + 2$

3. $m - \dfrac{1}{2} \leqslant \dfrac{2}{3}m + 2$

4. $\dfrac{2x - 3}{4} > \dfrac{x - 1}{3}$

5. $10 - y \geqslant 8 - (y - 2)$

6. $5(c + 3) + 4 < c - 1$

7. $\dfrac{3}{2}w + \dfrac{1}{6} \leqslant \dfrac{7}{6}$

8. $3(y - 4) - (6y + 2) \geqslant 5 - 3y$

9. $5(4x + 3) - 7(3x - 4) \leqslant 10$

10. $-5[2 - (1 + b)] \geqslant 25 + 7(2b - 3)$

11. $|x| \leqslant 2$

12. $|x| > 5$

13. $|x| - 2 = 6$

14. $|x| - 4 > 3$

Use graph paper. Graph these inequalities.

15. $x + y > -1$ **16.** $2x - y \leqslant 4$

17. $2x + y \geqslant 5$ **18.** $x < 2$

19. $y < 3x - 2$ **20.** $y \geqslant 2x$

21. $2x - y < 3$ **22.** $y \leqslant 4$

23. $2x + 3y < 12$ **24.** $y < 3x$

25. $y < |x|$ **26.** $y \geqslant |x|$

27. $2x + 3y \leqslant 12$ **28.** $y > 2x$
       $3x - 2y < 6$ $y < 2x + 4$

NAME_____

CHAPTER 7 TEST

CLASS_____SCORE_____GRADE_____

ANSWERS

Before taking the test *be sure* to allow yourself a day or so for review. Use the objectives listed in the margins to guide your study. The test will evaluate your progress and aid your preparation for a possible classroom test. Allow about an hour for the test. Remove the test from the book. When you finish read the test analysis on the answer page at the end of the book.

1.

Make the following statements true by inserting the correct inequality symbol.

1. If $a > b$ and c is negative, then ac _____ bc.

_____

2. If $a \leqslant b$ and c is positive, then ac _____ bc.

Solve the inequalities.

2.

3. $3x + 4 < 2x + 7$

4. $\dfrac{2}{3} - \dfrac{3}{4}x < \dfrac{1}{3}x - \dfrac{5}{12}$

_____

3.

5. $3 + 2(3n - 5) \geqslant 2n - 3(1 - 2n)$

_____

4.

Graph the inequalities.

6. $6x - 3 < x + 2$

7. $-3x + 1 > -8$

8. $|x| \leqslant 2$

9. $|x| > 1$

5.

_____

Graph the inequalities.

10. $y \geqslant -2$

11. $y < -5x + 3$

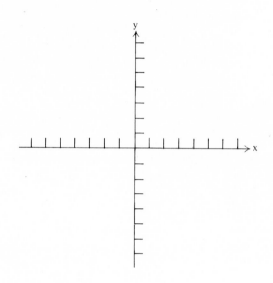

12. $x \leqslant y$

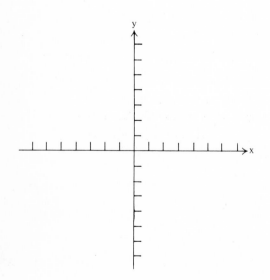

13. $3x - 4y > 12$

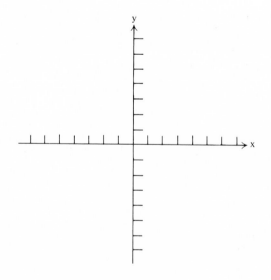

8 POLYNOMIALS IN SEVERAL VARIABLES

8.1 WHAT ARE POLYNOMIALS?

Most of the polynomials you have studied so far have had only one variable. A polynomial in several variables is an expression like those you have already seen, but we allow that there can be more than one variable. Here are some examples.

$$3x + xy^2 + 5y + 4$$

$$8xy^2z - 2x^3z - 13x^4y^2 + 5$$

Each term has a _coefficient, which is a number._ The variables can be raised to various powers.

DEGREES

The _degree_ of a term is the sum of the exponents of the variables.

Examples

Term	Degree		Coefficient
$5x^2$	2		5
$9x^2y^3$	5		9
$7xy$	2		7
$-14xy^2z^3$	6		-14
7	0	(think: $7 = 7x^0$)	7
$4y$	1	(think: $4y = 4y^1$)	4

The _degree_ of a polynomial is the same as that of its term of highest degree.

Do exercises 1 and 2 at the right.

SIMILAR TERMS

In order that terms be similar (or be "like" terms) they must have exactly the same variables and exactly the same exponents.

Examples

Similar terms:

a) $3x^2y^3$ and $-7x^2y^3$

b) $9y^4z^7$ and $12y^4z^7$

Terms not similar:

c) $13xy^2$ and $-2x^2y$

d) $3xyz^2$ and $4xy$

You should be able to:

a) Identify the coefficients of the terms of a polynomial.

b) Determine the degree of a term of a polynomial in several variables.

c) Arrange a polynomial in ascending or descending powers of a given variable.

d) Simplify by combining like terms.

1. What is the coefficient of each term?

$$-3xy^2 + 3x^2y - 2y^3 + xy + 2$$

2. What is the degree of each term?

$$4xy^2 + 7x^2y^3z^2 - 5x + 2y + 4$$

263

Combine like terms.

3. $4x^2y + 3xy - 2x^2y$

4. $-3pq - 5pqr^3 + 8pq + 5pqr^3 + 4$

5. Arrange in descending powers of y.

$3xy - 7xy^2 + 5xy^4 - 3xy^3$

6. Arrange in ascending powers of x.

$2x^2yz + 5xy^2z + 5x^3yz^2 - 2$

Collecting like terms (or "combining similar terms") is based on the distributive law as it was for polynomials in one variable. Similar terms are combined in just the same way.

Examples. Combine like terms

a) $5x^2y + 3xy^2 - 5x^2y - xy^2 = (5 - 5)x^2y + (3 - 1)xy^2$
$$= 2xy^2$$

b) $3xy - 5xy^2 + 3xy^2 + 9xy = -2xy^2 + 12xy$

c) $4ab^2 - 7a^2b^2 + 9a^2b^2 - 4a^2b = 4ab^2 + 2a^2b^2 - 4a^2b$

d) $3pq + 5pqr^3 - 8pq - 5pqr^3 - 4 = -5pq - 4$

Do exercises 3 and 4 at the left.

ASCENDING AND DESCENDING ORDER

We usually arrange polynomials in one variable so that the exponents decrease (descending order) or so that they increase (ascending order). For polynomials in several variables we choose one of the variables and arrange the terms with respect to it.

Examples

a) This polynomial is arranged in descending powers of x.
$3x^5y - 5x^3 + 7xy + y^4 + 2$

b) This polynomial is arranged in ascending powers of y.
$3 + 13xy - x^3y^2 + 4xy^3$

c) This polynomial is arranged in ascending powers of p.
$3pq - 2p^2qr + 7p^3q^5r^2 + 5p^4q^2r^4 - 3p^5r^2$

d) This polynomial is arranged in descending powers of t.
$5s^5t^7 + 2s^6t^4 - 7s^6t^2 + 4s^3t + 6s^4$

Do exercises 5 and 6 at the left.

Do exercise set 8.1, p. 275.

8.2 OPERATIONS WITH POLYNOMIALS

Calculations with polynomials in several variables are very much like those for polynomials in one variable. Additional practice is needed, however, and you should strive for as much speed as possible without losing accuracy.

ADDITION

To add polynomials in several variables, we combine like terms. The use of columns is often helpful.

Example 1. Add.

$3ax^2 + 4bx - 2$ and $2ax^2 + 5bx + 4$

We can first write one polynomial under the other, keeping similar terms in columns, like this. Then add.

$3ax^2 + 4bx - 2$
$2ax^2 + 5bx + 4$
$\overline{5ax^2 + 9bx + 2}$

Although the use of columns is helpful for complicated examples, you should attempt to write only the answer in simple cases like the example above.

Example 2. Add.

$(5xy^2 - 4x^2y + 5x^3 + 2) + (3xy^2 - 2x^2y + 3x^3y - 5)$

We look at the two polynomials, seeking similar terms. The first terms are similar, $5xy^2$ and $3xy^2$. We combine them to get $8xy^2$. We write this term in the answer. In the same way we find that the second terms of the two polynomials are similar. We add them to obtain $-6x^2y$. We write this term in our answer. There are no other two terms similar, and no more combining can be done. We finish writing the answer.

$8xy^2 - 6x^2y + 5x^3 + 3x^3y - 3$

Do exercises 7 and 8 at the right.

SUBTRACTION

Remember, to subtract one polynomial from another, we add the inverse. The additive inverse of a polynomial can be obtained by changing the sign of each term.

OBJECTIVES

You should be able to:

Add, subtract and multiply polynomials in several variables.

You should strive for speed as well as accuracy. Whenever possible you should write only the answer.

Add.

7. $(13x^3y + 3x^2y - 5y)$
     $+ (x^3y + 4x^2y - 3xy + 3y)$

8. $(-5p^2q^4 + 2p^2q^2 + 3q)$
     $+ (6pq^2 + 3p^2q + 5)$

Subtract.

9. $(-4s^4t + s^3t^2 + 2s^2t^3)$

   $- (4s^4t - 5s^3t^2 + s^2t^2)$

10. $(-5p^4q + 5p^3q^2 - 3p^2q^3 - 7q^4)$

   $- (4p^4q - 5p^3q^2 + p^2q^3 + 2q^4)$

Multiply.

11. $(x^2y^3 + 2x)(x^3y^2 + 3x)$

12. $(p^4q - 2p^3q^2 + 3q^3)(p + 2q)$

Example 3. Subtract:

$$(4x^2y + x^3y^2 + 3x^2y^3 + 6y) - (4x^2y - 6x^3y^2 + x^2y^2 - 5y)$$

We may first write the polynomial to be subtracted under the other, keeping similar terms in columns.

$$\frac{\begin{array}{l} 4x^2y + x^3y^2 + 3x^2y^3 \qquad\quad + 6y \\ 4x^2y - 6x^3y^2 \qquad\quad + x^2y^2 - 5y \end{array}}{7x^3y^2 + 3x^2y^3 - x^2y^2 + 11y}$$

As with addition, you should avoid the use of columns wherever possible. Rather, attempt to write only the answer when subtracting.

Do exercises 9 and 10 at the left.

MULTIPLICATION

Multiplication of polynomials is based upon the distributive laws. You will recall that this means we can multiply each term of one polynomial by every term of the other. Then we collect like terms. For most polynomials in several variables having three or more terms you will probably want to use columns.

Example 4. Multiply:

$$(3x^2y - 2xy + 3y)(xy + 2y)$$

It is helpful to first write one polynomial under the other.

$$\frac{\begin{array}{l} 3x^2y - 2xy + 3y \\ xy + 2y \\ \hline 3x^3y^2 - 2x^2y^2 + 3xy^2 \\ \qquad\quad 6x^2y^2 - 4xy^2 + 6y^2 \end{array}}{3x^3y^2 + 4x^2y^2 - xy^2 + 6y^2}$$

In multiplying polynomials, you should write one under the other, except in simple cases.

Do exercises 11 and 12 at the left.

Do exercise set 8.2, p. 277.

8.3 SPECIAL PRODUCTS OF POLYNOMIALS

Multiplying polynomials is based on the distributive law. The methods of handling the special kinds of products you learned for polynomials in one variable are the same for polynomials in several variables.

PRODUCTS OF TWO BINOMIALS

To multiply two binomials, we multiply the first terms. We multiply the last terms. Then we multiply the inside terms and finally the outside terms. If there are similar terms in the result we combine them.

Example 1

$(x^2y + 2x)(xy^2 + y^2) = x^3y^3 + x^2y^3 + 2x^2y^2 + 2xy^2$

Example 2

$$(p + 5q)(2p - 3q) = 2p^2 - 3pq + 10pq - 15q^2$$
$$= 2p^2 + 7pq - 15q^2$$

Do exercises 13 and 14 at the right.

SQUARES OF BINOMIALS

To square a binomial we square the first term. We take twice the product of the two terms. Then we square the second term.

Example 3

$$(3x + 2y)^2 = (3x)^2 + 2(3x)(2y) + (2y)^2$$
$$= 9x^2 + 12xy + 4y^2$$

Example 4

$$(2y^2 - 5x^2y)^2 = (2y^2)^2 - 2(2y^2)(5x^2y) + (5x^2y)^2$$
$$= 4y^4 - 20x^2y^3 + 25x^4y^2$$

Do exercises 15 and 16 at the right.

Multiply.

13. $(3xy + 2x)(x^2 + 2xy^2)$

14. $(x - 3y)(2x - 5y)$

Multiply.

15. $(4x + 5y)^2$

16. $(3x^2 - 2xy^2)^2$

Multiply

17. $(2xy^2 + 3x)(2xy^2 - 3x)$

18. $(3y + 4 - 3x)(3y + 4 + 3x)$

19. $(3xy^2 + 4y)(-3xy^2 + 4y)$

PRODUCTS OF SUMS AND DIFFERENCES

The product of the sum and difference of the same two terms is the difference of two squares. To find such a product we square the first term, and square the second term. Then write a minus sign between them.

Example 5

$$(3x^2y + 2y)(3x^2y - 2y) = (3x^2y)^2 - (2y)^2$$
$$= 9x^4y^2 - 4y^2$$

Example 6

$$(2x + 3 - 2y)(2x + 3 + 2y) = (2x + 3)^2 - (2y)^2$$
$$= 4x^2 + 12x + 9 - 4y^2$$

Example 7

$$(-2x^3y^2 + 5t)(2x^3y^2 + 5t) = (5t - 2x^3y^2)(5t + 2x^3y^2)$$
$$= (5t)^2 - (2x^3y^2)^2,$$
$$= 25t^2 - 4x^6y^4$$

Example 8

$$(4x + 3y + 2)(4x - 3y - 2) =$$
$$(4x + 3y + 2)(4x - [3y + 2]) =$$
$$16x^2 - (3y + 2)^2 = 16x^2 - (9y^2 + 12y + 4)$$
$$= 16x^2 - 9y^2 - 12y - 4$$

Do exercises 17 through 19 at the left.

Do exercise set 8.3, p. 279.

8.4 FACTORING

Factoring polynomials in several variables is quite similar to factoring polynomials in one variable.

TERMS WITH COMMON FACTORS

Whenever you factor polynomials you should look for common factors in the terms before trying any other kind of factoring.

Example 1

$$20x^3y + 12x^2y = (4x^2y)(5x) + (4x^2y) \cdot 3$$
$$= 4x^2y(5x + 3)$$

You should attempt to write the answer directly, as follows.

Example 2

$$6x^2y - 21x^3y^2 + 3x^2y^3 = 3x^2y(2 - 7xy + y^2)$$

Do exercises 20 and 21 at the right.

FACTORING BY GROUPING

Sometimes a common factor is itself a binomial, or pairs of terms have a common factor which can be removed, as in the following examples.

Example 3

$$(p + q)(x + 2) + (p + q)(x + y) = (p + q)[(x + 2) + (x + y)]$$
$$= (p + q)(2x + y + 2)$$

Example 4

$$px + py + qx + qy = p(x + y) + q(x + y)$$
$$= (p + q)(x + y)$$

Do exercises 22 and 23 at the right.

OBJECTIVES

You should be able to factor polynomials in several variables:

a) Where the terms have a common factor.
b) By grouping.
c) Which are differences of squares.

You should work up speed as much as you can, and when possible write only the answer.

Factor.

20. $x^4y^2 + 2x^3y + 3x^2y$

21. $10p^6q^2 - 4p^5q^3 + 2p^4q^4$

Factor.

22. $(a - b)(x + 5) + (a - b)(x + y^2)$

23. $ax^2 + ay + bx^2 + by$

Factor.

24. $9a^2 - 16x^4$

DIFFERENCES OF SQUARES

Differences of squares are factored as before. This process is the reverse of multiplying a sum and difference of the same two terms.

Example 5

$$36x^2 - 25y^6 = (6x)^2 - (5y^3)^2$$
$$= (6x + 5y^3)(6x - 5y^3)$$

25. $25x^2y^4 - 4a^2$

Do exercises 24 and 25 at the left.

Whenever terms have a common factor it should be removed first.

Example 6

$$32x^4y^4 - 50 = 2(16x^4y^4 - 25)$$
$$= 2(4x^2y^2 + 5)(4x^2y^2 - 5)$$

Whenever a factor can be factored again, this should be done.

Factor completely.

26. $5 - 5x^2y^6$

Example 7

$$5x^4 - 5y^4 = 5(x^4 - y^4)$$
$$= 5(x^2 + y^2)(x^2 - y^2)$$
$$= 5(x^2 + y^2)(x + y)(x - y)$$

Whenever you factor you should do so completely. Look first for a common factor. If any factor can be factored further, this should be done.

Example 8. Factor completely.

$$2x^2 + 24ax + 72a^2 - 98 = 2(x^2 + 12ax + 36a^2 - 49)$$
$$= 2[(x + 6a)^2 - 49]$$
$$= 2(x + 6a + 7)(x + 6a - 7)$$

27. $16y^2 - 81x^4y^2$

Do exercises 26 and 27 at the left.

Do exercise set 8.4, p. 281.

8.5 FACTORING TRINOMIALS

If a trinomial is a square then it is easy to factor. Whenever you have a trinomial to factor you should therefore check to see if it is a perfect square.

Example 1

Find whether $25x^2 + 20xy + 4y^2$ is a square.

If so, factor it. The first term and the last are squares:

$25x^2 = (5x)^2$ and $4y^2 = (2y)^2$

Twice the product of $5x$ and $2y$ should be the other term.

$2(5x)(2y) = 20xy$

Hence the trinomial is a square.

We factor by writing the square roots of the square terms:

$25x^2 + 20xy + 4y^2 = (5x + 2y)^2$

We can check this by squaring $5x + 2y$.

Do exercises 28 and 29 at the right.

If a trinomial is not a square we use trial and error to factor.

Example 2

$$p^2q^2 + 7pq + 12 = (pq)^2 + (3 + 4)pq + 3 \cdot 4$$
$$= (pq + 3)(pq + 4)$$

Remember always to look first for common factors. Remember, too, that some of the factors obtained may be factorable, and factor completely. Further, not all trinomials can be factored.

Do exercises 30 and 31 at the right.

Do exercise set 8.5, p. 283.

Do exercise set 8.5, p. 283.

OBJECTIVES

You should be able to:

a) Tell whether a trinomial is a square.

b) Factor trinomial squares.

c) Factor trinomials that are not squares, if they are factorable.

Factor.

28. $x^4 + 2x^2y^2 + y^4$

29. $-4x^2 + 12xy - 9y^2$

Hint: first factor out -1.

Factor.

30. $x^2y^2 + 5xy + 4$

31. $2x^4y^6 + 6x^2y^3 - 20$

You should be able to:

a) Solve simple equations, such as 32 and 33 below, where some letters are used as constants.

b) Solve equations with parentheses, where some letters are used as constants.

Solve.

32. $ay - b = 3$

33. $2abx - 6a = abx$

8.6 SOLVING EQUATIONS

CONSTANTS AND VARIABLES

Sometimes we use a letter to represent a specific number, although we may not know what number it is. When we do this the letter is not a variable. It is called a *constant*. It is not always easy to tell whether a letter is a variable or a constant. In much of algebra, however, we make an agreement. In this lesson we shall use the agreement. It is as follows: Letters near the end of the alphabet, x, y, z, will be variables. Letters near the first of the alphabet, a, b, c, will be constants. Thus when we solve equations with letters for constants in them we know to solve for one of the later letters, such as x, y or z.

EQUATIONS WITH LETTERS FOR CONSTANTS

When we solve an equation with unknown constants we treat them as if they were known. The laws of numbers apply, whether we know what the numbers are or not.

Example 1

Solve $cx + b^2 = 2$

By our agreement, x is a variable and b and c are constants. Hence we solve for x.

$$cx = 2 - b^2 \quad \text{(adding } -b^2\text{)}$$

$$x = \frac{2 - b^2}{c} \quad \left(\text{multiplying by } \frac{1}{c}\right)$$

Note that we treated c and b as if they were known. In this sense they are much the same as the 2 in the equation.

Example 2

Solve $\dfrac{3}{4}x + \dfrac{b}{4} = 2b$

$$\frac{3}{4} \cdot \frac{x}{1} + \frac{b}{4} = 2b$$

$$\frac{3x}{4} + \frac{b}{4} = 2b$$

$$\frac{3x + b}{4} = 2b$$

$$3x + b = 4 \cdot 2b = 8b$$

$$3x = 7b$$

$$x = \frac{7b}{3} \quad \text{or} \quad \frac{7}{3}b$$

Do exercises 32 and 33 at the left.

EQUATIONS WITH PARENTHESES

To solve an equation we ordinarily try to get the variable alone on one side. If the variable appears in several terms we usually collect like terms, or factor it out. It also happens that we must multiply some polynomials to remove parentheses so that we can collect the terms containing the variable.

Example 3

Solve: $3(7 + 2x) = 30 + 7(x - 1)$

$21 + 6x = 30 + 7x - 7$ (multiplying)

$21 = 23 + 7x - 6x$ (adding $-6x$)

$-2 = x$ (adding -23 and collecting x-terms)

In Examples 4 and 5, remember that x, y, and z are used as variables and a, b, and c are used as constants.

Example 4

Solve: $(x - a)(x + a) = x^2 - x$

$x^2 - a^2 = x^2 - x$ (multiplying)

$-a^2 = -x$ (adding $-x^2$)

$a^2 = x$ (multiplying by -1)

Example 5

Solve: $(x + a)(x + 2b) = x^2 + a^2 + b^2$

$x^2 + ax + 2bx + 2ab = x^2 + a^2 + b^2$ (multiplying)

$ax + 2bx + 2ab = a^2 + b^2$ (adding $-x^2$)

$(a + 2b)x + 2ab = a^2 + b^2$ (collecting x-terms)

$(a + 2b)x = a^2 + b^2 - 2ab$ (adding $-2ab$)

$x = \dfrac{a^2 + b^2 - 2ab}{a + 2b}$ $\left(\text{multiplying by } \dfrac{1}{a + 2b}\right)$

$x = \dfrac{(a - b)^2}{a + 2b}$ (factoring numerator)

Do exercises 34 through 36 at the right.

Do exercise set 8.6, p. 285.

Solve.

34. $3(7 - 2y) + 7(2y + 1) = 36$

35. $(x + a)(x + b) = x^2 + 2$

36. $2(x + a) + 3(x + b) = 4(x - 2)$

OBJECTIVES

You should be able to solve a formula for a given letter.

37. Solve.

$$f = \frac{kQ_1Q_2}{r^2} \text{ for } Q_2$$

38. Solve $V = \frac{1}{3}\pi h(3R^2 - h^2)$ for R^2.

8.7 FORMULAS

The use of formulas is important in many applications of mathematics to science, engineering and technology. The ability to manipulate a formula, solving it for a given letter, is highly important in most of these applications.

It is difficult, if not impossible, to look at a formula and tell which symbols are variables and which are constants. In fact, given a single formula, a certain letter may be a variable at times and a constant at others, depending upon how the formula is used.

Example 1

Solve: $\quad f = \frac{kmM}{d^2}$ for m.

Note that m and M are different variables (or constants). Therefore be careful how you write them.

$fd^2 = kmM \quad$ (multiplying by d^2)

$\dfrac{fd^2}{kM} = m \qquad \left(\text{multiplying by } \dfrac{1}{kM}\right)$

Example 2

Solve: $\quad A = \frac{1}{2}(b_1 + b_2)h$ for b_2

Note that b_1 and b_2 are different variables (or constants).

Therefore, do not forget to write the subscripts (the 1 and 2).

$A = \dfrac{1}{2}b_1h + \dfrac{1}{2}b_2h \quad$ (multiplying)

$A - \dfrac{1}{2}b_1h = \dfrac{1}{2}b_2h \quad \left(\text{adding } -\dfrac{1}{2}b_1h\right)$

$\dfrac{2}{h}\left(A - \dfrac{1}{2}b_1h\right) = b_2 \quad \left(\text{multiplying by } \dfrac{2}{h}\right)$

In both of the above examples, the letter for which we solved was left on the right side of the equation. Ordinarily we put the letter for which we solve on the left. This, however, is a matter of choice, since all equations are reversible.

Do exercises 37 and 38 at the left.

Do exercise set 8.7, p. 287.

EXERCISE SET 8.1 (pp. 263–264)

What is the coefficient of each term of the following polynomials?

1. $x^3y - 2xy + 3x^2 - 5$ **2.** $5y^3 - y^2 + 15y + 1$

3. $17x^2y^3 - 3x^3yz - 7$ **4.** $6 - xy + 8x^2y^2 - y^5$

5–8. What is the degree of each term in the expressions above?

Simplify by combining like terms.

9. $a + b - 2a - 3b$ **10.** $y^2 - 1 + y - 6 - y^2$

11. $3x^2y - 2xy^2 + x^2$ **12.** $m^3 + 2m^2n - 3m^2 + 3mn^2$

13. $2u^2v - 3uv^2 + 6u^2v - 2uv^2$ **14.** $3(x + y)^2 - 5(x + y)^2$

ANSWERS

15.

_____

16.

_____

17.

_____

18.

_____

19.

_____

20.

_____

21.

_____

22.

_____

23.

_____

24.

_____

15. $3a(2u + v) + 7a(2u + v)$ **16.** $3x^2y - 2z^2y + 3xy^2 + 5z^2y$

Arrange in ascending powers of the first variable (using alphabetical order). Simplify if possible.

17. $-xy^2 - y^3 + 5x^2y$

18. $-2x^3 - xy^2 + x^2y + x^3 - y^3 + 2y^3$

19. $3p^4rt - 2r^2t + 2t^3 - 2p^4rt$

20. $7r^4s - 18s^3 + 32r^5 - 2r^2s^2$

Arrange in descending powers of the second variable (using alphabetical order). Simplify if possible.

21. $x^2 + 3y^2 + 2xy$

22. $4n^2 - 3mn + m^2 - mn + n^2 + 5m^2$

23. $5uv - 8uv^2 + 7u^2v - 3uv^2$

24. $6r^2s + 11 + 3r^2s - 5r + 3$

NAME

CLASS

ANSWERS

EXERCISE SET 8.2 (pp. 265–266)

Add.

1. $(2x^2 - xy + y^2) + (-x^2 - 3xy + 2y^2)$

2. $(2z - z^2 + 5) + (z^2 - 3z + 1)$

3. $(r - 2s + 3) + (2r + s) + (s + 4)$

4. $(b^3a^2 - 2b^2a^3 + 3ba + 4) + (b^2a^3 - 4b^3a^2 + 2ba - 1)$

5. $(2x^2 - 3xy + y^2) + (-4x^2 - 6xy - y^2) + (x^2 + xy - y^2)$

Subtract.

6. $(x^3 - y^3) - (-2x^3 + x^2y - xy^2 + 2y^3)$

7. $(xy - ab) - (xy - 3ab)$

8. $(3y^4x^2 + 2y^3x - 3y) - (2y^4x^2 + 2y^3x - 4y - 2x)$

9. $(-2a + 7b - c) - (-3b + 4c - 8d)$

10. Find the sum of $2a + b$ and $3a - 4b$. Then subtract $5a + 2b$.

Multiply.

11. $(3z - u)(2z + 3u)$

12. $(a - b)(a^2 + b^2 + 2ab)$

13. $(a^2b - 2)(a^2b - 5)$

14. $(xy + 7)(xy - 4)$

15. $(a^2 + a - 1)(a^2 - y + 1)$

16. $(tx + r)(vx + s)$

ANSWERS

1.

2.

3.

4.

5.

6.

7.

8.

9.

10.

11.

12.

13.

14.

15.

16.

17. $(a^3 + bc)(a^3 - bc)$ **18.** $(m^2 + n^2 - mn)(m^2 + mn + n^2)$

17. _____

_____

19. $(y^4x + y^2 + 1)(y^2 + 1)$ **20.** $(a - b)(a^2 + ab + b^2)$

18. _____

_____

19. _____

_____   Simplify.

20. _____   **21.** $(r + s)(r^2 + ry + s^2)$ **22.** $(a + 2b + 3c)(2a - b + 2c)$

_____

21. _____

_____   **23.** $(xy + 1)(2xy - 3) + (xy + 1)(3xy - 1)$

22. _____

_____   **24.** $(2a + b)(3a - b) - (a - b)(2a - 3b)$

23. _____

_____   **25.** $(yz^2 + 2)(yz^2 - 2) + (2yz^2 + 1)(2yz^2 + 1)$

24. _____

_____   **26.** $(3a - d)(2a + 3d) - (4a + d)(2a - d)$

25. _____

_____

26. _____   **27.** $(2r - s + t)(r + 2s - t)$

_____

27. _____

_____

NAME _____

CLASS _____

ANSWERS

1. _____

2. _____

3. _____

4. _____

5. _____

6. _____

7. _____

8. _____

9. _____

10. _____

11. _____

12. _____

13. _____

14. _____

15. _____

16. _____

17. _____

18. _____

19. _____

20. _____

21. _____

22. _____

EXERCISE SET 8.3 (pp. 267–268)

Multiply.

1. $(3xy - 1)(4xy + 2)$

2. $(rs + 3)(rs - 1)$

3. $(m^3n + 8)(m^3n - 6)$

4. $(3 - c^2d)(5 - 4c^2d)$

5. $(c^3 - 8d)(c^3 + 8d)$

6. $(6x - 2y)(5x + 3y)$

7. $(3a + 2b)^2$

8. $(c^2d^3 + 2)^2$

9. $(x - y^3)(x + 2y^3)$

10. $(5st^2 + 4t)(2st^2 - t)$

11. $(2a^3 - \frac{1}{2}b^3)^2$

12. $-5x(x + 3y)^2$

13. $(ab + cd^2)(ab - cd^2)$

14. $(c + d + 5)(c + d - 5)$

15. $(x + y + 3)(x + y + 3)$

16. $(a + b + c)(a - c + b)$

17. $[5 + (x + y)][4 - (x + y)]$

18. $(3a^2b - b^2)^2$

19. $(a^2 + b + 2)^2$

20. $[2m + n + p][2m - (n + p)]$

21. $[1 - (r - s)][1 + 2(r - s)]$

22. $(ab - .3)(.5ab + .2)$

ANSWERS

23. _____

24. _____

25. _____

26. _____

27. _____

28. _____

29. _____

30. _____

31. _____

32. _____

33. _____

34. _____

35. _____

36. _____

37. _____

38. _____

39. _____

40. _____

23. $3a(a - 2b)^2$

24. $(-m^2n^2 + p^4)^2$

25. $(h - k + 4)(h + k - 4)$

26. $(m + n)(m^2 - n^2)$

27. $(x^2 - xy + y^2)(x^2 + xy + y^2)$

28. $(3x^2y - 2x^3y)^2$

29. $(3x - 4y - 5z)(3x + 4y + 5z)$

30. $(3x - 2y)^2$

Perform the indicated operations and simplify.

31. $(a - b)^2 - (a + b)(a - b)$

32. $(a - b)^2 + (a + b)^2$

33. $(a + b)^2 - (a - b)^2$

34. $9(x - 1)^2 - 25(x + 1)^2$

35. $y^2 - (y - 2)^2$

36. $(c^2 - cd + d^2)(c^2 + cd + d^2)$

37. $(a - b)^2 - (b - a)$

38. $(4x - y)(y - 4x)$

39. $(a + b)(a + 3b)(a + 2b) - 12(a + b)$

40. $(2x^2y - 3x + y)^2$

EXERCISE SET 8.4 (pp. 269–270)

Factor.

1. $12n^2 + 24n^3$

2. $12a^4b^4 - 9a^3b^3 + 6a^2b^2$

3. $9x^2y^2 - 36xy$

4. $5c(a^3 + b) - (a^3 + b)$

5. $(x - 1)(x + 1) - y(x + 1)$

6. $(c + 3d)(2c) + (c + 3d)(3d)$

7. $x^2y - xy^2$

8. $a^2bc - ab^2$

9. $\pi r^2 + \pi rh$

10. $a^2b^2 - c^2$

11. $x^2 + x + xy + y$

12. $ax - bx + ay - by$

13. $a^4b^4 - 16$

14. $bx + 2b + cx + 2c$

15. $-12x^3y^3 + 18x^2y^4 + 27xy^5$

16. $n^2 + 2n + np + 2p$

17. $9s^4 - 9s^2$

18. $2x^2 - 4x + xz - 2z$

19. $3y^2 - 48$

20. $18m^4 + 12m^3 + 2m^2$

ANSWERS

21. _____

22. _____

23. _____

24. _____

25. _____

26. _____

27. _____

28. _____

29. _____

30. _____

31. _____

32. _____

33. _____

34. _____

35. _____

36. _____

37. _____

38. _____

39. _____

40. _____

21. $a^2 - 3a + ay - 3y$

22. $4xy - 2x - 3x^3y$

23. $9x^4y^2 - b^2$

24. $4x^2 + 16y^2$

25. $6y^2 - 3y + 2py - p$

26. $3ab - b^2 + 3a^2 - ab$

27. $4x + 8x^3 + 1 + 2x^2$

28. $(a + b)^2 - c^2$

29. $a^2 - (m - 1)^2$

30. $(x - a)^2 - (y - b)^2$

31. $ay^2 - a - y^2 + 1$

32. $c^2 - (a + b)^2$

33. $a^2x - x - 3a^2 + 3$

34. $15x^2y - 30xy + 35xy^2$

35. $9a^2 - (a + 3b)^2$

36. $(b - 4)^2 - a^6$

37. $(s - 2)^2 - (t - 2)^2$

38. $(x - 2y)^2 - (t - 2y)^2$

39. $c^2xy - c^3 - x^2y + cx$

40. $-18u^3v^2 + 12u^2v^2 - 6uv^2$

NAME

CLASS

ANSWERS

EXERCISE SET 8.5 (p. 271)

Which of the following are perfect square trinomials? Write yes or no.

1. $4x^2 + 12xy + 9y^2$

2. $9c^2 - 6cd + d^2$

3. $x^2 + 4xy - y^2$

4. $36x^2 - 12xy + 4y^2$

5. $16x^2 - 24xy + 9y^2$

6. $-4a^2 + 4ab - b^2$

7. $.01x^4 - .1x^2y + .25y^2$

8. $49m^2 - 56mn + 64n^2$

9. $\frac{1}{4}r^2 - \frac{1}{3}rs + \frac{1}{9}s^2$

Fill in the blank so that the trinomial is a perfect square.

10. $64x^2 + \underline{\quad} + 16y^2$

11. $9a^2 - 12ab + \underline{\quad}$

12. $\underline{\quad} + \frac{1}{2}yz + \frac{1}{4}z^2$

Factor.

13. $9q^2 + 6q + 1$

14. $4x^2y^2 - 12xyz + 9z^2$

15. $y^4 + 10y^2z + 25z^2$

16. $6a^4 - 12a^2b^2 + 6b^4$

17. $4p^2q + pq^2 + 4p^3$

18. $z^2 + 18zab + 81a^2b^2$

1.	
2.	
3.	
4.	
5.	
6.	
7.	
8.	
9.	
10.	
11.	
12.	
13.	
14.	
15.	
16.	
17.	
18.	

ANSWERS

19. _____

20. _____

21. _____

22. _____

23. _____

24. _____

25. _____

26. _____

27. _____

28. _____

29. _____

30. _____

31. _____

32. _____

33. _____

34. _____

35. _____

36. _____

19. $(r - s)^2 - 2(r - s) + 1$
(Hint: Think of this as
▮$^2 - 2$ ▮ $+ 1$.)

20. $4p^2 - q^2 - 6q - 9$
[Hint: Think of this as
$4p^2 - (q^2 + 6q + 9)$.]

21. $t^2 - 4t + 4 - s^2$

22. $r^2 + 6rs + 9s^2$

23. $-k^2 + 36kl - 324l^2$

24. $4a^2 - 4ab - 36 + b^2$

25. $1 - n^2 - 2nx - x^2$

26. $x^4 + x^3 + x^2$

27. $3b^2 - 17ab - 6a^2$

28. $6a^2 - 47ab - 63b^2$

29. $a^2c^2 - 4a^2 - b^2c^2 + 4b^2$

30. $-16a^2b - 10a^2br - a^2br^2$

31. $m^2 + 2mn - 360n^2$

32. $n^2 + 23nr - 420r^2$

33. $a^2 - b^2 + (a - b)^2$

34. $5ac - 5ad + 5bc - 5bd$

35. $k^2t^2 - 9s^2 - 9t^2 + k^2s^2$

36. $42m^2n - 24mn^2 - 18n^3$

NAME _____

CLASS _____

ANSWERS

EXERCISE SET 8.6 (pp. 272–273)

Solve for x, y, or z.

1. $2a - 3x = 12$

2. $z - k = 3k$

3. $3b - 2y = 1$

4. $5 = \dfrac{a + x + b}{3}$

5. $3acy - 9a = acy$

6. $(z + c)(z - d) = z^2 + 5$

7. $ax - bx = a^2 - 2ab + b^2$

8. $a(x + b) = c$

9. $3(z + c) - 5(z + b) = 4(z - 3)$

10. $(2x - 5)(3x + 6) = 6x^2 - 8$

11. $b(x - b) = x - (2 - b)$

12. $3z - a = b + 11$

13. $a(3 + 2x) = 5 + x$

14. $24a - 3x = 5x$

15. $4(x - a) + 4a = 7(b + 3) - 21$

16. $-3(c + y) + 3c = 6(c - 2) + 12$

17. $5(2a - z) + 6 = -2(z - 5a)$

18. $x(a + 2) - 3 = b(x + 7)$

19. $bx - b^2 = 3x + b - 12$

20. $\dfrac{5x - a}{7} = \dfrac{4}{3}$

ANSWERS
1.
2.
3.
4.
5.
6.
7.
8.
9.
10.
11.
12.
13.
14.
15.
16.
17.
18.
19.
20.

ANSWERS

21. $ax + b^2 = a^2 + bx$

22. $m(m - x) = n(n - x)$

21. _____

22. _____

23. $b^2(ay - 1) = a^2(1 - ay)$

24. $m^2x = m + 1$

23. _____

24. _____

25. $3(x - a) = x + 3a$

26. $cy - 1 = c^2 + 2c - y$

25. _____

26. _____

27. $dx + x = (d + 1)(d - 1)$

28. $a^2x = a^6$

27. _____

28. _____

29. $rxt = c$

30. $mx + 4 = x + 4m^2$

29. _____

30. _____

31. $ax + bx = 3ax - c$

32. $b^2x = 4b$

31. _____

32. _____

33. $3(m - x) = x - 7m$

34. $2(3 - x) + b = b(x - 5)$

33. _____

34. _____

NAME

CLASS

ANSWERS

EXERCISE SET 8.7 (p. 274)

Solve for the variable indicated.

1. $S = 2\pi rk$; r

2. $A = P(1 + rt)$; t

3. $A = \frac{1}{2}bk$; b

4. $s = \frac{1}{2}gt^2$; g

5. $S = 180(n - 2)$; n

6. $S = \frac{n}{2}(a + l)$; a

7. $V = \frac{1}{3}k(B + b + 4M)$; b

8. $A = p + prt$; p
(Hint: Factor the right hand side.)

9. $S(r - 1) = rl - a$; r

10. $T = mg - mf$; m
(Hint: Factor the right hand side.)

11. $A = \frac{1}{2}h(b_1 + b_2)$; h

12. $S = 2\pi r(r + k)$; k

13. $r = \frac{v^2 pL}{a}$; a

14. $L = \frac{Mt - g}{t}$; M

1.

2.

3.

4.

5.

6.

7.

8.

9.

10.

11.

12.

13.

14.

15. $A = \dfrac{1}{2}h(b_1 + b_2)$; b_2

16. $l = a + (n - 1)d$; n

17. $A = \dfrac{\pi r^2 E}{180}$; E

18. $R = \dfrac{WL - x}{L}$; W

19. $V = -h(B + c + 4M)$; M

20. $W = I^2R$; R

21. $y = \dfrac{v^2 pL}{a}$; L

22. $V = \dfrac{1}{3}bh$; b

23. $r = \dfrac{v^2 pL}{a}$; p

24. $P = 2(l + w)$; l

25. $\dfrac{a}{c} = n + bn$; n

26. $C = \dfrac{Ka - b}{a}$; K

(Hint: Factor the right hand side.)

27. $C = \dfrac{5}{9}(F - 32)$; F

28. $V = \dfrac{4}{3}\pi r^3$; π

29. $f = \dfrac{gm - t}{m}$; g

30. $S = \dfrac{rl - a}{r - l}$; a

NAME_____

CLASS_____ SCORE_____ GRADE_____

CHAPTER 8 TEST

Before taking the test *be sure* to allow yourself a day or so for review. Use the objectives listed in the margins to guide your study. The test will evaluate your progress and aid your preparation for a possible classroom test. Allow about an hour for the test. Remove the test from the book. When you finish read the test analysis on the answer page at the end of the book.

ANSWERS

1. Identify the coefficient of each term of the following polynomials:

a) $5x^4y^3 + x^3 - 3xy^2$ b) $8b^2 + 7a^2b^6 - 3ab^3$

1.

a)_____ _____ _____

b)_____ _____ _____

2. Determine the degree of each term of the following polynomials.

a) $a^8b^8 + 15ab^3 + 6a^5 + 2$ b) $5c^2 + 4c^2d^5 - 2cd^3 + c$

2.

a)_____

b)_____

3. Write in descending powers of a: $b^3 + 3a^2b + 3ab^2 + a^3$

3.

Combine like terms.

4. $8(7j + 4m) + 5(2m - 5j) - 6(m + j)$

4.

5. $3x^2yz^3 - 2xy^2 + x^2yz^3 + x^2y^2 + 2xy^2$

5.

Perform the indicated operations and simplify.

6. $(a^3b - 6a^2b^2 - ab^3 + 3) + (a^3b - 4a^2b^2 + 8) + (12 - ab + a^2b^2)$

6.

7. $(y^3z^2 - 2yz - 3)(2yz - 1)$

7.

8. From $3x - 7y + z$ subtract $7x - 4y + 6z$

8.

9. $(18s^2t^3 + 12st^2 + 4) - (6st^2 - s^2t^2 - 3)$

9.

10. $(c^2 + d)(c^2 - d)$

10.

11. $(a + 2b - c)^2$

11.

ANSWERS

Factor.

12.

12. $16x^2y^2 - 1$ **13.** $9p^2 + 42pq + 49q^2$

13.

14. $35x^2 - 22xy + 3y^2$ **15.** $(x + y)^2 - 4$

14.

16. $c^2 + 6c + 9 - 16x^2$ **17.** $(m - n)^2 + 8(m - n) + 15$

15.

16.

18. $2uw - 6ux - 3vx + vw$

17.

19. Solve for b: $c^2 - 6b = 3c - 2bc$

18.

20. Solve for x: $r(x - r) + s(x - s) = s(s + x) - (r + s)$

19.

21. Solve for p: $A - p = prt$

20.

22. Solve for x: $ax + 6a = -2x - 3a^2$

21.

22.

23. Factor: $ax^4 + 2ax^2 - 24a$

23.

24. Factor: $16c^4 - d^8$

24.

25. Multiply: $[(x - y) - 6][(x - y) + 2]$

25.

9 FRACTIONAL EXPRESSIONS AND EQUATIONS

9.1 MULTIPLYING AND SIMPLIFYING FRACTIONAL EXPRESSIONS

Expressions like the following are called *fractional expressions*.

$$\frac{3}{4} \qquad \frac{5}{x+2} \qquad \frac{(x+5)(x-2)}{7(x^2-1)}$$

Fractional expressions indicate division. For example,

$$\frac{3}{4} \text{ means } 3 \div 4 \text{ and } \frac{5}{x+2} \text{ means } 5 \div (x+2).$$

$$\frac{3x^2-5x+2}{7y-4x^2} \text{ means } (3x^2-5x+2) \div (7y-4x^2).$$

MULTIPLYING

For fractional expressions, multiplication is done as in arithmetic. That is, we <u>multiply numerators</u> and <u>multiply denominators</u>.

Examples

a) Multiply: $\dfrac{x-2}{3} \cdot \dfrac{x+2}{x+3}$

$\dfrac{(x-2)(x+2)}{3(x+3)}$ (multiplying numerators and multiplying denominators)

$\dfrac{x^2-4}{3(x+3)}$ (simplifying numerator)

b) Multiply: $\dfrac{-2}{2y+3} \cdot \dfrac{3}{2y-3}$

$\dfrac{-2 \cdot 3}{(2y+3)(2y-3)}$ (multiplying numerators and denominators)

$\dfrac{-6}{4y^2-9}$ (simplifying)

Do exercises 1 and 2 at the right.

OBJECTIVES

You should be able to:

a) Multiply fractional expressions.

b) Simplify fractional expressions by factoring numerator and denominator and removing factors of 1.

Multiply.

1. $\dfrac{x+3}{5} \cdot \dfrac{x+2}{x+4}$

2. $\dfrac{-3}{2x+1} \cdot \dfrac{4}{2x-1}$

Multiply.

3. a) $\dfrac{2x+1}{3x-2}\cdot\dfrac{x}{x}$

b) $\dfrac{x+1}{x-2}\cdot\dfrac{x+2}{x+2}$

4. $\dfrac{x-8}{x-y}\cdot\dfrac{-1}{-1}$

MULTIPLYING BY 1

When we multiply any number by 1 we get that same number. Any fractional expression with the same numerator and denominator will name the number 1.*

Example

$\dfrac{x+2}{x+2},\ \dfrac{3x^2-4}{3x^2-4},\ \dfrac{-1}{-1}$ all name the number 1 for all sensible replacements.

In the following examples we are really multiplying by 1.

Examples. Multiply:

a) $\dfrac{3x+2}{x+1}\cdot\dfrac{2x}{2x}=\dfrac{(3x+2)2x}{(x+1)2x}=\dfrac{6x^2+4x}{2x^2+2x}$

b) $\dfrac{x+2}{x-1}\cdot\dfrac{x+1}{x+1}=\dfrac{(x+2)(x+1)}{(x-1)(x+1)}=\dfrac{x^2+3x+2}{x^2-1}$

c) $\dfrac{x-2}{x+2}\cdot\dfrac{-1}{-1}=\dfrac{(x-2)(-1)}{(x+2)(-1)}=\dfrac{-x+2}{-x-2}=\dfrac{2-x}{-x-2}$

Do exercises 3 and 4 at the left.

SIMPLIFYING FRACTIONAL EXPRESSIONS

We can simplify fractional expressions by reversing the previous procedure. That is, we factor numerator and denominator and try to "remove" a factor of 1.

Example. Simplify:

$\dfrac{3x}{x}=\dfrac{3x}{1\cdot x}=\dfrac{3}{1}\cdot\dfrac{x}{x}=\dfrac{3}{1}\cdot 1=3$

In this example we supplied a 1 in the denominator. This can always be done, but is not necessary.

---

* In $\dfrac{x+2}{x+2}$ we could not substitute -2 for x.

If we did we would get 0 for a denominator. Since we cannot divide by 0, this would not be a sensible replacement. In a fractional expression the nonsensible replacements are those that make the denominator 0.

Example. Simplify:

$$\frac{7xy}{x} = \frac{7y \cdot x}{x} = 7y \cdot \frac{x}{x} = 7y$$

In this example, as in the previous one, we could have multiplied by 1 in the denominator. However, this is not necessary.

Examples. Simplify.

a) $\dfrac{5(a+2)}{8(a+2)} = \dfrac{5}{8} \cdot \dfrac{a+2}{a+2} = \dfrac{5}{8}$

b) $\dfrac{6a+12}{7(a+2)} = \dfrac{6(a+2)}{7(a+2)}$ (factoring)

   $= \dfrac{6}{7} \cdot \dfrac{a+2}{a+2} = \dfrac{6}{7}$ ("removing" a factor of 1)

c) $\dfrac{6x^2+4x}{2x^2+2x} = \dfrac{2x(3x+2)}{2x(x+1)}$ (factoring)

   $= \dfrac{2x}{2x} \cdot \dfrac{3x+2}{x+1}$

   $= \dfrac{3x+2}{x+1}$ ("removing" a factor of 1)

d) $\dfrac{x^2+3x+2}{x^2-1} = \dfrac{(x+2)(x+1)}{(x+1)(x-1)}$

   $= \dfrac{x+1}{x+1} \cdot \dfrac{x+2}{x-1}$

   $= \dfrac{x+2}{x-1}$

e) $\dfrac{5a+15}{10} = \dfrac{5(a+3)}{10}$

   $= \dfrac{5}{10}(a+3) = \dfrac{1}{2}(a+3)$

   $= \dfrac{a+3}{2}$

Do exercises 5 through 7 at the right.

Simplify:

5. a) $\dfrac{5y}{y}$ b) $\dfrac{13xy}{y}$

6. a) $\dfrac{(x+2)(3x+1)}{(x+2)(x-1)}$

b) $\dfrac{2x^2+x}{3x^2+2x}$

c) $\dfrac{x^2-1}{2x^2-x-1}$

7. a) $\dfrac{7x+14}{7}$

b) $\dfrac{12y+24}{48}$

8. Multiply and simplify.

$$\frac{a^2 - 4a + 4}{a^2 - 9} \cdot \frac{a + 3}{a - 2}$$

a. DO NOT CANCEL.

b. You may cancel.

When we multiply, we usually simplify if possible.

Example

Multiply and simplify.

$$\frac{x^2 + 6x + 9}{x^2 - 4} \cdot \frac{x - 2}{x + 3} = \frac{(x^2 + 6x + 9)(x - 2)}{(x^2 - 4)(x + 3)} \quad \text{(multiplying)}$$

$$= \frac{(x + 3)(x + 3)(x - 2)}{(x + 2)(x - 2)(x + 3)} \quad \text{(factoring)}$$

$$= \frac{(x + 3)(x - 2)}{(x + 3)(x - 2)} \cdot \frac{(x + 3)}{(x + 2)}$$

$$= \frac{x + 3}{x + 2} \quad \text{("removing" a factor of 1)}$$

CANCELING

Canceling is a shortcut for part of the procedure in the preceding examples. The use of canceling saves a step in simplifying.

Example. Simplify:

$$\frac{x^2 + x - 2}{2x^2 - 3x + 1} = \frac{(x + 2)(x - 1)}{(x - 1)(2x - 1)}$$

$$= \frac{x + 2}{2x - 1}$$

Note: If you can't factor, you can't cancel!

Do exercise 8 at the left.

The use of canceling causes a great many errors, because it is used mechanically and without understanding. If you find that you can cancel without making errors, we suggest that you do it.

If you make errors by canceling, you should not do it.

Do exercise set 9.1, p. 317.

9.2 DIVISION AND RECIPROCALS

Let us review division with fractional numerals in arithmetic.

Example. Divide:

$$\frac{2}{3} \div \frac{5}{7} = \frac{\dfrac{2}{3}}{\dfrac{5}{7}} = \frac{\dfrac{2}{3} \cdot \dfrac{7}{5}}{\dfrac{5}{7} \cdot \dfrac{7}{5}} = \frac{\dfrac{2}{3} \cdot \dfrac{7}{5}}{1} = \frac{14}{15}$$

In this example we accomplished the division by multiplying by 1. We used the name $\dfrac{\frac{7}{5}}{\frac{7}{5}}$ for the number 1.

Do exercises 9 and 10 at the right.

In the above example, we found that the answer could be obtained by doing the multiplication $\frac{2}{3} \cdot \frac{7}{5}$. In other words we multiplied by the reciprocal of the divisor. We can always do this.

Example. Divide by multiplying by a reciprocal:

$$\frac{3}{10} \div \frac{5}{7} = \frac{3}{10} \cdot \frac{7}{5} = \frac{21}{50}$$

Do exercises 11 and 12 at the right.

The reciprocal of a fractional expression is easy to find. We interchange numerator and denominator.

Examples

a) The reciprocal of $\dfrac{2x^2 - 3}{x + 4}$ is $\dfrac{x + 4}{2x^2 - 3}$

b) The reciprocal of $x + 2$ is $\dfrac{1}{x + 2}$.

Do exercise 13 at the right.

OBJECTIVES

You should be able to:

a) Find the reciprocal of a fractional expression.

b) Divide fractional expressions by multiplying by a reciprocal and simplify the result.

Divide, by multiplying by 1.

9. $\dfrac{3}{4} \div \dfrac{7}{5}$

10. $\dfrac{5}{9} \div \dfrac{3}{4}$

Divide, by multiplying by a reciprocal.

11. $\dfrac{4}{13} \div \dfrac{7}{2}$

12. $\dfrac{5}{9} \div \dfrac{3}{5}$

13. Find the reciprocal of

$$\dfrac{x^3 + 5x}{2x^2 - 3}$$

Divide and simplify.

14. $\dfrac{9x^2 + 3x}{6} \div \dfrac{9x^2 + 6x + 1}{2}$

15. $\dfrac{y^2 - 1}{y + 1} \div \dfrac{y^2 - 2y + 1}{y + 1}$

We divide fractional expressions the same way that we divide in arithmetic with fractional numerals. That is, we multiply by the reciprocal of the divisor.

Examples.

Divide:

a) $\dfrac{x + 1}{x + 2} \div \dfrac{x - 1}{x + 3} = \dfrac{x + 1}{x + 2} \cdot \dfrac{x + 3}{x - 1}$

$\qquad\qquad = \dfrac{x^2 + 4x + 3}{x^2 + x - 2}$

b) $\dfrac{y - 2}{2x + 1} \div \dfrac{x - 2}{y + 1} = \dfrac{y - 2}{2x + 1} \cdot \dfrac{y + 1}{x - 2}$

$\qquad\qquad = \dfrac{y^2 - y - 2}{2x^2 - 3x - 2}$

We can also simplify our answers when dividing. In fact, this should be done.

Example. Divide and simplify.

$\dfrac{x^2 + 7x + 10}{y} \div \dfrac{x^2 - 3x - 10}{y} =$

$\dfrac{x^2 + 7x + 10}{y} \cdot \dfrac{y}{x^2 - 3x - 10} =$

$\dfrac{(x + 2)(x + 5)}{y} \cdot \dfrac{y}{(x - 5)(x + 2)} =$

$\dfrac{(x + 2)y}{(x + 2)y} \cdot \dfrac{x + 5}{(x - 5)} = \dfrac{x + 5}{x - 5}$

Do exercises 14 and 15 at the left.

Do exercise set 9.2, p. 319.

9.3 ADDITION

When denominators are the same it is easy to add fractional expressions. As in arithmetic, we merely add the numerators and keep the common denominator. The new numerator should be simplified when possible.

Examples. Add:

a) $\dfrac{x}{x+1} + \dfrac{2}{x+1} = \dfrac{x+2}{x+1}$

b) $\dfrac{2x^2+3x-7}{2x+1} + \dfrac{x^2+x-8}{2x+1} = \dfrac{3x^2+4x-15}{2x+1}$

Do exercises 16 through 18 at the right.

It is also easy to add when one denominator is the additive inverse of the other. We first multiply one expression by $\dfrac{-1}{-1}$.

Examples. Add:

a) $\dfrac{x}{2} + \dfrac{3}{-2} = \dfrac{x}{2} + \dfrac{-1}{-1}\cdot\dfrac{3}{-2} = \dfrac{x}{2} + \dfrac{-3}{2} = \dfrac{x-3}{2}$

b) $\dfrac{3x+4}{x-2} + \dfrac{x-7}{2-x} = \dfrac{3x+4}{x-2} + \dfrac{-1}{-1}\cdot\dfrac{x-7}{2-x}$

$\qquad = \dfrac{3x+4}{x-2} + \dfrac{-1(x-7)}{-1(2-x)}$

$\qquad = \dfrac{3x+4}{x-2} + \dfrac{7-x}{x-2} = \dfrac{2x+11}{x-2}$

Do exercises 19 through 21 at the right.

Do exercise set 9.3, p. 321.

OBJECTIVES

You should be able to:

a) Add fractional expressions having the same denominator.

b) Add fractional expressions whose denominators are additive inverses of each other.

Add.

16. $\dfrac{5}{9} + \dfrac{2}{9}$

17. $\dfrac{3}{x-2} + \dfrac{x}{x-2}$

18. $\dfrac{4x+5}{x-1} + \dfrac{2x-1}{x-1}$

Add.

19. $\dfrac{y}{3} + \dfrac{y}{-3}$

20. $\dfrac{2x+1}{x-3} + \dfrac{x+2}{3-x}$

21. $\dfrac{x+2}{2x-1} + \dfrac{x-3}{1-2x}$

You should be able to:

a) Subtract fractional expressions when denominators are the same.

b) Subtract fractional expressions when one denominator is the additive inverse of the other.

Subtract.

22. $\dfrac{7}{11} - \dfrac{3}{11}$

23. $\dfrac{3}{x+2} - \dfrac{x}{x+2}$

24. $\dfrac{4x+5}{x+1} - \dfrac{2x-1}{x+1}$

25. $\dfrac{2x^2+3x-7}{2x+1} - \dfrac{x^2+x-8}{2x+1}$

Subtract.

26. $\dfrac{x}{3} - \dfrac{2x-1}{-3}$

27. $\dfrac{3x}{x-2} - \dfrac{x-3}{2-x}$

9.4 SUBTRACTION

To subtract fractional expressions we proceed as in arithmetic. When denominators are the same, we subtract numerators and keep the common denominator.

Example. Subtract:

$$\frac{3x}{x+2} - \frac{x-2}{x+2} = \frac{3x-(x-2)}{x+2} = \frac{3x-x+2}{x+2}$$

$$= \frac{2x+2}{x+2}$$

In the example above note the use of parentheses. This is important, to make sure you subtract the entire numerator and not just part of it.

Do exercises 22 through 25 at the left.

When one denominator is the additive inverse of the other we multiply by $\dfrac{-1}{-1}$.

Example. Subtract: $\dfrac{5y}{y-5} - \dfrac{2y-3}{5-y}$

The denominators are additive inverses of each other (their sum is 0). Thus we shall multiply one fraction by $-1/-1$.

$$\frac{5y}{y-5} - \frac{-1}{-1} \cdot \frac{2y-3}{5-y} = \frac{5y}{y-5} - \frac{-2y+3}{-5+y}$$

$$= \frac{5y}{y-5} - \frac{3-2y}{y-5}$$

$$= \frac{5y-(3-2y)}{y-5} \quad \text{(Note the parentheses here, to make sure we subtract the entire numerator, not just part of it.)}$$

$$= \frac{5y-3+2y}{y-5} = \frac{7y-3}{y-5}$$

Do exercises 26 and 27 at the left.

Do exercise set 9.4, p. 323.

9.5 LEAST COMMON MULTIPLES

In arithmetic, to add when denominators are different, we first find a common denominator. We shall do the same thing for fractional expressions. Let us review the procedure in arithmetic first. To do the addition

$$\frac{5}{12} + \frac{7}{30}$$

we first find a common denominator. We look for a number that is a multiple of both 12 and 30 (a common multiple). Actually, we prefer to have the smallest such number, or the *Least Common Multiple* (L.C.M.). To find the L.C.M. we factor both numbers.

$12 = 2 \cdot 2 \cdot 3$

$30 = 2 \cdot 3 \cdot 5$

The L.C.M. is the number that has 2 as a factor twice, 3 as a factor once, and 5 as a factor once. The L.C.M. is $2 \cdot 2 \cdot 3 \cdot 5$, or 60.

To find the L.C.M., we use each factor the greatest number of times it appears in any one factorization.

Example. Find the L.C.M. of 24 and 36.

$24 = 2 \cdot 2 \cdot 2 \cdot 3$ ⎫
$36 = 2 \cdot 2 \cdot 3 \cdot 3$ ⎭   The L.C.M. is $2 \cdot 2 \cdot 2 \cdot 3 \cdot 3$, or 72.

Do exercises 28 and 29 at the right.

Now let us return to the problem of adding $\frac{5}{12}$ and $\frac{7}{30}$.

$$\frac{5}{12} + \frac{7}{30} = \frac{5}{2 \cdot 2 \cdot 3} + \frac{7}{2 \cdot 3 \cdot 5}$$

The L.C.M. is $2 \cdot 2 \cdot 3 \cdot 5$. To have this in the first denominator we need a 5. In the second denominator we need another 2. To accomplish this, we multiply by 1.

$$\frac{5}{2 \cdot 2 \cdot 3} \cdot \frac{5}{5} + \frac{7}{2 \cdot 3 \cdot 5} \cdot \frac{2}{2} = \frac{25}{2 \cdot 2 \cdot 3 \cdot 5} + \frac{14}{2 \cdot 2 \cdot 3 \cdot 5}$$

$$= \frac{39}{2 \cdot 2 \cdot 3 \cdot 5}$$

$$= \frac{39}{60}$$

Example. Add:

$$\frac{5}{12} + \frac{11}{18}$$

$12 = 2 \cdot 2 \cdot 3$ ⎫
$18 = 2 \cdot 3 \cdot 3$ ⎭   L.C.M. $= 2 \cdot 2 \cdot 3 \cdot 3$, or 36.

$$\frac{5}{12} + \frac{11}{18} = \frac{5}{2 \cdot 2 \cdot 3} \cdot \frac{3}{3} + \frac{11}{2 \cdot 3 \cdot 3} \cdot \frac{2}{2}$$

$$= \frac{15 + 22}{2 \cdot 2 \cdot 3 \cdot 3}$$

$$= \frac{37}{36}$$

Do exercises 30 and 31 at the right.

OBJECTIVES

You should be able to:

a) Find the L.C.M. of several numbers by factoring.

b) Add numbers with fractional notation, first finding the L.C.M. of the denominators.

c) Find the L.C.M. of several algebraic expressions by factoring.

Find the L.C.M. by factoring.

28. 12, 18

29. 24, 30, 20

Add, first finding the L.C.M. of the denominators.

30. $\dfrac{5}{24} + \dfrac{7}{18}$

31. $\dfrac{3}{28} + \dfrac{5}{21} + \dfrac{4}{49}$

Find the L.C.M.

32. $12xy^2$, $15x^3y$

33. $y^2 + 5y + 4$, $y^2 + 2y + 1$

34. $x^2 + 2x + 1$, $3x - 3x^2$, $x^2 - 1$

L.C.M.'s OF ALGEBRAIC EXPRESSIONS

To find the L.C.M. of several algebraic expressions we proceed as we do for numbers. We factor them and then use each factor the greatest number of times it occurs in any one place.

Example 1. Find the L.C.M. of $12x$, $16y$, and $8xyz$.

$$\left.\begin{array}{l} 12x = 2 \cdot 2 \cdot 3 \cdot x \\ 16y = 2 \cdot 2 \cdot 2 \cdot 2 \cdot y \\ 8xyz = 2 \cdot 2 \cdot 2 \cdot x \cdot y \cdot z \end{array}\right\} \quad \begin{array}{l} \text{L.C.M.} = 2 \cdot 2 \cdot 2 \cdot 2 \cdot 3 \cdot x \cdot y \cdot z \\ \qquad\quad = 48xyz \end{array}$$

Example 2. Find the L.C.M. of $x^2 + 5x - 6$ and $x^2 - 1$.

$$\left.\begin{array}{r} x^2 + 5x - 6 = (x + 6)(x - 1) \\ x^2 - 1 = (x + 1)(x - 1) \end{array}\right\} \quad \text{L.C.M.} = (x + 6)(x + 1)(x - 1)$$

Example 3. Find the L.C.M. of $x^2 + 4$, $x + 1$, and 5.

These expressions are not factorable, so the L.C.M. is their product, $5(x^2 + 4)(x + 1)$

If an L.C.M. is multiplied by -1 we still consider the answer to be an L.C.M. For example, the L.C.M. of 5 and -5 is either 5 or -5. The L.C.M. of $x - 2$ and $2 - x$ is either $x - 2$ or $2 - x$.

Example 4. Find the L.C.M. of $x^2 - y^2$ and $2y - 2x$.

$$\left.\begin{array}{l} x^2 - y^2 = (x + y)(x - y) \\ 2y - 2x = 2(y - x) \end{array}\right\} \quad \begin{array}{l} \text{L.C.M.} = 2(x + y)(x - y) \\ \qquad\quad \text{or } 2(x + y)(y - x) \end{array}$$

Example 5. Find the L.C.M. of $x - 2y$, $x^2 - 4y^2$, and $x^2 - 4xy + 4y^2$.

$x - 2y = x - 2y$

$x^2 - 4y^2 = (x - 2y)(x + 2y)$

$x^2 - 4xy + 4y^2 = (x - 2y)(x - 2y)$

L.C.M. $= (x - 2y)^2(x + 2y)$

Do exercises 32 through 34 at the left.

Do exercise set 9.5, p. 325.

9.6 ADDITION WITH DIFFERENT DENOMINATORS

Now that we know how to find L.C.M.s we can add fractional expressions with different denominators. The procedure is as in arithmetic. We first find a common denominator and then add.

Example 1. Add: $\dfrac{3}{x+1} + \dfrac{5}{x-1}$

The denominators do not factor, so the L.C.M. is their product. We multiply by 1 to get the L.C.M. in each expression.

$$\frac{3}{x+1} \cdot \frac{x-1}{x-1} + \frac{5}{x-1} \cdot \frac{x+1}{x+1} = \frac{3(x-1) + 5(x+1)}{(x-1)(x+1)}$$

$$= \frac{3x - 3 + 5x + 5}{x^2 - 1}$$

$$= \frac{8x + 2}{x^2 - 1}$$

Example 2. Add: $\dfrac{5}{x^2 + x} + \dfrac{4}{2x + 2}$

We first find the L.C.M. of the denominators.

$x^2 + x = x(x + 1)$

$2x + 2 = 2(x + 1)$

L.C.M. $= 2x(x + 1)$

Now we multiply by 1 to get the L.C.M. in each expression. Then we add and simplify.

$$\frac{5}{x(x+1)} \cdot \frac{2}{2} + \frac{4}{2(x+1)} \cdot \frac{x}{x} = \frac{10}{2x(x+1)} + \frac{4x}{2x(x+1)} =$$

$$\frac{10 + 4x}{2x(x+1)} = \frac{2(5 + 2x)}{2x(x+1)}$$

$$= \frac{2}{2} \cdot \frac{5 + 2x}{x(x+1)}$$

$$= \frac{5 + 2x}{x(x+1)}$$

Do exercises 35 and 36 at the right.

Do exercise set 9.6, p. 327.

OBJECTIVES

You should be able to add fractional expressions with different denominators and simplify the result.

Add and simplify.

35. $\dfrac{3}{x^3 - x} + \dfrac{4}{x^2 + 2x + 1}$

36. $\dfrac{5}{x^2 + 17x + 16} + \dfrac{3}{x^2 + 9x + 8}$

You should be able to add and subtract fractional expressions with different denominators and simplify the result.

Subtract.

37. $\dfrac{x-2}{3x} - \dfrac{2x-1}{5x}$

38. $\dfrac{2y+1}{y^2-7y+6} - \dfrac{y+3}{y^2-5y-6}$

Simplify.

39. $\dfrac{1}{x} - \dfrac{5}{3x} + \dfrac{2x}{x+1}$

9.7 ADDITION AND SUBTRACTION WITH DIFFERENT DENOMINATORS

Subtraction is like addition, except that we subtract numerators instead of adding them.

Examples

a) Subtract:

$$\frac{x+1}{x-1} - \frac{2x-3}{x-1} = \frac{x+1-(2x-3)}{x-1}$$

(Note the use of parentheses here, to make sure we subtract the entire numerator.)

$$= \frac{x+1-2x+3}{x-1} = \frac{-x+4}{x-1}$$

b) Subtract:

$$\frac{1}{a-2} - \frac{1}{2} = \frac{1}{a-2}\cdot\frac{2}{2} - \frac{1}{2}\cdot\frac{a-2}{a-2}$$

$$= \frac{2}{2(a-2)} - \frac{a-2}{2(a-2)}$$

$$= \frac{2-(a-2)}{2(a-2)}$$

(Note the use of parentheses here, to make sure we subtract the entire numerator.)

$$= \frac{2-a+2}{2(a-2)} = \frac{4-a}{2(a-2)}$$

c) Simplify:

$$\frac{1}{x} - \frac{1}{x^2} + \frac{2}{x+1} \qquad \text{L.C.M.} = x^2(x+1)$$

$$\frac{1}{x}\cdot\frac{x(x+1)}{x(x+1)} - \frac{1}{x^2}\cdot\frac{x+1}{x+1} + \frac{2}{x+1}\cdot\frac{x^2}{x^2} =$$

$$\frac{x(x+1)-(x+1)+2x^2}{x^2(x+1)} = \frac{x^2+x-x-1+2x^2}{x^2(x+1)}$$

$$= \frac{3x^2-1}{x^2(x+1)}$$

Do exercises 37 through 39 at the left.

Do exercise set 9.7, p. 329.

9.8 COMPLEX FRACTIONAL EXPRESSIONS

You should be able to simplify complex fractional expressions.

A *complex fractional expression* is one which has a fractional expression in its numerator or its denominator, or both. Here are some examples.

$$\frac{1 + \frac{2}{x}}{3}, \quad \frac{\frac{x+y}{2}}{\frac{2x}{x+1}}, \quad \frac{\frac{1}{3} + \frac{1}{5}}{\frac{2}{x} - \frac{x}{y}}$$

To simplify a complex fractional expression we first add or subtract if necessary, to get a single fractional expression in both numerator and denominator. Then we divide, by multiplying by the reciprocal of the denominator.

Simplify.

40. $\dfrac{\frac{2}{7} + \frac{3}{7}}{\frac{3}{4}}$

Example 1. Simplify.

$$\frac{\frac{1}{5} + \frac{2}{5}}{\frac{7}{3}} = \frac{\frac{3}{5}}{\frac{7}{3}} \qquad \text{(adding in the numerator)}$$

$$= \frac{3}{5} \cdot \frac{3}{7} \qquad \text{(multiplying by the reciprocal of the denominator)}$$

$$= \frac{9}{35}$$

41. $\dfrac{3 + \frac{x}{2}}{\frac{5}{4}}$

Example 2. Simplify.

$$\frac{1 + \frac{2}{x}}{\frac{3}{4}} = \frac{1 \cdot \frac{x}{x} + \frac{2}{x}}{\frac{3}{4}} = \frac{\frac{x+2}{x}}{\frac{3}{4}}$$

$$= \frac{x+2}{x} \cdot \frac{4}{3} = \frac{4(x+2)}{3x}$$

Do exercises 40 and 41 at the right.

Simplify.

42. $\dfrac{\dfrac{x}{2} + \dfrac{2x}{3}}{\dfrac{1}{x} - \dfrac{x}{2}}$

Example 3. Simplify.

$$\frac{\dfrac{3}{x} + \dfrac{1}{2x}}{\dfrac{1}{3x} - \dfrac{3}{4x}} = \frac{\dfrac{3}{x} \cdot \dfrac{2}{2} + \dfrac{1}{2x}}{\dfrac{1}{3x} \cdot \dfrac{4}{4} - \dfrac{3}{4x} \cdot \dfrac{3}{3}}$$

$$= \frac{\dfrac{6}{2x} + \dfrac{1}{2x}}{\dfrac{4}{12x} - \dfrac{9}{12x}} = \frac{\dfrac{7}{2x}}{\dfrac{-5}{12x}}$$

$$= \frac{7}{2x} \cdot \frac{12x}{-5} = \frac{2x}{2x} \cdot \frac{7 \cdot 6}{-5}$$

$$= \frac{42}{-5} = -\frac{42}{5}$$

43. $\dfrac{b - \dfrac{x^2}{b}}{\dfrac{x}{b} + 1}$

Example 4. Simplify.

$$\frac{1 - \dfrac{x}{a}}{a - \dfrac{x^2}{a}} = \frac{\dfrac{a}{a} - \dfrac{x}{a}}{a \cdot \dfrac{a}{a} - \dfrac{x^2}{a}} = \frac{\dfrac{a - x}{a}}{\dfrac{a^2 - x^2}{a}}$$

$$= \frac{a - x}{a} \cdot \frac{a}{a^2 - x^2} = \frac{a}{a} \cdot \frac{a - x}{a^2 - x^2}$$

$$= \frac{a - x}{(a - x)(a + x)} = \frac{a - x}{a - x} \cdot \frac{1}{a + x} = \frac{1}{a + x}$$

Do exercises 42 and 43 at the left.

Do exercise set 9.8, p. 331.

9.9 DIVISION OF POLYNOMIALS

DIVISOR A MONOMIAL

In a division of polynomials where the divisor is a monomial, we can see how to proceed by considering fractional expressions. Recall that fractional expressions indicate division.

Example 1

Divide $x^3 + 10x^2 + 8x$ by $2x$

$$\frac{x^3 + 10x^2 + 8x}{2x} = \frac{x^3}{2x} + \frac{10x^2}{2x} + \frac{8x}{2x}$$

Here we have expressed the division by a fractional expression and done what amounts to the opposite of addition.*
Next we do the three indicated divisions.

$$\frac{x^3}{2x} + \frac{10x^2}{2x} + \frac{8x}{2x} = \frac{1}{2}x^2 + 5x + 4$$

Do exercises 44 and 45 at the right.

Example 2

Divide and check: $(5y^2 - 2y + 4) \div 2$

$$\frac{5y^2 - 2y + 4}{2} = \frac{5y^2}{2} - \frac{2y}{2} + \frac{4}{2}$$

$$= \frac{5}{2}y^2 - y + 2$$

Check: $\frac{5}{2}y^2 -\ \ y + 2$

$$\underline{\hspace{5cm} 2 \quad \text{(We multiply)}}$$

$$5y^2 - 2y + 4$$

When doing divisions like this you should write only the answer.

Do exercises 46 and 47 at the right.

_____

* To see this, do the addition $\dfrac{x^3}{2x} + \dfrac{10x^2}{2x} + \dfrac{8x}{2x}$ and you will get the fractional expression $\dfrac{x^3 + 10x^2 + 8x}{2x}$.

OBJECTIVES

You should be able to:

a) Divide a polynomial by a monomial and check the result.

b) Divide a polynomial by a binomial and check the result. If there is a remainder, express the result two ways.

Divide.

44. $\dfrac{2x^3 + 6x^2 + 4x}{2x}$

45. $(6x^2 + 3x - 2) \div 3$

Divide and check.

46. $(8x^2 - 3x + 1) \div 2$

47. $\dfrac{(2x^4 - 3x^3 + 5x^2)}{x^2}$

DIVISOR NOT A MONOMIAL

When the divisor is not a monomial we use a procedure very much like long division in arithmetic.

Example 3

Divide $x^2 + 5x + 6$ by $x + 2$

$$
\begin{array}{r}
x \\
x + 2 \overline{) x^2 + 5x + 6} \\
\underline{x^2 + 2x} \\
3x
\end{array}
$$

Divide first term by first term, to get x.

Multiply x by divisor

Subtract

Below we have rewritten the problem and "brought down" the next term of the dividend, 6.

$$
\begin{array}{r}
x + 3 \\
x + 2 \overline{) x^2 + 5x + 6} \\
\underline{x^2 + 2x} \\
3x + 6 \\
\underline{3x + 6} \\
0
\end{array}
$$

Divide first term by first term to get 3.

Multiply 3 by divisor

Subtract

To check, we multiply quotient by divisor and add the remainder, if any, to see if we get the dividend.

$(x + 2)(x + 3) = x^2 + 5x + 6$, so this division checks.

Example 4

Divide and check:

$$
\begin{array}{r}
x + 5 \\
x - 3 \overline{) x^2 + 2x - 12} \\
\underline{x^2 - 3x} \\
5x - 12 \\
\underline{5x - 15} \\
3
\end{array}
$$

Check:

$(x - 3)(x + 5) + 3 = x^2 + 2x - 15 + 3$

$= x^2 + 2x - 12$

Quotient

Remainder

We can also express the result this way:

$$\frac{x^2 + 2x - 12}{x - 3} = x + 5 + \frac{3}{x - 3}$$

Example 5. Divide:

$$
\begin{array}{r}
x^2 - x\ \ + 1 \\
x + 1\ \overline{)\ x^3\ \qquad + 1} \\
x^3 + x^2 \\
\hline
-x^2 \\
-x^2 - x \\
\hline
x + 1 \\
x + 1 \\
\hline
\end{array}
$$

When there are missing terms space must be left for them

Example 6. Divide:

$$
\begin{array}{r}
x^3 + 4x^2 + 13x + 52 \\
x - 4\ \overline{)\ x^4\qquad\ - 3x^2\qquad\quad + 1} \\
x^4 - 4x^3 \\
\hline
4x^3 - 3x^2 \\
4x^3 - 16x^2 \\
\hline
13x^2 \\
13x^2 - 52x \\
\hline
52x + 1 \\
52x - 208 \\
\hline
209 \\
\end{array}
$$

The quotient is $x^3 + 4x^2 + 13x + 52$ and the remainder is 209. Or, we can express the answer this way:

$$x^3 + 4x^2 + 13x + 52 + \frac{209}{x - 4}$$

Do exercises 48 through 51 at the right.

Do exercise set 9.9, p. 333.

Divide and check.

48. $x + 3\ \overline{)\ x^2 + x - 6}$

49. $x - 2\ \overline{)\ x^2 + 2x - 8}$

Divide. Express your answer two ways.

50. $x + 3\ \overline{)\ x^2 + 7x + 10}$

Divide.

51. $(x^3 - 1) \div (x - 1)$

OBJECTIVE

You should be able to solve fractional equations.

Solve.

52. $\dfrac{1}{x} = \dfrac{1}{6 - x}$

53. $\dfrac{1}{3x} = -\dfrac{1}{x} - 9$

9.10 FRACTIONAL EQUATIONS

A fractional equation is an equation containing one or more fractional expressions. To solve such an equation, we first eliminate the fractional expressions. We can do this by multiplying on both sides of the equation by the L.C.M. of all the denominators. We call this *clearing of fractions*.

Example 1. Clear of fractions.

$$\frac{2}{3} + \frac{5}{6} = \frac{x}{9}$$

The L.C.M. of all denominators is 18, or $2 \cdot 3 \cdot 3$. We multiply by this.

$$2 \cdot 3 \cdot 3 \left[\frac{2}{3} + \frac{5}{6} \right] = 2 \cdot 3 \cdot 3 \cdot \frac{x}{9}$$

$$2 \cdot 3 \cdot 3 \cdot \frac{2}{3} + 2 \cdot 3 \cdot 3 \cdot \frac{5}{6} = 2 \cdot 3 \cdot 3 \cdot \frac{x}{9} \qquad \text{(multiplying, to remove the parentheses)}$$

$$2 \cdot 3 \cdot 2 + 3 \cdot 5 = 2 \cdot x$$

$$12 + 15 = 2x \qquad \text{(simplifying)}$$

Note that when we *clear of fractions* all the denominators disappear. Thus we have an equation without fractional expressions, which is easier to solve.

When clearing of fractions, always be sure to multiply *all* terms in the equation by the L.C.M.

Example 2. Clear of fractions and solve.

$$\frac{1}{x} = \frac{1}{4 - x}$$

The L.C.M. is $x(4 - x)$. We multiply by this.

$$x(4 - x) \cdot \frac{1}{x} = x(4 - x) \cdot \frac{1}{4 - x}$$

$$4 - x = x \qquad \text{(simplifying)}$$

$$4 = 2x$$

$$x = 2$$

Check: $\qquad \dfrac{1}{x} = \dfrac{1}{4 - x}$

$$\frac{\dfrac{1}{2} \quad \bigg| \quad \dfrac{1}{4 - 2}}{\qquad \qquad \dfrac{1}{2}}$$

This checks, so 2 is the solution.

Do exercises 52 and 53 at the left.

The next example shows how important it is to multiply all terms in an equation by the L.C.M.

Example 3. Solve.

$$x + \frac{6}{x} = -5$$

The L.C.M is x. We multiply by this.

$$x\left(x + \frac{6}{x}\right) = -5x \qquad \text{(multiplying on both sides by } x)$$

$$x^2 + x \cdot \frac{6}{x} = -5x \qquad \text{(notice that each term on the left is now multiplied by } x)$$

$$x^2 + 5x + 6 = 0$$

$$(x + 3)(x + 2) = 0 \qquad \text{(factoring)}$$

$$x = -3 \quad \text{or} \quad x = -2. \qquad \text{(principle of zero products)}$$

Both of these check, so there are two solutions, -3 and -2.

It is important always to check when solving fractional equations.

Example 4. Solve.

$$\frac{x^2}{x - 1} = \frac{1}{x - 1}$$

The L.C.M. is $x - 1$.

$$(x - 1) \cdot \frac{x^2}{x - 1} = (x - 1) \cdot \frac{1}{x - 1}$$

$$x^2 = 1, \text{ or } x^2 - 1 = 0$$

$$(x + 1)(x - 1) = 0 \qquad \text{(factoring)}$$

Possible solutions are 1 and -1.

Check:
$$\frac{x^2}{x - 1} = \frac{1}{x - 1} \qquad\qquad \frac{x^2}{x - 1} = \frac{1}{x - 1}$$

$$\begin{array}{c|c} \dfrac{1^2}{1 - 1} & \dfrac{1}{1 - 1} \\[2ex] \dfrac{1}{0} & \dfrac{1}{0} \end{array} \qquad\qquad \begin{array}{c|c} \dfrac{(-1)^2}{-1 - 1} & \dfrac{1}{-1 - 1} \\[2ex] -\dfrac{1}{2} & -\dfrac{1}{2} \end{array}$$

The number -1 is a solution, but 1 is not because it makes a denominator zero.

Do exercises 54 through 56 at the right.

Do exercise set 9.10, p. 335.

Solve.

54. $x + \dfrac{1}{x} = 2$

55. $\dfrac{x^2}{x + 2} = \dfrac{4}{x + 2}$

56. $\dfrac{1}{2x} + \dfrac{1}{x} = -12$

OBJECTIVES

You should be able to:

a) Solve applied problems involving fractional equations.

b) Solve formulas for a given letter when fractional expressions occur in the formula.

9.11 FORMULAS AND APPLIED PROBLEMS

Now that you have learned to solve fractional equations you can use that skill in handling formulas and in solving applied problems.

Example 1

The reciprocal of 2 less than a certain number is twice the reciprocal of the number itself. What is the number?

First, we translate to an equation. We shall use x for the number. Then 2 less than the number is $x - 2$, and the reciprocal of the number is $\frac{1}{x}$.

Reciprocal of 2 less than number	is $\downarrow$	Twice the reciprocal of the number
$\frac{1}{x-2}$	$=$	$2 \cdot \frac{1}{x}$

Now we solve.

$$\frac{1}{x - 2} = \frac{2}{x} \qquad \text{(The L.C.M. is } x(x - 2)).$$

$$\frac{x(x - 2)}{x - 2} = \frac{2x(x - 2)}{x} \qquad \text{(Multiplying by L.C.M.)}$$

$$x = 2(x - 2) \qquad \text{(Simplifying)}$$

$$x = 2x - 4$$

$$x = 4$$

Check:

We go to the original problem. The number we are checking is 4. Two less than this is 2. The reciprocal of 2 is $\frac{1}{2}$. The reciprocal of the number itself is $\frac{1}{4}$. Now $\frac{1}{2}$ is twice $\frac{1}{4}$, so the conditions are satisfied.

Thus 4 is the solution.

Example 2

One car travels 20 km/hr faster than another. While one of them goes 240 km the other goes 160 km. Find their speeds.

To translate, we recall that distance, rate and time are related as follows: $d = rt$. Also, we note that both cars travel the same time. Thus.

$$t_1 = t_2 \quad \text{or} \quad \frac{d_1}{r_1} = \frac{d_2}{r_2}$$

Now, if we let $d_1 = 160$, $d_2 = 240$ and note that $r_2 = r_1 + 20$, we get our equation:

$$\frac{240}{r_1 + 20} = \frac{160}{r_1}$$

We solve:

$$\frac{240}{r_1 + 20} = \frac{160}{r_1}$$

$$\frac{240r_1(r_1 + 20)}{r_1 + 20} = \frac{160r_1(r_1 + 20)}{r_1} \qquad \text{(Multiplying by L.C.M.)}$$

$$240r_1 = 160(r_1 + 20) \qquad \text{(Simplifying)}$$

$$240r_1 = 160r_1 + 3200$$

$$80r_1 = 3200$$

$$r_1 = 40 \text{ km/hr}$$

$$r_2 = r_1 + 20 = 60 \text{ km/hr}$$

These answers check.

Do exercises 57 and 58 at the right.

57. The reciprocal of 2 more than a number is three times the reciprocal of the number. Find the number.

58. One car goes 10 km/hr faster than another. While one car goes 120 km the other goes 150 km. How fast is each car?

59. Solve for p.

$$\frac{n}{p} = 2 - m$$

FORMULAS

The following examples show how formulas involving fractional expressions can be solved for a given letter.

Example 3. Solve for t.

$$\frac{b}{t} = 1 + c$$

$$\frac{bt}{t} = t(1 + c)$$

$$b = t(1 + c) \qquad \text{(Simplifying left side)}$$

$$\frac{b}{1 + c} = \frac{t(1 + c)}{1 + c} \qquad \text{(Dividing by } 1 + c\text{)}$$

$$\frac{b}{1 + c} = t \qquad \text{(Simplifying)}$$

We note that this will not hold for $c = -1$ or $t = 0$.

60. Solve for s.

$$\frac{s}{r} + \frac{s}{t} = 1$$

Example 4. Solve for s.

$$\frac{ms}{n} + \frac{ns}{m} = 1 \qquad \text{(L.C.M. is } mn\text{)}$$

$$\left[\frac{ms}{n} + \frac{ns}{m}\right]mn = 1 \cdot mn \qquad \text{(Multiplying by L.C.M.)}$$

$$\frac{ms \cdot mn}{n} + \frac{ns \cdot mn}{m} = mn$$

$$m^2s + n^2s = mn$$

$$(m^2 + n^2)s = mn \qquad \text{(Factoring out } s\text{)}$$

$$\frac{(m^2 + n^2)s}{m^2 + n^2} = \frac{mn}{m^2 + n^2} \qquad \text{(Dividing by } m^2 + n^2\text{)}$$

$$s = \frac{mn}{m^2 + n^2} \qquad \text{(This holds if } mn \neq 0 \text{ and if } m^2 + n^2 \neq 0.\text{)}$$

Do exercises 59 and 60 at the left.

Do exercise set 9.11, p. 337.

9.12 RATIO, PROPORTION, AND VARIATION

RATIOS

When we speak of the ratio of two quantities, we mean their quotient. For example, the ratio of the age of a 27-year old man to that of his 3-year old son is $27 \div 3$ or 9. In the rectangle below the ratio of width to length is $\frac{2}{3}$. The ratio of length to width is $\frac{3}{2}$. An older notation for ratios is 2 : 3 (read 2 is to 3). This notation is still used occasionally.

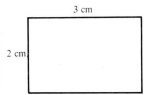

PROPORTIONS

In applied problems we often find a single ratio expressed in two ways. For example, suppose it takes 9 gallons of gas to drive 110 miles, and we wish to find how much will be required to go 550 miles. We can set up ratios.

$$\frac{9 \text{ gal}}{110 \text{ mi}} \qquad \frac{x \text{ gal}}{550 \text{ mi}}$$

Since the car should use gas at the same rate throughout the trip, these ratios are the same. Thus we have an equation

$$\frac{9}{110} = \frac{x}{550}$$

An equation like this is sometimes read "9 is to 110 as x is to 550". We solve this equation, first multiplying by 550.

$$\frac{9}{110} \cdot 550 = x$$

$$45 = x$$

The answer is, since it checks, that 45 gallons will be required.

An equality of two ratios, $\dfrac{a}{b} = \dfrac{c}{d}$ is called a *proportion*.

The numbers named in a true proportion are said to be *proportional*.

Many applied problems can be solved rather simply using proportions.

Use proportions to solve.

61. In the post office a machine can cancel stamps at the rate of 300 in 20 minutes. How many can it cancel in 45 minutes?

62. In a rectangle the ratio of width to length is $\frac{5}{7}$. The length is 8 cm greater than the width. Find the length and width.
(Hint: Let w = width and $w + 8$ = length.)

Example 1. A secretary can address envelopes at the rate of 100 every 35 minutes. How many can she address in 55 minutes.

To solve this we think of ratios and set up a proportion. We can think "100 envelopes is to 35 minutes as x envelopes is to 55 minutes". Then we write

$$\frac{100}{35} = \frac{x}{55}$$

Now we solve the proportion, first multiplying by 55.

$$\frac{100}{35} \cdot 55 = x$$

Computing, we get $x = 157\frac{1}{7}$, which we round to 157.

Do exercises 61 and 62 at the left.

VARIATION

Suppose a bicycle is traveling at 10 km/hr. In one hour it goes 10 km. In two hours it goes 20 km; in three hours 30 km, and so on. This situation gives rise to a set of pairs of numbers, all having the same ratio, (1, 10), (2,20), (3,30), (4, 40) and so on. The ratio of distance to time is always $\frac{10}{1}$, or 10. Whenever a situation gives rise to pairs of numbers like this, in which the ratio is constant, we say that there is *direct variation*. In this case we might say that the distance *varies (directly) as the time*.
Notice that we have

$$\frac{d}{t} = 10 \text{ (a constant) or } d = 10t$$

Whenever a situation gives rise to a relation among variables $y = kx$, where k is a constant, we say that there is *direct variation*. The number k is called the *variation constant*.

Whenever we know an equation of variation $y = kx$, we can find the variation constant if we know one set of values for x and y. When we know the variation constant we can find either x or y when the other is given. Thus we can solve applied problems involving variation.

Example 2. A man's paycheck varies as the number of hours he works. If he works 15 hours he gets paid $22.50. Find the variation constant. Then find his pay for 35 hours work.

We first write an equation of variation

$P = kh$, where P is the pay in dollars and h is the number of hours.

Now we substitute the known set of values and compute k.

$22.50 = k \cdot 15$

$$k = \frac{22.50}{15} = 1.5$$

Thus the variation equation is

$P = 1.5h$

Now we can use this equation to find his pay for working 35 hours.

$P = 1.5 \times 35$

$\quad = \$52.50$

Do exercises 63 and 64 at the right.

You should notice that we can solve the same kinds of problems with variation equations as with proportions. Sometimes one approach is simpler or more convenient than the other. Then, of course, one should use the one that makes the work easier.

Do exercise set 9.12, p. 339.

To solve, first find a variation constant and then use an equation of variation.

63. The number of screws a machine can make varies as the time it operates. It can make 1000 screws in 2.5 hours. How many can it make in 4 hours?

64. The electric current, in amperes, in a circuit varies as the voltage. When 12 volts are applied the current is 4 amperes. What is the current when 15 volts are applied?

EXERCISE SET 9.1 (pp. 291–294)

Multiply.

1. $\dfrac{2x+3}{4} \cdot \dfrac{x+1}{x-5}$

2. $\dfrac{x-1}{x+2} \cdot \dfrac{x+1}{x+2}$

3. $\dfrac{-5}{3x-4} \cdot \dfrac{-6}{5x+6}$

4. $\dfrac{a+2b}{a+b} \cdot \dfrac{a-2b}{a-b}$

5. $\dfrac{3x-1}{2x+y} \cdot \dfrac{y}{y}$

6. $\dfrac{3x-2}{x+7} \cdot \dfrac{2x+5}{2x+5}$

Simplify.

7. $\dfrac{(3x+y)(x+1)}{(x+1)(2x-3)}$

8. $\dfrac{x(3x+2)(5x-7)}{x(3x-2)(5x+7)}$

9. $\dfrac{3(x-2)(5x+4)}{2(5x+4)(x+2)}$

10. $\dfrac{6x-6y}{6x+6y}$

11. $\dfrac{ab}{ax-ay}$

12. $\dfrac{a^2-9}{a^2+5a+6}$

13. $\dfrac{2x^2+6x+4}{4x^2-12x-16}$

14. $\dfrac{x^2-3x-4}{2x^2+10x+8}$

ANSWERS

15. _____

16. _____

17. _____

18. _____

19. _____

20. _____

21. _____

22. _____

23. _____

24. _____

25. _____

26. _____

15. $\dfrac{x^2 - 25}{x^2 - 10x + 25}$

16. $\dfrac{6(x^2 - 9)}{4(x + 3)(x - 3)}$

17. $\dfrac{6x + 12}{x^2 - x - 6}$

18. $\dfrac{5y + 5}{y^2 + 7y + 6}$

19. $\dfrac{a^2 + 1}{a + 1}$

20. $\dfrac{x^2 - 3x - 18}{x^2 - 2x - 15}$

Multiply and simplify.

21. $\dfrac{-1}{-1} \cdot \dfrac{b - 3}{b + 4}$

22. $\dfrac{x^2 - 3x - 10}{(x - 2)^2} \cdot \dfrac{x - 2}{x - 5}$

23. $\dfrac{24a^2}{3(a^2 - 4a + 4)} \cdot \dfrac{3a - 6}{2a}$

24. $\dfrac{5v + 5}{v - 2} \cdot \dfrac{v^2 - 4v + 4}{v^2 - 1}$

25. $\dfrac{ab - b^2}{2a} \cdot \dfrac{2a + 2b}{a^2b - b^3}$

26. $\dfrac{c^2 - 6c}{c - 6} \cdot \dfrac{c + 3}{c}$

EXERCISE SET 9.2 (pp. 295–296)

Divide by multiplying by 1.

1. $\dfrac{2}{7} \div \dfrac{1}{2}$

2. $\dfrac{2}{5} \div \dfrac{3}{4}$

3. $\dfrac{24}{25} \div \dfrac{8}{5}$

4. $\dfrac{3}{x} \div \dfrac{6}{x}$

5. $\dfrac{4}{3} \div 12$

6. $\dfrac{x}{y} \div \dfrac{2}{3}$

Divide by multiplying by a reciprocal.

7. $\dfrac{3}{5} \div \dfrac{7}{8}$

8. $4 \div \dfrac{2}{3}$

9. $\dfrac{2}{3} \div 12$

10. $\dfrac{2c}{3d} \div \dfrac{6c}{9d}$

11. $\dfrac{x}{x+1} \div \dfrac{y}{x+1}$

12. $\dfrac{3x+6}{5} \div \dfrac{x+2}{10}$

Divide and simplify.

13. $\dfrac{a+2}{a-3} \div \dfrac{a-1}{a+3}$

14. $\dfrac{y+2}{4} \div \dfrac{y}{2}$

15. $\dfrac{x^2-1}{x} \div \dfrac{x+1}{x-1}$

16. $\dfrac{4y-8}{y+2} \div \dfrac{y-2}{y^2-4}$

17. $\dfrac{x+1}{6} \div \dfrac{x+1}{3}$

18. $\dfrac{a}{a-b} \div \dfrac{a}{b-a}$

19. $\dfrac{x^2-9}{4x+12} \div \dfrac{x-3}{6}$

20. $\dfrac{c^2+3c}{c^2+2c-3} \div \dfrac{c}{c+1}$

1.

2.

3.

4.

5.

6.

7.

8.

9.

10.

11.

12.

13.

14.

15.

16.

17.

18.

19.

20.

21. $\dfrac{x+y}{x-y} \div \dfrac{x^2+y}{x^2-y^2}$

22. $\dfrac{x-b}{2x} \div \dfrac{x^2-b^2}{5x^2}$

23. $\dfrac{x^2-x-20}{x^2+7x+12} \div \dfrac{x^2-10x+25}{x^2+6x+9}$

24. $\dfrac{1}{d(d-1)} \div \dfrac{d+1}{(1-d)^2}$

25. $\dfrac{c^2+10c+21}{c^2-2c-15} \div (c^2+2c-35)$

26. $(1-z) \div \dfrac{1-z}{1+2z-z^2}$

27. $\dfrac{2y^2-7y+3}{2y^2+3y-2} \div \dfrac{6y^2-5y+1}{3y^2+5y-2}$

Perform the indicated operations and simplify:

28. $\left[\dfrac{r^2-4s^2}{r+2s} \div (r+2s)\right] \cdot \dfrac{2s}{r-2s}$

29. $\left[\dfrac{15-13a+2a^2}{4a^2-9} \cdot \dfrac{2a+1}{1-2a}\right] \div \dfrac{5-a}{2a-1}$

30. $\left[\dfrac{d^2-d}{d^2-6d+8} \cdot \dfrac{d-2}{-d^2-5d}\right] \div \dfrac{5d}{d^2+d-20}$

EXERCISE SET 9.3 (p. 297)

1. Name the additive inverse of each of the following. Use as few minus signs as possible.

a) 3 b) -4 c) $x + 1$ d) $x - 1$

e) $1 - 2x$ f) $3 - x$ g) $-x - 5$ h) $2x - 6$

Add. Simplify if possible.

2. $\dfrac{5}{12} + \dfrac{7}{12}$ **3.** $\dfrac{3}{17} + \dfrac{5}{17}$ **4.** $\dfrac{1}{3 + x} + \dfrac{5}{3 + x}$

5. $\dfrac{7}{2x + 1} + \dfrac{3}{1 + 2x}$ **6.** $\dfrac{x}{x - 6} + \dfrac{x + 1}{x - 6}$ **7.** $\dfrac{3x + 2}{3x + 4} + \dfrac{x - 5}{4 + 3x}$

8. $\dfrac{2x + 7}{x - 6} + \dfrac{3x}{6 - x}$ **9.** $\dfrac{2x + 5}{3x - 2} + \dfrac{x + 5}{2 - 3x}$ **10.** $\dfrac{4x + 1}{6x + 5} + \dfrac{3x - 7}{5 + 6x}$

11. $\dfrac{3x - 2}{4x - 3} + \dfrac{2x - 5}{3 - 4x}$ **12.** $\dfrac{2x + 5}{5x - 4} + \dfrac{9 - 3x}{4 - 5x}$ **13.** $\dfrac{a}{x + y} + \dfrac{b}{x + y}$

14. $\dfrac{x^2}{x + 4} + \dfrac{16}{x + 4}$ **15.** $\dfrac{z}{(y + z)(y - z)} + \dfrac{y}{(y + z)(y - z)}$

ANSWERS

1.

2.

3.

4.

5.

6.

7.

8.

9.

10.

11.

12.

13.

14.

15.

ANSWERS

16. $\dfrac{x+4}{x^2-x+2} + \dfrac{2x-3}{x^2-x+2}$ 17. $\dfrac{a-5}{a^2-25} + \dfrac{a-5}{25-a^2}$

16.

_____

17.

18. $\dfrac{3(2a+5)}{a-1} + \dfrac{6a-1}{a-1}$ 19. $\dfrac{a^2}{a-b} + \dfrac{b^2}{b-a}$

_____

18.

20. $\dfrac{x+3}{x-5} + \dfrac{2x-1}{5-x} + \dfrac{2(3x-1)}{x-5}$

_____

19.

_____

21. $\dfrac{3(x-2)}{2x-3} + \dfrac{5(2x+1)}{2x-3} + \dfrac{3(x-1)}{3-2x}$

20.

_____

22. $\dfrac{2(4x+1)}{5x-7} + \dfrac{3(x-2)}{7-5x} + \dfrac{-10x-1}{5x-7}$

21.

_____

22.

23. $\dfrac{5(x-2)}{3x-4} + \dfrac{2(x-3)}{4-3x} + \dfrac{3(5x+1)}{4-3x}$

_____

23.

24. $\dfrac{x+1}{(x+3)(x-3)} + \dfrac{4(x-3)}{(x-3)(x+3)} + \dfrac{(x-1)(x-3)}{(3-x)(x+3)}$

_____

24.

_____

25. $\dfrac{2(x+5)}{(2x-3)(x-1)} + \dfrac{3x+4}{(2x-3)(1-x)} + \dfrac{x-5}{(3-2x)(x-1)} + \dfrac{2(x+3)}{(3-2x)(1-x)}$

25.

_____

EXERCISE SET 9.4 (p. 298)

Subtract. Simplify if possible.

1. $\dfrac{7}{11} - \dfrac{3}{11}$

2. $\dfrac{5}{y} - \dfrac{7}{y}$

3. $\dfrac{x-2}{3} - \dfrac{x-7}{3}$

4. $\dfrac{4x}{4x-y} - \dfrac{2x}{4x-y}$

5. $\dfrac{3x-1}{x-2} - \dfrac{2x-5}{x-2}$

6. $\dfrac{5x-6}{3-x} - \dfrac{2(x-1)}{3-x}$

7. $\dfrac{a}{(a+b)(a-b)} - \dfrac{b}{(a+b)(a-b)}$

8. $\dfrac{5x^2-3x+2}{2x-1} - \dfrac{3x^2+3x-2}{2x-1}$

9. $\dfrac{x^2}{x+4} - \dfrac{16}{x+4}$

10. $\dfrac{x}{x-1} - \dfrac{1}{x-1}$

11. $\dfrac{2}{x-1} - \dfrac{2}{1-x}$

12. $\dfrac{x}{2x+4} - \dfrac{2-x}{2x+4}$

13. $\dfrac{3}{x^2+2x+1} - \dfrac{2-x}{x^2+2x+1}$

14. $\dfrac{x+1}{x^2-2x+1} - \dfrac{5-3x}{x^2-2x+1}$

15. $\dfrac{2a+1}{a^2-a-6} - \dfrac{1-a}{a^2-a-6}$

16. $\dfrac{x}{6x-3y} - \dfrac{y}{3(y-2x)}$

17. $\dfrac{2x-3}{x^2+3x-4} - \dfrac{x-7}{x^2+3x-4}$

NAME

CLASS

ANSWERS

1.

2.

3.

4.

5.

6.

7.

8.

9.

10.

11.

12.

13.

14.

15.

16.

17.

ANSWERS

18.

_____

19.

_____

20.

_____

21.

_____

22.

_____

23.

_____

24.

_____

25.

_____

26.

_____

18. $\dfrac{3 - x}{x - 7} - \dfrac{2x - 5}{7 - x}$

19. $\dfrac{2(x - 1)}{2x - 3} - \dfrac{3(x + 2)}{2x - 3} - \dfrac{x - 1}{3 - 2x}$

20. $\dfrac{(x + 1)(2x - 1)}{(x - 2)(x - 3)} - \dfrac{(x + 2)(x - 1)}{(x - 2)(3 - x)} - \dfrac{(x + 5)(2x + 1)}{(3 - x)(2 - x)}$

Perform the indicated operations.

21. $\dfrac{3(2x + 5)}{x - 1} - \dfrac{3(2x - 3)}{1 - x} + \dfrac{6x - 1}{x - 1}$

22. $\dfrac{2x - y}{x - y} + \dfrac{x - 2y}{y - x} - \dfrac{3x - 3y}{x - y}$

23. $\dfrac{x - y}{x^2 - y^2} + \dfrac{x + y}{x^2 - y^2} - \dfrac{2x}{x^2 - y^2}$

24. $\dfrac{x + y}{2(x - y)} - \dfrac{2x - 2y}{2(x - y)} + \dfrac{x - 3y}{2(y - x)}$

25. $\dfrac{10}{2y - 1} - \dfrac{6}{1 - 2y} + \dfrac{y}{2y - 1} + \dfrac{y - 4}{1 - 2y}$

26. $\dfrac{(x + 3)(2x - 1)}{(2x - 3)(x - 3)} - \dfrac{(x - 3)(x + 1)}{(3 - x)(3 - 2x)} + \dfrac{(2x + 1)(x + 3)}{(3 - 2x)(x - 3)}$

EXERCISE SET 9.5 (pp. 299–300)

Find the L.C.M.

1. 12,27   **2.** 10,15   **3.** 8,9

4. 12,15   **5.** 6,9,21   **6.** 8,36,40

7. 24,36,40   **8.** 3,4,5   **9.** 28,42,60

Find the L.C.M. of the denominators, then add.

10. $\dfrac{7}{24} + \dfrac{11}{18}$ **11.** $\dfrac{7}{60} + \dfrac{6}{75}$ **12.** $\dfrac{1}{6} + \dfrac{3}{40} + \dfrac{2}{75}$

13. $\dfrac{5}{24} + \dfrac{3}{20} + \dfrac{7}{30}$ **14.** $\dfrac{2}{15} + \dfrac{5}{9} + \dfrac{3}{20}$ **15.** $\dfrac{1}{20} + \dfrac{1}{30} + \dfrac{2}{45}$

1.

2.

3.

4.

5.

6.

7.

8.

9.

10.

11.

12.

13.

14.

15.

ANSWERS

Find the L.C.M.

16. _____

17. _____

18. _____

19. _____

20. _____

21. _____

22. _____

23. _____

24. _____

25. _____

26. _____

27. _____

16. $6x^2, 12x^3$ **17.** $2a^2b, 8ab^2$ **18.** $2x^2, 6xy, 18y^2$

19. c^2d, cd^2, c^3d **20.** $2(y-3), 6(3-y)$

21. $8xy, 12x^2, 18y^2$ **22.** $x^2 - y^2, 2x + 2y, x^2 + 2xy + y^2$

23. $a + 1; (a-1)^2; a^2 - 1$ **24.** $m^2 - 5m + 6; m^2 - 4m + 4$

25. $2 + 3k; 9k^2 - 4; 2 - 3k$ **26.** $10v^2 + 30v; -5v^2 - 35v - 60$

27. $9x^3 - 9x^2 - 18x; 6x^5 - 24x^4 + 24x^3$

NAME

CLASS

ANSWERS

EXERCISE SET 9.6 (p. 301)

Add, and simplify if possible.

1. $\dfrac{2}{x} + \dfrac{5}{x^2}$

2. $\dfrac{5}{6r} + \dfrac{7}{8r}$

3. $\dfrac{x+y}{xy^2} + \dfrac{3x+y}{x^2y}$

4. $\dfrac{2c-d}{c^2d} + \dfrac{c+d}{cd^2}$

5. $\dfrac{2}{x-1} + \dfrac{2}{x+1}$

6. $\dfrac{3}{x+1} + \dfrac{2}{3x}$

7. $\dfrac{2x}{x^2-16} + \dfrac{x}{x-4}$

8. $\dfrac{5}{z+4} + \dfrac{3}{3z+12}$

9. $\dfrac{3}{x-1} + \dfrac{2}{(x-1)^2}$

10. $\dfrac{4a}{5a-10} + \dfrac{3a}{10a-20}$

1.

2.

3.

4.

5.

6.

7.

8.

9.

10.

ANSWERS

11.

_____

12.

_____

13.

_____

14.

_____

15.

_____

16.

_____

17.

_____

18.

_____

19.

_____

20.

_____

11. $\dfrac{x}{x^2 + 2x + 1} + \dfrac{1}{x^2 + 5x + 4}$ 12. $\dfrac{7}{a^2 + a - 2} + \dfrac{5}{a^2 - 4a + 3}$

13. $\dfrac{x + 3}{x - 5} + \dfrac{x - 5}{x + 3}$ 14. $\dfrac{3x}{2y - 3} + \dfrac{2x}{3y - 2}$

15. $\dfrac{a}{a^2 - 1} + \dfrac{2a}{a^2 - a}$ 16. $\dfrac{3x + 2}{3x + 6} + \dfrac{x - 2}{x^2 - 4}$

17. $\dfrac{6}{x - y} + \dfrac{4x}{y^2 - x^2}$ 18. $\dfrac{a - 2}{3 - a} + \dfrac{4 - a^2}{a^2 - 9}$

19. $\dfrac{10}{x^2 + x - 6} + \dfrac{3x}{x^2 - 4x + 4}$ 20. $\dfrac{2}{z^2 - z - 6} + \dfrac{3}{z^2 - 9}$

NAME _____

CLASS _____

ANSWERS

EXERCISE SET 9.7 (p. 302)

Subtract, and simplify if possible.

1. $\dfrac{x-2}{6} - \dfrac{x+1}{3}$

2. $\dfrac{a+2}{2} - \dfrac{a-4}{4}$

3. $\dfrac{4z-9}{3z} - \dfrac{3z-8}{4z}$

4. $\dfrac{x-1}{4x} - \dfrac{2x+3}{x}$

5. $\dfrac{y-5}{y} - \dfrac{3y-1}{4y}$

6. $\dfrac{5x+3y}{2x^2y} - \dfrac{3x+4y}{xy^2}$

7. $\dfrac{5}{x+5} - \dfrac{3}{x-5}$

8. $\dfrac{2z}{z-1} - \dfrac{3z}{z+1}$

9. $\dfrac{3}{2t^2-2t} - \dfrac{5}{2t-2}$

10. $\dfrac{8}{x^2-4} - \dfrac{3}{x+2}$

1. _____

2. _____

3. _____

4. _____

5. _____

6. _____

7. _____

8. _____

9. _____

10. _____

ANSWERS

11.

_____

12.

_____

13.

_____

14.

_____

15.

_____

16.

_____

17.

_____

18.

_____

19.

_____

20.

_____

11. $\dfrac{2s}{t^2 - s^2} - \dfrac{s}{t - s}$

12. $\dfrac{3}{12 + x - x^2} - \dfrac{2}{x^2 - 9}$

Perform the indicated operations. Simplify when possible.

13. $\dfrac{4y}{y^2 - 1} - \dfrac{2}{y} - \dfrac{2}{y + 1}$

14. $\dfrac{x + 6}{4 - x^2} - \dfrac{x + 3}{x + 2} + \dfrac{x - 3}{2 - x}$

15. $\dfrac{2z}{1 - 2z} + \dfrac{3z}{2z + 1} - \dfrac{3}{4z^2 - 1}$

16. $\dfrac{1}{x + y} + \dfrac{1}{x - y} - \dfrac{2x}{x^2 - y^2}$

17. $\dfrac{5}{3 - 2x} + \dfrac{3}{2x - 3} - \dfrac{x - 3}{2x^2 - x - 3}$

18. $\dfrac{2r}{r^2 - s^2} + \dfrac{1}{r + s} - \dfrac{1}{r - s}$

19. $\dfrac{3}{2c - 1} - \dfrac{1}{c + 2} - \dfrac{5}{2c^2 + 3c - 2}$

20. $\dfrac{3y - 1}{2y^2 + y - 3} - \dfrac{2 - y}{y - 1}$

NAME

CLASS

ANSWERS

EXERCISE SET 9.8 (pp. 303–304)

Simplify.

1. $\dfrac{1 + \dfrac{9}{16}}{1 - \dfrac{3}{4}}$

2. $\dfrac{9 - \dfrac{1}{4}}{3 + \dfrac{1}{2}}$

3. $\dfrac{1 - \dfrac{3}{5}}{1 + \dfrac{1}{5}}$

4. $\dfrac{\dfrac{5}{27} - 5}{\dfrac{1}{3} + 1}$

5. $\dfrac{\dfrac{1}{x} + 3}{\dfrac{1}{x} - 5}$

6. $\dfrac{\dfrac{3}{s} + s}{\dfrac{s}{3} + s}$

7. $\dfrac{1 - \dfrac{x}{y}}{\dfrac{x - y}{y}}$

8. $\dfrac{a - 1}{a - \dfrac{1}{a}}$

9. $\dfrac{\dfrac{2}{y} + \dfrac{1}{2y}}{y + \dfrac{y}{2}}$

1. _____

2. _____

3. _____

4. _____

5. _____

6. _____

7. _____

8. _____

9. _____

ANSWERS

10. _____

11. _____

12. _____

13. _____

14. _____

15. _____

16. _____

17. _____

18. _____

10. $\dfrac{4 - \dfrac{1}{x^2}}{2 - \dfrac{1}{x}}$

11. $\dfrac{c + \dfrac{c}{d}}{1 + \dfrac{1}{d}}$

12. $\dfrac{2 - \dfrac{a}{b}}{2 - \dfrac{b}{a}}$

13. $\dfrac{\dfrac{1}{b} - \dfrac{1}{a}}{\dfrac{b - a}{5}}$

14. $\dfrac{2 - \dfrac{1}{x}}{\dfrac{2}{x}}$

15. $\dfrac{\dfrac{x}{x - y}}{\dfrac{x^2}{x^2 - y^2}}$

16. $\dfrac{\dfrac{x}{y} - \dfrac{y}{x}}{\dfrac{1}{y} + \dfrac{1}{x}}$

17. $\dfrac{x - 3 + \dfrac{2}{x}}{x - 4 + \dfrac{3}{x}}$

18. $\dfrac{1 + \dfrac{a}{b - a}}{\dfrac{a}{a + b} - 1}$

NAME

CLASS

ANSWERS

EXERCISE SET 9.9 (pp. 305–307)

Divide. Check your answers by multiplying.

1. $\dfrac{u - 2u^2 - u^5}{u}$ **2.** $\dfrac{12a^4 - 3a^2 + a - 6}{6}$

3. $(15t^3 + 24t^2 - 6t) \div 3t$ **4.** $(25x^3 + 15x^2 - 30x) \div 5x$

5. $\dfrac{20x^6 - 20x^4 - 5x^2}{-5x^2}$ **6.** $\dfrac{9r^2s^2 + 3r^2s - 6rs^2}{-3rs}$

7. $\dfrac{4x^4y - 8x^6y^2 + 12x^8y^6}{4x^4y}$ **8.** $\dfrac{12a^3 - 9a^2 + 3a}{3a}$

9. $\dfrac{3(a - b) + n(a - b)}{a - b}$ **10.** $\dfrac{4(x - y) - 8(x - y)^3}{-2(x - y)}$

ANSWERS

1.

2.

3.

4.

5.

6.

7.

8.

9.

10.

ANSWERS

11. _____

12. _____

13. _____

14. _____

15. _____

16. _____

17. _____

18. _____

11. $(x^2 + 4x + 4) \div (x + 2)$

12. $(a^2 - 6a + 9) \div (a - 3)$

13. $\dfrac{x^6 - 13x^3 + 42}{x^3 - 7}$

14. $\dfrac{x^2 + 4x - 14}{x + 6}$

15. $\dfrac{x^5 + 1}{x + 1}$

16. $\dfrac{x^2 - 6}{x + 3}$

17. $\dfrac{8x^3 - 22x^2 - 5x + 12}{4x + 3}$

18. $\dfrac{2x^3 - 9x^2 + 11x - 3}{2x - 3}$

EXERCISE SET 9.10 (pp. 308–309)

Solve.

1. $\dfrac{5}{x} = \dfrac{6}{x} - \dfrac{1}{3}$ 2. $\dfrac{5}{3x} + \dfrac{3}{x} = 1$

3. $\dfrac{x-7}{x+2} = \dfrac{1}{4}$ 4. $\dfrac{a-2}{a+3} = \dfrac{3}{8}$

5. $\dfrac{2}{x+1} = \dfrac{1}{x-2}$ 6. $\dfrac{5}{x-1} = \dfrac{3}{x+2}$

7. $\dfrac{x}{8} - \dfrac{x}{12} = \dfrac{1}{8}$ 8. $\dfrac{x+1}{3} - \dfrac{x-1}{2} = 1$

9. $\dfrac{a-3}{3a+2} = \dfrac{1}{5}$ 10. $\dfrac{2}{x+3} = \dfrac{5}{x}$

ANSWERS

11.

_____

12.

_____

13.

_____

14.

_____

15.

_____

16.

_____

17.

_____

18.

_____

11. $\dfrac{x-2}{x-3} = \dfrac{x-1}{x+1}$

12. $\dfrac{2b-3}{3b+2} = \dfrac{2b+1}{3b-2}$

13. $\dfrac{6x-2}{2x-1} = \dfrac{9x}{3x+1}$

14. $\dfrac{2a}{a+1} = 2 - \dfrac{5}{2a}$

15. $\dfrac{1}{x+3} + \dfrac{1}{x-3} = \dfrac{1}{x^2-9}$

16. $\dfrac{4}{x-3} + \dfrac{2x}{x^2-9} = \dfrac{1}{x+3}$

17. $\dfrac{x}{x+4} - \dfrac{4}{x-4} = \dfrac{x^2+16}{x^2-16}$

18. $\dfrac{5}{y-3} - \dfrac{30}{y^2-9} = 1$

EXERCISE SET 9.11 (pp. 310–312)

Solve.

1. In a fractional numeral the denominator is 5 more than the numerator. If 3 is added to both numerator and denominator, the result equals $\frac{1}{2}$. Find the original fractional numeral.

1. _____

2. Gloria drove 120 miles. Driving back she doubled her speed, and took 3 hr less time. Find her speed going.

2. _____

3. One number is 4 more than another. The quotient of the larger divided by the smaller is $\frac{5}{2}$. Find the numbers.

3. _____

4. If a number is increased by 4 times its reciprocal, the sum is 5. Find the number.

4. _____

5. After making a trip of 126 kilometers, Jack found he could have made the trip in 1 hour less time if he had increased his average speed by 8 km/hr. What was his actual speed?

5. _____

ANSWERS

6.

_____

7.

_____

8.

_____

9.

_____

10.

_____

11.

_____

12.

_____

13.

_____

14.

_____

6. If 12 is subtracted from both numerator and denominator of a certain fractional numeral the result equals $\frac{3}{17}$. If 5 is added only to the numerator the result equals $\frac{1}{2}$. Find the original fractional numeral.

Solve each formula for the indicated variable:

7. $\dfrac{1}{p} + \dfrac{1}{q} = \dfrac{1}{f}$; p

8. $a = \dfrac{v - v_0}{t}$; v_0

9. $E = RI + \dfrac{rI}{n}$; I

10. $\dfrac{E}{e} = \dfrac{R + r}{r}$; e

11. $V = \dfrac{Q}{r_1} - \dfrac{Q}{r_2}$; Q

12. $u = -F\left(E - \dfrac{P}{T}\right)$; T

13. $h_1 = q\left(1 + \dfrac{h_2}{p}\right)$; h_2

14. $S = \dfrac{a - ar^n}{1 - r}$; a

NAME

CLASS

ANSWERS

EXERCISE SET 9.12 (pp. 313–315)

Solve the problems.

1. If, on a map, 10 miles is represented by $\frac{3}{4}$ inch, how many miles does 6 inches represent?

1.

2. S. T. Broker owns 300 shares of stock and receives $180 per year in dividends. How much does I. Makit receive for an annual dividend if he owns 450 shares in the same company?

2.

3. A sample of 184 light bulbs proved to contain 6 defective bulbs. How many would you expect in 1288 bulbs?

3.

4. If 10 cubic centimeters of a normal specimen of human blood contains 1.2 grams of hemoglobin, how many grams would 16 cubic centimeters of the same blood contain.

4.

5. A car uses 9 gallons of gas to travel 125 miles. How many gallons would be required to drive 500 miles?

5.

6.

7.

8.

9.

10.

11.

6. What is the variation constant in problem 5?

7. In an election for freshman class president the successful candidate won by a vote of 3 to 2. He received 324 votes. How many did the loser receive?

8. How many kilograms of coffee will be required to make 2000 cups of coffee if 1.2 kilograms will make 150 cups?

9. If 90 feet of wire weighs 18 pounds, what will 110 feet of the same kind of wire weigh?

10. Separate 72 into two parts that are in the ratio of 4 to 5.

11. The ratio of the weight of an object on the moon to the weight of an object on earth is 0.16 to 1.

a) How much would a 12-ton rocket weigh on the moon?

b) How much would a 90-kilogram astronaut weigh on the moon?

NAME_____

CHAPTER 9 TEST   CLASS_____SCORE_____GRADE_____

Before taking the test *be sure* to allow yourself a day or so for review.
Use the objectives to guide your study. The test will evaluate your progress
and aid your preparation for a possible classroom test. Allow about an
hour for the test. Remove the test from the book. When you finish read
the test analysis on the answer page at the end of the book.

ANSWERS

Simplify.

1. $\dfrac{8xy^2}{4x^2y - 8xy^2}$ **2.** $\dfrac{14x^2 - x - 3}{2x^2 - 7x + 3}$

1. _____

2. _____

3. _____

Perform the indicated operations and simplify.

3. $\dfrac{x - 3}{x + 4} \cdot \dfrac{x - 2}{x + 1}$ **4.** $\dfrac{a^2 - 36}{10a} \cdot \dfrac{2a}{a + 6}$

4. _____

5. _____

5. $\dfrac{4x^4}{x^2 - 1} \div \dfrac{2x^3}{x^2 - 2x + 1}$ **6.** $\dfrac{4x^2 - 1}{6x^2 - 9x} \div \dfrac{4x^2 + 8x + 3}{2x^2 - 5x + 3}$

6. _____

7. _____

7. $\dfrac{3}{3x - 9} + \dfrac{x - 2}{3 - x}$ **8.** $\dfrac{5}{4z - 1} + \dfrac{5}{4z}$

8. _____

9. _____

9. $\dfrac{1}{x^2 - 25} - \dfrac{x - 5}{x^2 - 4x - 5}$ **10.** $\dfrac{\dfrac{1}{a} - 2}{\dfrac{1}{2a} - 1}$

10. _____

11. _____

11. $\dfrac{12x^2 + 8x - 4}{-2}$ **12.** $\dfrac{6k^3 - 5k^2 - 13k + 13}{2k + 3}$

12. _____

ANSWERS

13.

_____

14.

_____

15.

_____

16.

_____

17.

_____

18.

_____

19.

_____

Solve.

13. $\dfrac{4x-2}{3} - \dfrac{3x+5}{4} = 1$ **14.** $\dfrac{15}{x} - \dfrac{15}{x+2} = 2$

15. Solve for n: $E = RI + \dfrac{rI}{n}$

16. Solve for R: $\dfrac{E}{e} = \dfrac{R+r}{r}$

17. In a fractional numeral, the denominator is 5 less than the numerator. If the numerator is increased by 9, the result will be equal to $\frac{3}{2}$. Find the original fractional numeral.

18 One car travels 90 miles in the same time that a car traveling 10 mph slower travels 60 miles. Find the speed of each.

19. Several friends wanted to rent a cottage for $320. When 2 more people joined the group, each person's share was reduced by $8. How many were in the original group?

10 RADICAL NOTATION

10.1 RADICAL EXPRESSIONS

Every positive number has two square roots. The square roots of 25 are 5 and -5, because

$5^2 = 25$ and also $(-5)^2 = 25$.

Zero has only one square root, 0 itself.

We use the radical symbol $\sqrt{}$ to name the non-negative square root of a number. This is called its *principal square root*. To name the negative square root we write $-\sqrt{}$. Thus $\sqrt{36} = 6$ and $-\sqrt{36} = -6$.

The symbol $\sqrt{}$ is called a *radical*.* When an expression is written under it we have a *radical expression*. The expression written under the radical is called the *radicand*. Here are some examples of radical expressions.

$$\sqrt{14}, \quad \sqrt{x}, \quad \sqrt{x^2 + 4}, \quad \sqrt{\frac{x^2 - 5}{2}}$$

Do exercises 1 through 5 at the right.

Negative numbers do not have square roots in the number system we are using, because when we square any negative number the result is positive.† Thus certain radical expressions are meaningless. The following expressions are meaningless.

$$\sqrt{-25}, \quad -\sqrt{-10}$$

If a radicand has a variable we can make various replacements. If a replacement makes a radicand negative it is not sensible. Thus radical expressions may have nonsensible replacements.

Examples

a) In $\sqrt{x}$ any replacement less than 0 is nonsensible.

b) In $\sqrt{x + 2}$ any replacement less than -2 is nonsensible, because it make $x + 2$ negative.

---

* There are other kinds of radicals that we will not consider in this book. For example $\sqrt[3]{}$ is a radical denoting cube root. $\sqrt[3]{8}$, for example, is 2 because $2^3 = 8$.
† In a more advanced course you may study the system of *complex numbers*, in which all numbers have square roots.

You should be able to:

a) Identify the radicand in a radical expression.

b) Find the positive and negative square roots of numbers which are perfect squares.

c) Evaluate expressions such as $\sqrt{25}$ and $-\sqrt{25}$.

d) Identify meaningless radical expressions, such as $\sqrt{-4}$.

e) Determine the nonsensible replacements in a radical expression.

f) Simplify radical expressions having a perfect square for a radicand, using

$$\sqrt{E^2} = |E|.$$

Find these square roots.

1. $\sqrt{16}$

2. $\sqrt{49}$

3. $\sqrt{100}$

In each radical expression what is the radicand?

4. $\sqrt{45 + x}$

5. $\sqrt{\dfrac{x}{x + 2}}$

Find two square roots of each number.

6. 36

7. 81

8. 100

Find the following.

9. $\sqrt{100}$

10. $-\sqrt{100}$

11. $-\sqrt{36}$

12. $\sqrt{36}$

13. $\sqrt{0}$

14. Which of these expressions are meaningless?

a) $\sqrt{-25}$ b) $-\sqrt{25}$

c) $-\sqrt{-36}$ d) $-\sqrt{36}$

15. What are the nonsensible replacements for the following?

a) $\sqrt{x-3}$ b) $\sqrt{x^2}$

Simplify.

16. $\sqrt{(xy)^2}$

17. $\sqrt{x^2 y^2}$

18. $\sqrt{(x-1)^2}$

19. $\sqrt{x^2 + 2xy + y^2}$

20. $\sqrt{25x^2}$

Do exercises 6 through 15 at the left.

EXPRESSIONS OF THE TYPE $\sqrt{E^2}$

Let us consider radical expressions in which the radicand is a perfect square, such as $\sqrt{x^2}$. Since squares of numbers are never negative there are no nonsensible replacements in such cases.

Suppose $x = 3$. Then we have $\sqrt{3^2}$, which is $\sqrt{9}$, or 3.

Suppose $x = -3$. Then we have $\sqrt{(-3)^2}$, which is $\sqrt{9}$, or 3.

In any case, $\sqrt{x^2} = |x|$. This is always true.

Any radical expression $\sqrt{E^2}$ can be simplified to $|E|$.

Examples

a) $\sqrt{(3x)^2} = |3x|$

b) $\sqrt{a^2 b^2} = \sqrt{(ab)^2} = |ab|$

c) $\sqrt{x^2 + 2x + 1} = \sqrt{(x+1)^2} = |x+1|$

We can sometimes simplify absolute value notation. In example (a) above, $|3x|$ simplifies to $|3| \cdot |x|$ or $3|x|$. In (b) we can simplify $|ab|$ to $|a| \cdot |b|$ if we wish. The absolute value of a product is always the product of the absolute values.

For any numbers a and b, $|a \cdot b| = |a| \cdot |b|$.

Do exercises 16 through 20 at the left.

Do exercise set 10.1, p. 359.

10.2 IRRATIONAL NUMBERS AND REAL NUMBERS

Recall that all rational numbers can be named by fractional notation, $\frac{a}{b}$, where both numerator and denominator are integers. They can of course be named other ways, but they are all nameable this way. Suppose we try to find such a number for $\sqrt{2}$. That is, we look for a number $\frac{a}{b}$ for which $\frac{a}{b} \cdot \frac{a}{b} = 2$. We can find numbers whose square is quite close to 2, but we never can find one whose square is exactly 2. This statement can be proved, but we shall not do it here. Since $\sqrt{2}$ is not a rational number we call it an *irrational number*.

There are many irrational numbers. Unless a whole number is a perfect square its square root is irrational. Here, then, are examples of other irrational numbers.

$$\sqrt{5}, \quad \sqrt{7}, \quad \sqrt{35}$$

There are many irrational numbers. There are many of them that we do not obtain by taking square roots. For example, the number π is an irrational number.

DECIMALS FOR IRRATIONAL NUMBERS

Recall that the decimal for a rational number either ends or repeats.

Examples

a) $\frac{1}{4} = .25$

b) $\frac{1}{3} = .3333\ldots$

Thus the decimal numeral for an irrational number will never end and it will never repeat.

The number π is an example.

$\pi = 3.1415926535\ldots$

Its decimal numeral never ends and never repeats.

Do exercises 21 through 28 at the right.

Which of these are rational? Which are irrational!?

21. $\frac{-3}{5}$ **22.** $\frac{95}{37}$

23. 6.12 **24.** $.0353535\ldots$ (numeral repeats)

25. $3.01001000100001\ldots$ (numeral does not repeat)

26. $\sqrt{17}$

27. $-\sqrt{81}$ **28.** $-\sqrt{8}$

29. Find the number halfway between

$$\frac{3}{64} \text{ and } \frac{4}{64}$$

Use the table on p. 423 to approximate these.

30. $\sqrt{7}$

31. $\sqrt{72}$

REAL NUMBERS

Consider the number line. The rational numbers are very close together on the line. No matter how close together two rational numbers are, we can find another one between them. We can do this by averaging them.

Example. The number halfway between $\frac{1}{32}$ and $\frac{2}{32}$ is their average. We add and divide by 2.

$$\frac{\frac{1}{32} + \frac{2}{32}}{2} = \frac{\frac{3}{32}}{2} = \frac{3}{64}$$

Do exercise 29 at the left.

In spite of the fact that the rational numbers seem to fill up the number line, they do not. There are many points of the line for which there is no rational number. For these points there are irrational numbers. Together the rational numbers and the irrational numbers make up the set known as the *real numbers*.

The *real numbers* consist of the rational numbers and the irrational numbers. There is a real number for each point of a number line.

How do we denote irrational numbers? We most often find an approximation which is rational. For example, on p. 423 is a table of rational approximations to square roots. You will find, for example, that $\sqrt{10} \approx 3.162$. This is rounded to three decimal places. The symbol $\approx$ means "is approximately equal to."

Do exercises 30 and 31 at the left.

Do exercise 10.2, p. 361.

10.3 MULTIPLICATION

You should be able to:

a) State the basic principle used in multiplying with radical notation.

b) Multiply with radical notation.

To see how we can multiply with radical notation, look at the following examples.

Example 1

$$\sqrt{9}\cdot\sqrt{4} = 3\cdot2 = 6$$
$$\sqrt{9\cdot4} = \sqrt{36} = 6$$

Example 2

$$\sqrt{4}\cdot\sqrt{25} = 2\cdot5 = 10$$
$$\sqrt{4\cdot25} = \sqrt{100} = 10$$

Example 3

$$\sqrt{-9}\cdot\sqrt{-4} \quad \text{meaningless}$$

$$\sqrt{(-9)(-4)} = \sqrt{36} = 6$$

Do exercise 32 at the right.

The above examples illustrate that we can multiply with radical notation by multiplying the radicands.* However, we cannot allow any radicand to be negative.

For any non-negative radicands A and B, $\sqrt{A}\cdot\sqrt{B} = \sqrt{A\cdot B}$

This basic principle allows us to multiply with radical notation.

Examples

a) $\sqrt{5}\cdot\sqrt{7} = \sqrt{5\cdot7} = \sqrt{35}$

b) $\sqrt{\dfrac{2}{3}}\cdot\sqrt{\dfrac{4}{5}} = \sqrt{\dfrac{2}{3}\cdot\dfrac{4}{5}} = \sqrt{\dfrac{8}{15}}$

c) $\sqrt{2x}\,\sqrt{3x-1} = \sqrt{2x(3x-1)} = \sqrt{6x^2 - 2x}$

Do exercises 33 through 35 at the right.

Do exercise set 10.3, p. 363.

32. Simplify.

a) $\sqrt{4}\cdot\sqrt{16}$ b) $\sqrt{4\cdot16}$

Multiply. Assume all radicands non-negative.

33. $\sqrt{3}\cdot\sqrt{7}$

34. $\sqrt{x}\ \ \sqrt{x+1}$

35. $\sqrt{x+1}\ \ \sqrt{x-1}$

---

* Here is a proof (optional). We consider a product $\sqrt{a}\cdot\sqrt{b}$, where a and b are not negative. We square this product, to show that we get $a\cdot b$, and thus the product is the square root of $a\cdot b$ (or $\sqrt{a\cdot b}$).

$(\sqrt{a}\cdot\sqrt{b})^2 = (\sqrt{a}\cdot\sqrt{b})(\sqrt{a}\cdot\sqrt{b}) = (\sqrt{a}\cdot\sqrt{a})(\sqrt{b}\cdot\sqrt{b}) = a\cdot b$

OBJECTIVES

You should be able to:

a) Factor radical expressions, and where possible simplify.

b) Use factoring and the square root table to find approximate square roots not in the table.

Factor. Simplify where possible.

36. $\sqrt{32}$

37. $\sqrt{56}$

38. $\sqrt{16y^2}$

39. $\sqrt{25x^2}$

40. $\sqrt{x^2 - 1}$

Approximate these square roots. Round to two decimal places.

41. $\sqrt{275}$

42. $\sqrt{102}$

10.4 FACTORING AND APPROXIMATING SQUARE ROOTS

We know that for non-negative radicands, $\sqrt{a} \cdot \sqrt{b} = \sqrt{a \cdot b}$. Remember that all equations are reversible. Thus we can just as well read this equation as $\sqrt{a \cdot b} = \sqrt{a} \cdot \sqrt{b}$. Therefore we can reverse the multiplication process to factor radical expressions. The following examples illustrate. Notice also that it allows simplification in some cases.

Example 1

$$\sqrt{18} = \sqrt{9 \cdot 2} = \sqrt{9} \cdot \sqrt{2} = 3\sqrt{2}$$

Example 2

$$\sqrt{25x} = \sqrt{25}\sqrt{x} = 5\sqrt{x}$$

Example 3

$$\sqrt{36x^2} = \sqrt{36}\sqrt{x^2} = 6|x|$$

Example 4

$$\sqrt{x^2 - 4} = \sqrt{(x-2)(x+2)} = \sqrt{x-2}\sqrt{x+2}$$

Do exercises 36 through 40 at the left.

The table on page 423 goes only to 100. We can use it to find approximate square roots for other numbers.

Example 5

$$\sqrt{160} = \sqrt{16} \cdot \sqrt{10} = 4\sqrt{10}$$
$$\approx 4 \times 3.162 = 12.648$$

Example 6

$$\sqrt{341} = \sqrt{11 \cdot 31} = \sqrt{11} \cdot \sqrt{31}$$
$$\approx 3.317 \times 5.568$$
$$\approx 18.469 \text{ (to 3 decimal places)}$$

Do exercises 41 and 42 at the left.

Do exercise set 10.4, p. 365.

10.5 FACTORING AND SIMPLIFYING

To simplify radical expressions we usually try to factor out as many perfect square factors as possible. Consider, for example

$$\sqrt{50} = \sqrt{10 \cdot 5} = \sqrt{10} \cdot \sqrt{5}$$

Compare this with

$$\sqrt{50} = \sqrt{25 \cdot 2} = \sqrt{25} \cdot \sqrt{2} = 5\sqrt{2}$$

In the second case we factored out the perfect square 25. The expression $5\sqrt{2}$ is usually considered simpler than $\sqrt{50}$ or $\sqrt{10} \cdot \sqrt{5}$. Further examples of simplification follow.

Example 1. Simplify

$$\sqrt{48} = \sqrt{16 \cdot 3} = \sqrt{16} \cdot \sqrt{3} = 4\sqrt{3}$$

Example 2. Simplify

$$\sqrt{3x^2 + 6x + 3} = \sqrt{3(x^2 + 2x + 1)} = \sqrt{3}\sqrt{(x + 1)^2}$$

$$= \sqrt{3}\,|x + 1|$$

Whenever a variable is involved, as in example 2, you should remember to use absolute values.

Do exercises 43 through 45 at the right.

SQUARE ROOTS OF POWERS

In order to see how to take square roots of powers, note the following.

a) $x^3 \cdot x^3 = x^6$, so $\sqrt{x^6} = |x^3|$

b) $x^5 \cdot x^5 = x^{10}$, so $\sqrt{x^{10}} = |x^5|$

To take a square root of a power such as x^6 the exponent must be even. We then take half the exponent. We should also remember to use absolute value.

OBJECTIVES

You should be able to:

a) Simplify radical expressions by factoring out perfect square factors.

b) Multiply radical expressions and then simplify.

Simplify.

43. $\sqrt{32}$

44. $\sqrt{16a}$

45. $\sqrt{3x^2 - 6x + 3}$

Simplify.

46. $\sqrt{x^8}$

47. $\sqrt{(x+2)^{14}}$

Multiply and simplify.

48. $\sqrt{2x^3}\sqrt{8x^3y^4}$

49. $\sqrt{10xy^2}\sqrt{5x^2y^3}$

Example 3. Simplify

$$\sqrt{x^{12}} = |x^6|$$

Do exercises 46 and 47 at the left.

Sometimes a simplification can be done after multiplying. When this is possible, it should be done.

Example 4. Multiply and simplify

$$\sqrt{2}\cdot\sqrt{14} = \sqrt{2\cdot14} = \sqrt{2\cdot2\cdot7}$$
$$= \sqrt{4}\cdot\sqrt{7}$$
$$= 2\cdot\sqrt{7}$$

Note from Example 4 that we factor the radicands and look for perfect square factors.

Example 5. Multiply and simplify

$$\sqrt{3x^2}\sqrt{9x^3} = \sqrt{3\cdot9x^5} = \sqrt{3\cdot9\cdot x^4\cdot x}$$
$$= 3|x^2|\sqrt{3x}$$

The absolute value signs could be omitted in this case, because x^2 cannot be negative. However, it is not wrong to write them.

Do exercises 48 and 49 at the left.

Do exercise set 10.5, p. 367.

10.6 ADDITION AND SUBTRACTION

You should be able to add or subtract with radical notation, using the distributive laws to simplify.

We can add any two real numbers. Thus there is some number which is the sum of 5 and $\sqrt{2}$. For this sum we can write $5 + \sqrt{2}$, but there is no way to simplify it unless we begin to use rational approximations. Sometimes, however, after naming a sum in this way we can use the distributive laws to do some simplifying.

Example 1. Add $3\sqrt{5}$ and $4\sqrt{5}$ and simplify.

$3\sqrt{5} + 4\sqrt{5} = (3 + 4)\sqrt{5}$ (using distributive law)

$\qquad\qquad\quad = 7\sqrt{5}$

Example 2. Subtract $\sqrt{8}$ from $\sqrt{2}$ and simplify

$\sqrt{2} - \sqrt{8} = \sqrt{2} - \sqrt{4 \cdot 2}$

$\qquad\quad = \sqrt{2} - 2\sqrt{2}$

$\qquad\quad = 1 \cdot \sqrt{2} - 2\sqrt{2}$

$\qquad\quad = (1 - 2)\sqrt{2}$ (using distributive law)

$\qquad\quad = -1 \cdot \sqrt{2}$

$\qquad\quad = -\sqrt{2}$

Example 3. Simplify

$\sqrt{x^3 - x^2} + \sqrt{4x - 4}$

$\sqrt{x^2(x - 1)} + \sqrt{4(x - 1)}$

$|x|\sqrt{x - 1} + 2\sqrt{x - 1}$

$\qquad (|x| + 2)\sqrt{x - 1}$

When variables occur, as in Example 3, you should remember to use absolute values correctly. Otherwise adding and subtracting depend upon the distributive laws.*

Add or subtract and simplify.

50. $3\sqrt{2} + 9\sqrt{2}$

51. $8\sqrt{5} - 3\sqrt{5}$

52. $2\sqrt{10} - 7\sqrt{40}$

53. $\sqrt{24} + \sqrt{54}$

Do exercises 50 through 54 at the right.

Do exercise set 10.6, p. 369.

54. $\sqrt{9x + 9} - \sqrt{4x + 4}$

---

* Be sure not to make the mistake of thinking that the square root of a sum is the sum of the square roots. For example, $\sqrt{9} + \sqrt{16} = 7$. But $\sqrt{9 + 16} = 5$.

You should be able to:

a) Simplify radical expressions with fractional radicands, when numerator and denominator are perfect squares or can be so simplified.

b) When numerator and denominator are not perfect squares, simplify so that there will be only a whole number radicand.

c) Use a table to approximate square roots of fractions.

d) Use the skill of b) when adding or subtracting with radicals.

Simplify.

55. $\sqrt{\dfrac{16}{9}}$

56. $\sqrt{\dfrac{1}{25}}$

57. $\sqrt{\dfrac{1}{9}}$

58. $-\sqrt{\dfrac{16}{25}}$

59. $\sqrt{\dfrac{18}{32}}$

10.7 FRACTIONAL RADICANDS

When a radicand is fractional, how do we proceed in simplifying? If the radicand is already simplified, we can take the square root of the numerator and denominator separately, as the following examples show.

Example 1

$$\sqrt{\frac{25}{9}} = \frac{5}{3} \text{ because } \frac{5}{3} \cdot \frac{5}{3} = \frac{25}{9}$$

Example 2

$$\sqrt{\frac{1}{16}} = \frac{1}{4} \text{ because } \frac{1}{4} \cdot \frac{1}{4} = \frac{1}{16}$$

Example 3

$$\sqrt{\frac{18}{50}} = \sqrt{\frac{9 \cdot 2}{25 \cdot 2}} = \sqrt{\frac{9}{25} \cdot 1} = \sqrt{\frac{9}{25}}$$

$$\sqrt{\frac{9}{25}} = \frac{3}{5} \text{ because } \frac{3}{5} \cdot \frac{3}{5} = \frac{9}{25}$$

Do exercises 55 through 59 at the left.

In the above examples, the numerators and denominators were perfect squares, or became perfect squares upon simplifying. Can we simplify when this is not the case? We can at least simplify to an expression which has a whole number radicand, as in the following examples.

Example 4. Simplify $\sqrt{\dfrac{2}{3}}$.

$$\sqrt{\frac{2}{3}} = \sqrt{\frac{2}{3} \cdot \frac{3}{3}}$$ Multiplying by 1. We chose $\frac{3}{3}$ for 1 to make the denominator a perfect square.

$$= \sqrt{\frac{6}{9}}$$

$$= \sqrt{\frac{1}{9} \cdot 6} = \sqrt{\frac{1}{9}}\sqrt{6}$$

$$= \frac{1}{3}\sqrt{6}$$

As in example 4, we can always multiply the radicand by 1, choosing the name for 1 in such a way as to make the denominator a perfect square. Finally, we get a radicand which is a whole number.

Example 5. Simplify $\sqrt{\dfrac{5}{12}}$.

$$\sqrt{\frac{5}{12}} = \sqrt{\frac{5}{12} \cdot \frac{3}{3}} = \sqrt{\frac{15}{36}} = \sqrt{\frac{1}{36} \cdot 15}$$

$$= \frac{1}{6}\sqrt{15}$$

In Example 5, notice that we did not multiply by $\frac{12}{12}$. This would have given us a correct answer, but not the simplest. We chose $\frac{3}{3}$ to get for our denominator the smallest multiple of 12 which is a perfect square.

Do exercises 60 and 61 at the right.

USING THE SQUARE ROOT TABLE

Now that we know how to simplify radicals with fractional radicands, we can use the square root table to find square roots of fractions.

Example 6

Find $\sqrt{\dfrac{3}{5}}$.

$$\sqrt{\frac{3}{5}} = \sqrt{\frac{3}{5} \cdot \frac{5}{5}} = \sqrt{\frac{15}{25}} = \frac{1}{5}\sqrt{15}$$

From the table, $\sqrt{15} \approx 3.873$.

We divide by 5:

$$\begin{array}{r} .7746 \\ 5\overline{)3.8730} \\ \underline{35} \\ 37 \\ \underline{35} \\ 23 \\ \underline{20} \\ 30 \end{array}$$ $\sqrt{\dfrac{3}{5}} \approx .775$, to 3 decimal places.

Simplify.

60. $\sqrt{\dfrac{3}{5}}$

61. $\sqrt{\dfrac{5}{8}}$ Hint: multiply the radicand by $\dfrac{2}{2}$.

62. Use the table, p. 423, to approximate

$$\sqrt{\frac{2}{5}}.$$

Simplify.

63. $2\sqrt{3} + \sqrt{3}$

64. $\sqrt{2} + \sqrt{\frac{1}{2}}$

65. $\sqrt{\frac{5}{3}} - \sqrt{\frac{3}{5}}$

Example 7. Find $\sqrt{\frac{2}{3}}$.

$$\sqrt{\frac{2}{3}} = \sqrt{\frac{2}{3} \cdot \frac{3}{3}} = \sqrt{\frac{6}{9}} = \frac{1}{3}\sqrt{6}$$

From the table, $\sqrt{6} \approx 2.450$

We divide by 3 and get .817

Do exercise 62 at the left.

ADDING AND SUBTRACTING

When the same radical occurs in more than one term, we can use the distributive law and factor.

Example 8. Simplify.

$$\sqrt{2} + 5\sqrt{2} = 1 \cdot \sqrt{2} + 5\sqrt{2}$$
$$= (1 + 5)\sqrt{2} = 6\sqrt{2}$$

Note that this is just like factoring $x + 5x$, where $x = \sqrt{2}$.

Sometimes eliminating fractional radicands will enable us to factor expressions.

Example 9. Simplify.

$$\sqrt{3} + \sqrt{\frac{1}{3}} = \sqrt{3} + \sqrt{\frac{1}{3} \cdot \frac{3}{3}} \qquad \text{(multiplying by 1)}$$

$$= \sqrt{3} + \frac{1}{3}\sqrt{3} \qquad \text{(simplifying)}$$

$$= \left(1 + \frac{1}{3}\right)\sqrt{3} = \frac{4}{3}\sqrt{3} \qquad \text{(factoring and simplifying)}$$

Do exercises 63 through 65 at the left.

Do exercise set 10.7, p. 371.

10.8 DIVISION

To multiply with radical notation we multiply the radicands. It seems reasonable that to divide we would divide the radicands. This is actually the case, as can easily be proved.* Stated below is the basic principle we use in dividing.

For any non-negative radicands A and B, $\dfrac{\sqrt{A}}{\sqrt{B}} = \sqrt{\dfrac{A}{B}}$.

Example 1

$$\frac{\sqrt{27}}{\sqrt{3}} = \sqrt{\frac{27}{3}} = \sqrt{9} = 3$$

Example 2

$$\frac{\sqrt{7}}{\sqrt{14}} = \sqrt{\frac{7}{14}} = \sqrt{\frac{1}{2}} = \sqrt{\frac{1}{2}\cdot\frac{2}{2}} = \sqrt{\frac{2}{4}} = \frac{1}{2}\sqrt{2}$$

Example 3

$$\frac{\sqrt{30a^3}}{\sqrt{6a^2}} = \sqrt{\frac{30a^3}{6a^2}} = \sqrt{5a}$$

Do exercises 66 and 67 at the right.

RATIONALIZING DENOMINATORS

Expressions with radicals are generally simpler if there are no radicals in denominators. We can eliminate them by multiplying by 1† as follows.

Example 4. Rationalize the denominator of $\dfrac{\sqrt{2}}{\sqrt{3}}$

$$\frac{\sqrt{2}}{\sqrt{3}} = \frac{\sqrt{2}}{\sqrt{3}}\cdot 1 = \frac{\sqrt{2}}{\sqrt{3}}\cdot\frac{\sqrt{3}}{\sqrt{3}} = \frac{\sqrt{6}}{3}$$

Do exercises 68 and 69 at the right.

Do exercise set 10.8, p. 373.

---

* Here is a proof (optional). We consider a quotient $\dfrac{\sqrt{a}}{\sqrt{b}}$, where a and b are not negative. We shall also assume that $b \neq 0$. Why? We square this quotient, to show that we get $\dfrac{a}{b}$, and thus that the quotient is the square root of $\dfrac{a}{b}$ $\left(\text{or } \sqrt{\dfrac{a}{b}}\right)$.

$$\left(\frac{\sqrt{a}}{\sqrt{b}}\right)^2 = \frac{\sqrt{a}}{\sqrt{b}}\cdot\frac{\sqrt{a}}{\sqrt{b}} = \frac{\sqrt{a}\sqrt{a}}{\sqrt{b}\sqrt{b}} = \frac{a}{b}$$

† Denominators can also be rationalized by multiplying by 1 under a radical, as in the preceding lesson.

OBJECTIVES

You should be able to:

a) State the basic principle used in dividing with radical notation.

b) Divide with radical notation.

c) Rationalize denominators.

Divide and simplify.

66. $\dfrac{\sqrt{5}}{\sqrt{45}}$

67. $\dfrac{\sqrt{2}}{\sqrt{6}}$

Rationalize the denominators.

68. $\dfrac{\sqrt{5}}{\sqrt{7}}$

69. $\dfrac{\sqrt{x}}{\sqrt{y}}$

OBJECTIVES

You should be able to:

a) Identify the legs and hypotenuse of a right triangle.

b) State the Pythagorean property of right triangles.

c) Given lengths of any two sides of a right triangle find the length of the third side.

d) Solve applied problems involving right triangles.

70. Find the length of the hypotenuse in this right triangle.

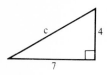

71. Find the length of the leg of this right triangle.

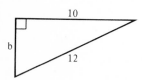

10.9 RADICALS AND RIGHT TRIANGLES

Many interesting problems can be solved if we remember an important property of right triangles. In a right triangle, we call the longest side the *hypotenuse*. The other two sides are called the *legs*. We usually use the letters a and b for the lengths of the legs and c for the length of the hypotenuse.

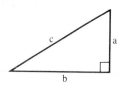

They are related as follows.

The Pythagorean Property of Right Triangles: **In any right triangle if a and b are the lengths of the legs and c is the length of the hypotenuse, then $a^2 + b^2 = c^2$.**

If we know the lengths of any two sides we can find the length of the third side.

Example 1. Find the length of the hypotenuse of this right triangle.

$$4^2 + 5^2 = c^2$$
$$16 + 25 = c^2$$
$$41 = c^2$$
$$c = \sqrt{41} \approx 6.403$$

Example 2. Find the length of the leg of this right triangle.

$$a^2 + 11^2 = 14^2$$
$$a^2 = 14^2 - 11^2 = 196 - 121$$
$$a = \sqrt{75}$$
$$a \approx 8.660$$

Do exercises 70 and 71 at the left.

Example 3. Find the length of the leg of this right triangle.

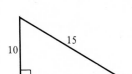

$$a^2 + 10^2 = 15^2$$

$$a^2 = 15^2 - 10^2$$

$$a^2 = 225 - 100$$

$$a^2 = 125$$

$$a = \sqrt{125} = \sqrt{25 \cdot 5} = 5\sqrt{5}$$

$$5\sqrt{5} \approx 5 \times 2.236$$

$$= 11.180$$

72. How long is a guy wire reaching from the top of a 15 ft pole to a point on the ground 10 ft from the pole?

Example 4. A 12 foot ladder is leaning against a house. The bottom of the ladder is 7 ft. from the house. How high is the top of the ladder?

We first make a drawing.

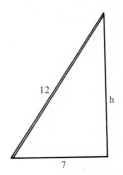

Now $7^2 + h^2 = 12^2$. We solve this equation.

$$h^2 = 12^2 - 7^2 = 144 - 49 = 95$$

$$h^2 = 95$$

$$h = \sqrt{95} \approx 9.747$$

Do exercise 72 at the right.

Do exercise set 10.9, p. 375.

You should be able to:

a) State the principle of squaring.
b) Solve simple equations with radicals.

Solve.

73. $\sqrt{3x} - 5 = 3$

10.10 EQUATIONS WITH RADICALS

To solve equations with radicals we first convert them to equations without radicals. To do this we use an equation solving principle that we have not yet considered. We can get rid of radicals by squaring both sides of an equation. This new principle guarantees that this will work.

The Principle of Squaring: **If an equation $a = b$ is true, then the equation $a^2 = b^2$ is also true.**

The following examples show how this principle is applied.

Example 1.

Solve: $\sqrt{2x} - 4 = 7$

$$\sqrt{2x} = 11 \qquad \text{(Adding 4, to get the radical alone on one side.)}$$
$$2x = 121 \qquad \text{(Squaring both sides.)}$$
$$x = \tfrac{121}{2}$$

Check: $\sqrt{2x} - 4 = 7$

$$
\begin{array}{c|c}
\sqrt{2 \tfrac{121}{2}} - 4 & 7 \\
\sqrt{121} - 4 & \\
11 - 4 & \\
7 & \\
\end{array}
$$

When we use the principle of squaring the checking of a solution is very important, because we may get extra numbers that are not solutions when we square both sides of an equation.*

74. $\sqrt{x - 2} - 5 = 3$
(Hint: First get $\sqrt{x - 2}$ alone on one side of the equation.)

Example 2

Solve: $\sqrt{x + 1} = \sqrt{2x - 5}$

$$x + 1 = 2x - 5 \quad \text{(Squaring both sides.)}$$
$$x = 6$$

The solution is 6, because it checks.

Do exercises 73 and 74 at the left.

Do exercise set 10.10, p. 377.

---

* To see this consider the equation $x = 1$. It has just one solution, the number 1. When we square both sides we get $x^2 = 1$, which has two solutions, **1** and -1.

NAME _____

CLASS _____

EXERCISE SET 10.1 (pp. 343–344)

In each expression what is the radicand?

1. $\sqrt{3x}$

2. $5\sqrt{x^2 - 6}$

3. $x^2y\sqrt{\dfrac{3}{x+2}}$

Which of these expressions are meaningless? Write yes or no.

4. $\sqrt{-16}$

5. $-\sqrt{49}$

6. $\sqrt{-0}$

For which values of x is each expression meaningful?

7. $\sqrt{-x}$

8. $\sqrt{-x^2}$

Find two square roots of each number.

9. 81

10. 1

11. 324

12. 169

13. 729

Evaluate

14. $\sqrt{9}$

15. $-\sqrt{64}$

16. $-\sqrt{0}$

17. $\sqrt{-4}$

18. $\sqrt{361}$

19. $\sqrt{144}$

20. $-\sqrt{-1}$

21. $-\sqrt{289}$

ANSWERS

1. _____

2. _____

3. _____

4. _____

5. _____

6. _____

7. _____

8. _____

9. _____

10. _____

11. _____

12. _____

13. _____

14. _____

15. _____

16. _____

17. _____

18. _____

19. _____

20. _____

21. _____

What are the nonsensible replacements for the following?

22. $\sqrt{x-5}$ 23. $\sqrt{x+7}$ 24. $\sqrt{x}$

25. $\sqrt{x^2+1}$ 26. $\sqrt{x+20}$ 27. $\sqrt{x^2+5}$

28. $\sqrt{5x}$ 29. $\sqrt{x^2+4}$

Find these square roots.

30. $\sqrt{100}$ 31. $\sqrt{4}$ 32. $\sqrt{121}$

33. $\sqrt{169}$

Simplify.

34. $\sqrt{x^2}$ 35. $\sqrt{4a^2}$ 36. $\sqrt{(-3b)^2}$

37. $\sqrt{(-5)^2}$ 38. $\sqrt{(-7)^2c^2d^2}$ 39. $\sqrt{(-4d)^2}$

40. $\sqrt{-16^2}$ 41. $\sqrt{(x-7)^2}$ 42. $\sqrt{x^2+6x+9}$

43. $\sqrt{4x^2-20x+25}$

EXERCISE SET 10.2 (pp. 345–346)

State whether each number is rational or irrational.

1. $-\dfrac{2}{3}$

2. $\dfrac{136}{51}$

3. 23

4. $\sqrt{3}$

5. $\sqrt{5}$

6. $\dfrac{\frac{2}{7}}{5}$

7. $\sqrt{4}$

8. $-\sqrt{12}$

9. $-\sqrt{49}$

10. 0

11. $\dfrac{2.5}{7}$

12. $\sqrt{.64}$

13. $-\sqrt{7}$

14. $2.1313\ldots$ (numeral repeats)

15. 3.141617

16. $1.131131113\ldots$ (does not repeat)

17. Every _____ number can be named as the quotient of two _____ .

18. The real numbers consist of the rational numbers and the _____ numbers.

19. Which of the numbers in Exercises 1–16 inclusive are real numbers?

Find the number halfway between each of the following.

20. $\dfrac{2}{7}, \dfrac{5}{7}$ **21.** $1, \dfrac{1}{3}$ **22.** $.25, \dfrac{3}{8}$

23. $\dfrac{2}{3}, \dfrac{7}{10}$ **24.** $8\dfrac{1}{6}, 8\dfrac{1}{7}$ **25.** $\dfrac{15}{24}, \dfrac{16}{24}$

Use the table, p. 423, to approximate these numbers:

26. $\sqrt{5}$ **27.** $\sqrt{43}$ **28.** $\sqrt{81}$

29. $\sqrt{93}$ **30.** $\sqrt{17}$ **31.** $\sqrt{63}$

32. $\sqrt{50}$ **33.** $\sqrt{87}$ **34.** $\sqrt{2}$

NAME _____

CLASS _____

ANSWERS

EXERCISE SET 10.3 (p. 347)

Which of the following are true? Write T or F in the answer spaces.

1. $\sqrt{5} \cdot \sqrt{2} = \sqrt{5 \cdot 2}$

2. $\sqrt{3} \cdot \sqrt{5} = \sqrt{15}$

3. $-\sqrt{2} \cdot \sqrt{3} = -\sqrt{6}$

4. $\sqrt{-2} \cdot \sqrt{-6} = \sqrt{(-2)(-6)}$

5. $\sqrt{3}(-\sqrt{15}) = -\sqrt{15}$

6. $\sqrt{-7}\sqrt{-3} = \sqrt{21}$

Simplify:

7. $\sqrt{36} \cdot \sqrt{16}$

8. $\sqrt{25} \cdot \sqrt{81}$

9. $\sqrt{9 \cdot 64}$

10. $\sqrt{49 \cdot 4}$

11. $\sqrt{9} \cdot \sqrt{121}$

12. $\sqrt{144} \cdot \sqrt{16}$

Multiply:

13. $\sqrt{3} \cdot \sqrt{3}$

14. $\sqrt{17} \cdot \sqrt{17}$

15. $\sqrt{6} \cdot \sqrt{7}$

1. _____

2. _____

3. _____

4. _____

5. _____

6. _____

7. _____

8. _____

9. _____

10. _____

11. _____

12. _____

13. _____

14. _____

15. _____

ANSWERS

16.

_____

17.

_____

18.

_____

19.

_____

20.

_____

21.

_____

22.

_____

23.

_____

24.

_____

25.

_____

26.

_____

27.

_____

28.

_____

29.

_____

30.

_____

16. $\sqrt{\dfrac{7}{11}} \cdot \sqrt{\dfrac{3}{7}}$

17. $\sqrt{x} \cdot \sqrt{x+1}$

18. $\sqrt{25} \cdot \sqrt{3}$

19. $\sqrt{x} \cdot \sqrt{x-3}$

20. $\sqrt{x+2} \cdot \sqrt{x+1}$

21. $\sqrt{5} \cdot \sqrt{2x-1}$

22. $\sqrt{x-3}\sqrt{x+4}$

23. $\sqrt{x+3} \cdot \sqrt{2x+3}$

24. $\sqrt{-3} \cdot \sqrt{x-1}$

25. $\sqrt{x-2} \cdot \sqrt{x+2}$

26. $\sqrt{x+4} \cdot \sqrt{x-4}$

27. $\sqrt{(-2)^2}\sqrt{3x}$

28. $\sqrt{2x+5} \cdot \sqrt{3x}$

29. $\sqrt{x+y} \cdot \sqrt{x-y}$

30. $\sqrt{2x}\sqrt{x-1}$

EXERCISE SET 10.4 (p. 348)

Factor. Simplify where possible.

1. $\sqrt{12}$ 2. $\sqrt{3x}$ 3. $\sqrt{8}$

4. $\sqrt{5x^2}$ 5. $\sqrt{16x}$ 6. $\sqrt{9x^2}$

7. $\sqrt{6}$ 8. $\sqrt{7xy}$ 9. $\sqrt{10}$

10. $\sqrt{8x^2}$ 11. $\sqrt{14x}$ 12. $\sqrt{x^2 - 1}$

13. $\sqrt{3x - 3}$ 14. $\sqrt{2x^2 - 5x - 12}$ 15. $\sqrt{x^3 - 2x^2}$

16. $\sqrt{9x^2 - 6x + 1}$ 17. $\sqrt{4x^2 - 8x + 4}$ 18. $\sqrt{(-7x)^2}$

19. $\sqrt{4x^2}$ 20. $\sqrt{x^2 - x - 2}$ 21. $\sqrt{12x^2 - 36x + 27}$

ANSWERS

1.
2.
3.
4.
5.
6.
7.
8.
9.
10.
11.
12.
13.
14.
15.
16.
17.
18.
19.
20.
21.

Approximate these square roots. Round to two decimal places.

22. _____

22. $\sqrt{180}$ **23.** $\sqrt{3969}$ **24.** $\sqrt{768}$

23. _____

24. _____

25. $\sqrt{105}$ **26.** $\sqrt{765}$ **27.** $\sqrt{500}$

25. _____

26. _____

28. Which of the numbers in Exercises 22–27 are rational?

What are the sensible values for x in the following expressions?

27. _____

29. $\sqrt{-3x}$ **30.** $\sqrt{-4(x-2)}$ **31.** $\sqrt{-5(x+3)}$

28. _____

29. _____

30. _____

31. _____

NAME _____

CLASS _____

ANSWERS

EXERCISE SET 10.5 (pp. 349–350)

Simplify.

1. $\sqrt{24}$

2. $\sqrt{48x}$

3. $\sqrt{121m}$

4. $\sqrt{40}$

5. $\sqrt{125y}$

6. $\sqrt{20x^2}$

7. $\sqrt{98x^4}$

8. $\sqrt{28x^2}$

9. $\sqrt{36m^3}$

10. $\sqrt{250y^3}$

11. $\sqrt{8a^5}$

12. $\sqrt{27xy^2}$

13. $\sqrt{243a^6}$

14. $\sqrt{448x^6y^3}$

15. $\sqrt{8x^2 + 8x + 2}$

16. $\sqrt{27x^2 - 36x + 12}$

17. $\sqrt{x^2y - 2xy + y}$

18. $\sqrt{36y + 12y^2 + y^3}$

1. _____

2. _____

3. _____

4. _____

5. _____

6. _____

7. _____

8. _____

9. _____

10. _____

11. _____

12. _____

13. _____

14. _____

15. _____

16. _____

17. _____

18. _____

ANSWERS

19.

_____

20.

_____

21.

_____

22.

_____

23.

_____

24.

_____

25.

_____

26.

_____

27.

_____

28.

_____

29.

_____

30.

_____

19. $\sqrt{(y-2)^8}$

20. $\sqrt{x^3(x-7)^6}$

21. $\sqrt{4(x+5)^4}$

Multiply and simplify.

22. $\sqrt{3}\sqrt{18}$

23. $\sqrt{8}\cdot\sqrt{6}$

24. $\sqrt{5b}\cdot\sqrt{15b}$

25. $\sqrt{ab}\sqrt{ac}$

26. $\sqrt{2x^2y}\sqrt{4xy^2}$

27. $\sqrt{18x^2y^3}\sqrt{6xy^4}$

28. $\sqrt{50ab}\sqrt{10a^2b^4}$

29. $\sqrt{2x^3y}\sqrt{3xy^2}\sqrt{6xy}$

30. $\sqrt{27(x+1)}\sqrt{18y(x+1)^2}$

EXERCISE SET 10.6 (p. 351)

Evaluate for $a = 1$, $b = 3$, $c = 2$, $d = 4$

1. $\sqrt{a^2 + c^2}$; $\sqrt{a^2} + \sqrt{c^2}$

2. $\sqrt{b^2 + c^2}$; $\sqrt{b^2} + \sqrt{c^2}$

3. $\sqrt{a^2 + d^2}$; $\sqrt{a^2} + \sqrt{d^2}$

4. $\sqrt{b^2 + d^2}$; $\sqrt{b^2} + \sqrt{d^2}$

5. Can you guess any numbers that would make the following true?

$\sqrt{x^2 + y^2} = \sqrt{x^2} + \sqrt{y^2}$

Simplify.

6. $3\sqrt{2} + 4\sqrt{2}$

7. $8\sqrt{3} + 3\sqrt{3}$

8. $7\sqrt{5} - 3\sqrt{5}$

9. $9\sqrt{x} - 11\sqrt{x}$

10. $\sqrt{45} - \sqrt{20}$

11. $3\sqrt{12} + 2\sqrt{3}$

12. $2\sqrt{12} + \sqrt{27} + \sqrt{48}$

13. $9\sqrt{8} - \sqrt{72} + \sqrt{98}$

1.

2.

3.

4.

5.

6.

7.

8.

9.

10.

11.

12.

13.

ANSWERS

14.

_____

15.

_____

16.

_____

17.

_____

18.

_____

19.

_____

20.

_____

21.

_____

22.

_____

23.

_____

14. $3\sqrt{18} - 2\sqrt{32} - 5\sqrt{50}$

15. $\sqrt{18} - \sqrt{12} + \sqrt{50}$

16. $2\sqrt{27} - 3\sqrt{48} + 2\sqrt{18}$

17. $\sqrt{4x} + \sqrt{81x^3}$

18. $\sqrt{12x^2} + \sqrt{27}$

19. $\sqrt{8x + 8} + \sqrt{2x + 2}$

20. $\sqrt{x^5 - x^2} + \sqrt{9x^3 - 9}$

21. $\sqrt{x^2y + 6xy + 9y} + \sqrt{y^3}$

22. $3x\sqrt{y^3x} - x\sqrt{yx^3} + y\sqrt{y^3x}$

23. $\sqrt{8(a + b)^3} - \sqrt{2(a + b)} + \sqrt{32(a + b)^3}$

NAME

CLASS

ANSWERS

EXERCISE SET 10.7 (pp. 352–354)

Simplify.

1. $\sqrt{\dfrac{9}{49}}$ **2.** $\sqrt{\dfrac{16}{81}}$ **3.** $\sqrt{\dfrac{1}{144}}$

4. $\sqrt{.09}$ **5.** $\sqrt{\dfrac{36}{x^2}}$ **6.** $\sqrt{\dfrac{64}{289}}$

7. $\sqrt{\dfrac{9a^2}{625}}$ **8.** $\sqrt{\dfrac{x^4y^2}{256}}$ **9.** $\sqrt{\dfrac{2}{5}}$

10. $\sqrt{\dfrac{2}{3}}$ **11.** $\sqrt{\dfrac{3}{8}}$ **12.** $\sqrt{\dfrac{9}{15}}$

13. $\sqrt{\dfrac{3}{27}}$ **14.** $\sqrt{\dfrac{7}{12}}$ **15.** $\sqrt{\dfrac{1}{7}}$

16. $\sqrt{\dfrac{3}{20}}$ **17.** $\sqrt{\dfrac{8}{3}}$ **18.** $\sqrt{\dfrac{12}{5}}$

19. $\sqrt{\dfrac{2}{x}}$ **20.** $\sqrt{\dfrac{x}{y}}$

ANSWERS

1.

2.

3.

4.

5.

6.

7.

8.

9.

10.

11.

12.

13.

14.

15.

16.

17.

18.

19.

20.

ANSWERS

Use the table to approximate the following.

21. _____

22. _____

23. _____

24. _____

25. _____

26. _____

27. _____

28. _____

29. _____

30. _____

31. _____

32. _____

21. $\sqrt{\dfrac{2}{5}}$

22. $\sqrt{\dfrac{5}{12}}$

23. $\sqrt{\dfrac{8}{3}}$

24. $\sqrt{3\dfrac{1}{2}}$

Simplify.

25. $2\sqrt{5} + 7\sqrt{5}$

26. $8\sqrt{7} - 2\sqrt{7}$

27. $\sqrt{2} + \sqrt{8}$

28. $5\sqrt{2} - 3\sqrt{\dfrac{1}{2}}$

29. $7\sqrt{3} + 5\sqrt{\dfrac{1}{3}}$

30. $\sqrt{\dfrac{1}{2}} - \sqrt{\dfrac{1}{8}}$

31. $\sqrt{\dfrac{2}{3}} - \sqrt{\dfrac{1}{6}}$

32. $\sqrt{\dfrac{1}{12}} - \sqrt{\dfrac{1}{27}}$

EXERCISE SET 10.8 (p. 355)

Divide.

1. $\dfrac{\sqrt{27}}{\sqrt{3}}$

2. $\dfrac{\sqrt{18}}{\sqrt{3}}$

3. $\dfrac{\sqrt{60}}{\sqrt{15}}$

4. $\dfrac{\sqrt{2}}{\sqrt{8}}$

5. $\dfrac{\sqrt{108}}{\sqrt{3}}$

6. $\dfrac{\sqrt{75}}{\sqrt{15}}$

7. $\dfrac{\sqrt{\frac{3}{7}}}{\sqrt{\frac{3}{14}}}$

8. $\dfrac{\sqrt{\frac{5}{6}}}{\sqrt{\frac{2}{3}}}$

9. $\dfrac{\sqrt{8x}}{\sqrt{2x}}$

10. $\dfrac{\sqrt{15x^5}}{\sqrt{3x}}$

11. $\dfrac{\sqrt{63y^3}}{\sqrt{7y}}$

12. $\dfrac{\sqrt{3x}}{\sqrt{\frac{3x}{4}}}$

13. $\dfrac{\sqrt{2}}{\sqrt{5}}$

14. $\dfrac{\sqrt{7}}{\sqrt{3}}$

15. $\dfrac{\sqrt{9}}{\sqrt{8}}$

ANSWERS

1. _____

2. _____

3. _____

4. _____

5. _____

6. _____

7. _____

8. _____

9. _____

10. _____

11. _____

12. _____

13. _____

14. _____

15. _____

Rationalize the denominator and simplify.

16. _____

17. _____

16. $\dfrac{\sqrt{5}}{\sqrt{11}}$

17. $\dfrac{\sqrt{7}}{\sqrt{27}}$

18. $\dfrac{\sqrt{48}}{\sqrt{32}}$

18. _____

19. _____

20. _____

19. $\dfrac{\sqrt{56}}{\sqrt{40}}$

20. $\dfrac{\sqrt{45}}{\sqrt{56}}$

21. $\dfrac{\sqrt{343}}{\sqrt{14}}$

21. _____

22. _____

22. $\dfrac{\sqrt{27c}}{\sqrt{32c^3}}$

23. $\dfrac{\sqrt{45}}{\sqrt{8a}}$

24. $\dfrac{\sqrt{a^3}}{\sqrt{8}}$

23. _____

24. _____

25. _____

25. $\dfrac{\sqrt{7x^3}}{\sqrt{12a}}$

26. $\dfrac{\sqrt{y^5}}{\sqrt{xy^2}}$

27. $\dfrac{\sqrt{x^3}}{\sqrt{xy}}$

26. _____

27. _____

28. $\dfrac{\sqrt{3y}}{\sqrt{2x^2}}$

29. $\dfrac{\sqrt{45mn^2}}{\sqrt{32m}}$

30. $\dfrac{\sqrt{16a^4b^6}}{\sqrt{128a^6b^6}}$

28. _____

29. _____

30. _____

ANSWERS

EXERCISE SET 10.9 (pp. 356–357)

Which side is the hypotenuse in each triangle?

1.

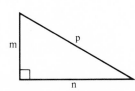

2.

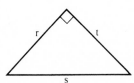

3.

Find the length of the hypotenuse of each triangle.

4.

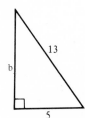

5.

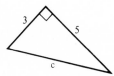

6.

Given the length of the hypotenuse and the length of one leg, find the length of the other leg.

7.

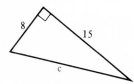

8.

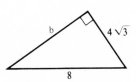

9.

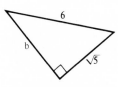

Given a right triangle with c as the length of the hypotenuse, use the Pythagorean Property to find the length of the side not given.

10. $a = 10; b = 24$

11. $a = 18; c = 30$

1.

_____

2.

_____

3.

_____

4.

_____

5.

_____

6.

_____

7.

_____

8.

_____

9.

_____

10.

_____

11.

_____

ANSWERS

12.

_____

13.

_____

14.

_____

15.

_____

16.

_____

17.

_____

18.

_____

19.

_____

20.

_____

12. $c = 10; b = 5\sqrt{3}$ **13.** $a = 7, b = 7\sqrt{3}$

14. $a = \sqrt{2}, b = \sqrt{3}$ **15.** $c = 6\sqrt{2}, a = 6$

16. $a = 5, b = 5$ **17.** $c = 26, b = 10$

Solve.

18. Find the length of a diagonal of a square whose sides are 3 centimeters long.

19. When the foot of a ladder 18 feet long is set 8 feet from the wall of a building, the top of the ladder just reaches the second-story window. How far from the ground is the window?

20. Find the length of a diagonal of a rectangle whose dimensions are 12 meters by 20 meters.

NAME _____

CLASS _____

EXERCISE SET 10.10 (p. 358)

Solve.

1. $\sqrt{x} = 5$

2. $3 + \sqrt{x-1} = 5$

3. $8 = \sqrt{x-9}$

4. $\sqrt{x+1} = -3$

5. $2\sqrt{x-5} = 7$

6. $\sqrt{x-5} = 4$

7. $^8\sqrt{x} = 4$

8. $\sqrt{x} = -6$

9. $\sqrt{5x+1} + 6 = 10$

10. $\sqrt{t} - 4 = 0$

11. $7 = 3\sqrt{x}$

12. $6 + 2\sqrt{3n} = 0$

13. $\sqrt{y+7} + 4 = 0$

14. $6 + \sqrt{x} = 13$

15. $7 = \dfrac{1}{\sqrt{y}}$

ANSWERS

1. _____

2. _____

3. _____

4. _____

5. _____

6. _____

7. _____

8. _____

9. _____

10. _____

11. _____

12. _____

13. _____

14. _____

15. _____

ANSWERS

16.

_____

17.

_____

18.

_____

19.

_____

20.

_____

21.

_____

22.

_____

23.

_____

16. $6 - \sqrt{x - 5} = 2$ **17.** $8 = \sqrt{5r + 1}$

18. $\sqrt{x^2 + 6} - x + 3 = 0$ **19.** $\sqrt{2y + 6} = \sqrt{2y - 5}$

20. $\sqrt{5x - 7} = \sqrt{x + 10}$ **21.** $2\sqrt{3x - 2} = \sqrt{2x - 3}$

22. How long must a wire be to reach from the top of a 40-foot telephone pole to a point on the ground 30 feet from the foot of the pole?

23. A number is 12 more than its positive square root. Find the number.

NAME_____

CHAPTER 10 TEST   CLASS_____SCORE_____GRADE_____

Before taking the test *be sure* to allow yourself a day or so for review. Use the objectives listed in the margins to guide your study. The test will evaluate your progress and aid your preparation for a possible classroom test. Allow about an hour for the test. Remove the test from the book. When you finish read the test analysis on the answer page at the end of the book.

1. What is the radicand in $3x\sqrt{\dfrac{x}{2+x}}$

Identify the following as rational or irrational.

2. $\sqrt{3}$

3. $\sqrt{36}$

4. $-\sqrt{12}$

5. $-\sqrt{4}$

Simplify.

6. $\sqrt{\dfrac{25}{64}}$

7. $\sqrt{27}$

8. $\sqrt{\dfrac{1}{2}}$

9. $\sqrt{\dfrac{3}{8}}$

10. $\sqrt{\dfrac{3}{x^2}}$

11. $\sqrt{\dfrac{a}{b}}$

12. $\sqrt{\dfrac{16m^3}{a^5}}$

13. $\sqrt{\dfrac{45x^3}{18y}}$

Multiply or divide as indicated. Simplify your answers.

14. $\sqrt{2}\cdot\sqrt{6}$

15. $-\sqrt{3}\cdot\sqrt{12}$

16. $\sqrt{-2}\cdot\sqrt{-3}$

17. $\sqrt{x-3}\cdot\sqrt{x-3}$

18. $\dfrac{\sqrt{27}}{\sqrt{45}}$

19. $\dfrac{2\sqrt{3}}{\sqrt{60}}$

ANSWERS

1. _____

2. _____

3. _____

4. _____

5. _____

6. _____

7. _____

8. _____

9. _____

10. _____

11. _____

12. _____

13. _____

14. _____

15. _____

16. _____

17. _____

18. _____

19. _____

ANSWERS

20. $\sqrt{x+4} \cdot \sqrt{x-4}$

21. $\dfrac{\sqrt{45x^2 y}}{\sqrt{54y}}$

20. _____

21. _____

Add or subtract as indicated. Simplify.

22. _____

22. $5\sqrt{12} - 3\sqrt{12}$

23. $\dfrac{2}{3}\sqrt{108} + \sqrt{192}$

23. _____

24. $10\sqrt{\dfrac{4}{5}} - 2\sqrt{180}$

25. $\dfrac{18}{\sqrt{12}} + 3\sqrt{27}$

24. _____

25. _____

Solve.

26. $3\sqrt{x} + 1 = 7$

27. $2 - k = \sqrt{k^2 + 10}$

26. _____

27. _____

28. $2 - y = 2\sqrt{y+1}$

29. Find a.

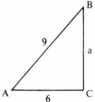

28. _____

29. _____

30. The length of a rectangle is 2 centimeters greater than the width. A diagonal is 10 centimeters long. Find the dimensions of the rectangle.

30. _____

11 QUADRATIC EQUATIONS

11.1 QUADRATIC EQUATIONS

The following are examples of quadratic equations. Each is of second degree.

$4x^2 + 7x - 5 = 0$, $3x^2 - \frac{1}{2}x = 9$, $5y^2 = -6y$

An equation of the type

$$ax^2 + bx + c = 0$$

where a, b, and c are real number constants and $a \neq 0$, is called the *standard form* of a *quadratic equation*.

Examples. Write each equation in standard form and determine a, b, and c.

a) $4x^2 + 7x - 5 = 0$; the equation is already in standard form.

  $a = 4, b = 7, c = -5$

b) $3x^2 - \frac{1}{2}x = 9$

  $3x^2 - \frac{1}{2}x - 9 = 0$

  $a = 3, b = -\frac{1}{2}, c = -9$

c) $5y^2 = -6y$

  $5y^2 + 6y = 0$

  $a = 5, b = 6, c = 0$

Do exercises 1 through 3 at the right.

EQUATIONS OF THE TYPE $ax^2 = k$

The following is in standard form, with b and c equal to 0.

$$7x^2 = 0$$

We obtain $x^2 = 0$ (multiplying by $\frac{1}{7}$)

  $x \cdot x = 0$

  $x = 0$ or $x = 0$ (principle of zero products)

The solution is 0.

Do exercise 4 at the right.

You should be able to:

a) Given a quadratic equation, write it in standard form,

$ax^2 + bx + c = 0$,

and determine a, b, and c.

b) Solve quadratic equations of the type $ax^2 = k$, $a \neq 0$.

c) Solve quadratic equations of the type $ax^2 + bx = 0$, $a \neq 0$, $b \neq 0$, by factoring.

d) Solve equations of the type $ax^2 + bx + c = 0$, $a \neq 0$, $b \neq 0$, $c \neq 0$, by factoring.

Write each equation in standard form and determine a, b, and c.

1. $x^2 = 7x$

2. $3 - x^2 = 9x$

3. $3x + 5x^2 = x^2 - 4 + x$

Solve.

4. $-\frac{1}{2}x^2 = 0$

5. Solve $x^2 - 5 = 0$.

Consider the equation $5x^2 = 15$ (where $b = 0$ and $c = -15$). Multiplying by $\frac{1}{5}$ we have $x^2 = \frac{15}{5}$, or 3.

Now we take the principal square root:

$$\sqrt{x^2} = \sqrt{3}$$
$$|x| = \sqrt{3}$$
$$x = \sqrt{3} \quad \text{or} \quad -x = \sqrt{3}$$
$$x = \sqrt{3} \quad \text{or} \quad x = -\sqrt{3}.$$

The solutions are $\sqrt{3}$ and $-\sqrt{3}$.

We could also solve the equation by recalling that every positive real number has two square roots, one positive and one negative.

Do exercise 5 at the left.

Example. Solve

$-3x^2 + 7 = 0$.
$$-3x^2 = -7 \qquad \text{(Adding } -7)$$
$$x^2 = \frac{-7}{-3}, \text{ or } \frac{7}{3} \quad \left(\text{Multiplying by } \frac{1}{-3}\right)$$

6. Solve $4x^2 - 9 = 0$.

$$|x| = \sqrt{\frac{7}{3}} \quad \text{(Taking square root)}$$

$$\left. \begin{array}{l} x = \sqrt{\frac{7}{3} \cdot \frac{3}{3}} \text{ or } -x = \sqrt{\frac{7}{3} \cdot \frac{3}{3}} \\[2mm] x = \frac{\sqrt{21}}{3} \text{ or } x = \frac{-\sqrt{21}}{3} \end{array} \right\} \text{(Rationalizing the denominator)}$$

Check: For $\dfrac{\sqrt{21}}{3}$:

$$\begin{array}{c|c} -3x^2 + 7 = 0 & \\ \hline -3\left(\dfrac{\sqrt{21}}{3}\right)^2 + 7 & 0 \\[3mm] -3 \cdot \dfrac{21}{9} + 7 & \\[3mm] -7 + 7 & \\[2mm] 0 & \end{array}$$

The solutions are $\dfrac{\sqrt{21}}{3}$ and $\dfrac{-\sqrt{21}}{3}$.

Do exercise 6 at the left.

SOLVING BY FACTORING

Let us first consider an equation of the type

$ax^2 + bx = 0 \quad (a \neq 0, b \neq 0)$

Example 1. Solve $4x^2 - 3x = 0$.

$x(4x - 3) = 0$ (factoring)

$x = 0$ or $4x - 3 = 0$ (principle of zero products)

$x = 0 \quad$ or $\quad x = \frac{3}{4}$

The solutions are 0 and $\frac{3}{4}$.

Do exercise 7 at the right.

Note that an equation of the above type will always have one solution 0 and the other nonzero.

Now let us consider an equation of the type

$ax^2 + bx + c = 0 \quad (a \neq 0, b \neq 0, c \neq 0)$

We will be developing several procedures for solving equations such as this. One procedure we have already considered is by factoring. Let us review it here.

Example 2. Solve $5x^2 - 8x + 3 = 0$

$(5x - 3)(x - 1) = 0$ (factoring)

$5x - 3 = 0$ or $x - 1 = 0$

$x = \frac{3}{5}$ or $\quad x = 1$

The solutions are $\frac{3}{5}$ and 1.

Example 3. Solve $(y - 3)(y - 2) = 6(y - 3)$

$y^2 - 5y + 6 = 6y - 18$

$y^2 - 11y + 24 = 0$ (This is standard form)

$(y - 8)(y - 3) = 0$

$y = 8$ or $y = 3$

The solutions are 8 and 3.

Do exercises 8 and 9 at the right.

Do exercise set 11.1, p. 397.

7. Solve $3x^2 + 5x = 0$.

8. Solve $3x^2 + x - 2 = 0$.

9. Solve $(x - 1)(x + 1) = 5(x - 1)$.

OBJECTIVES

You should be able to solve quadratic equations by completing the square.

10. Solve $x^2 + 6x + 8 = 0$ by factoring.

11. Solve $x^2 - 6x - 2 = 0$ by completing the square.

11.2 COMPLETING THE SQUARE

We used *completing the square* earlier to factor certain polynomials. We can also use *completing the square* to factor when solving quadratic equations.

Consider $x^2 + 6x + 8 = 0$. We see that this equation can be solved by factoring.

Do exercise 10 at the left.

Now let us consider another way to solve the equation. We consider the trinomial $x^2 + 6x + 8$, and complete the square. Since the trinomial $x^2 + 6x + 9$ is a square, the left side of the equation can be made a square by adding zero, using $9 - 9$. We get 9 by taking half of 6 and squaring it. Then

$$x^2 + 6x + 8 = 0$$

$$x^2 + 6x + 8 + (9 - 9) = 0$$

$$x^2 + 6x + 9 + (8 - 9) = 0$$

$$(x + 3)^2 - 1 = 0$$

$$(x + 3 - 1)(x + 3 + 1) = 0$$

$$(x + 2)(x + 4) = 0$$

$$x + 2 = 0 \text{ or } x + 4 = 0$$

$$x = -2 \text{ or } x = -4$$

The solutions are -2 and -4.

We can perform the above procedure faster if we first subtract 8 and then add 9:

$$x^2 + 6x + 8 = 0$$

$$x^2 + 6x = -8$$

$$x^2 + 6x + 9 = -8 + 9$$

$$(x + 3)^2 = 1$$

$$x + 3 = 1 \text{ or } x + 3 = -1$$

$$x = -2 \text{ or } x = -4$$

Example 1. Solve by completing the square.

$x^2 + 4x - 7 = 0$

$\quad x^2 + 4x = 7 \qquad$ (adding 7)

$x^2 + 4x + 4 = 7 + 4 \quad$ (taking half of 4, squaring it, adding)

$\quad (x + 2)^2 = 11$

$x + 2 = \sqrt{11} \text{ or } x + 2 = -\sqrt{11}$

$\qquad x = -2 + \sqrt{11} \text{ or } x = -2 - \sqrt{11}$

The solutions are $-2 \pm \sqrt{11}$ ($-2 \pm \sqrt{11}$ is read "-2 plus or minus $\sqrt{11}$").

Do exercise 11 on the previous page.

Do exercise 12 at the right.

When the coefficient of the x^2 term in the standard form is not 1 we first make this coefficient 1.

Example 2. Solve $2x^2 - 3x - 1 = 0$.

$x^2 - \frac{3}{2}x - \frac{1}{2} = 0 \qquad$ (multiplying by $\frac{1}{2}$)

$x^2 - \frac{3}{2}x \quad\quad = \frac{1}{2} \qquad$ (adding $\frac{1}{2}$)

$x^2 - \frac{3}{2}x + \frac{9}{16} = \frac{1}{2} + \frac{9}{16} \quad$ (taking half of $\frac{3}{2}$, squaring it, adding)

$(x - \frac{3}{4})^2 = \frac{8}{16} + \frac{9}{16} = \frac{17}{16}$

$x - \frac{3}{4} = \sqrt{\frac{17}{16}} \quad\text{ or }\quad x - \frac{3}{4} = -\sqrt{\frac{17}{16}}$

$x - \frac{3}{4} = \dfrac{\sqrt{17}}{4} \quad\text{ or }\quad x - \frac{3}{4} = \dfrac{-\sqrt{17}}{4}$

$x = \frac{3}{4} + \dfrac{\sqrt{17}}{4} \quad\text{ or }\quad x = \frac{3}{4} - \dfrac{\sqrt{17}}{4}$

The solutions are $\dfrac{3 \pm \sqrt{17}}{4}$.

Do exercise 13 at the right.

Do exercise set 11.2, p. 399.

12. Solve $x^2 - 4x - 3 = 0$ by completing the square.

13. Solve $2x^2 + 2x - 3 = 0$ by completing the square.

You should be able to solve quadratic equations, using the quadratic formula.

11.3 THE QUADRATIC FORMULA

You may have noticed while solving equations by completing the square that we do about the same thing each time. In situations like this in mathematics, when we do about the same kind of computation many times, we look for a formula so we can speed up our work.

We want to find the solutions of *any* quadratic equation by completing the square. Thus we consider the standard form

$$ax^2 + bx + c = 0, \ a \neq 0 \ (b \text{ or } c \text{ may be } 0).$$

and complete the square:

$$x^2 + \frac{b}{a}x + \frac{c}{a} = 0 \qquad \left(\text{multiplying by } \frac{1}{a}\right)$$

$$x^2 + \frac{b}{a}x \phantom{+ \frac{c}{a}} = -\frac{c}{a} \qquad \left(\text{adding } -\frac{c}{a}\right)$$

Taking half of $\frac{b}{a}$, we get $\frac{b}{2a}$. Squaring we get $\frac{b^2}{4a^2}$. Then we add $\frac{b^2}{4a^2}$:

$$x^2 + \frac{b}{a}x + \frac{b^2}{4a^2} = -\frac{c}{a} + \frac{b^2}{4a^2}$$

$$\left(x + \frac{b}{2a}\right)^2 = -\frac{4ac}{4a^2} + \frac{b^2}{4a^2} = \frac{b^2 - 4ac}{4a^2}$$

$$x + \frac{b}{2a} = \sqrt{\frac{b^2 - 4ac}{4a^2}} \quad \text{or} \quad x + \frac{b}{2a} = -\sqrt{\frac{b^2 - 4ac}{4a^2}}$$

$$x + \frac{b}{2a} = \frac{\sqrt{b^2 - 4ac}}{|2a|} \quad \text{or} \quad x + \frac{b}{2a} = \frac{-\sqrt{b^2 - 4ac}}{|2a|}$$

When $a > 0$, $|2a| = 2a$, so

$$x + \frac{b}{2a} = \frac{\sqrt{b^2 - 4ac}}{2a} \quad \text{or} \quad x + \frac{b}{2a} = \frac{-\sqrt{b^2 - 4ac}}{2a}$$

When $a < 0$, $|2a| = -2a$, so

$$x + \frac{b}{2a} = \frac{-\sqrt{b^2 - 4ac}}{2a} \quad \text{or} \quad x + \frac{b}{2a} = \frac{\sqrt{b^2 - 4ac}}{2a}$$

In either case $x + \frac{b}{2a} = \pm\frac{\sqrt{b^2 - 4ac}}{2a}$, so

$$x = -\frac{b}{2a} + \frac{\sqrt{b^2 - 4ac}}{2a} \quad \text{or} \quad x = -\frac{b}{2a} - \frac{\sqrt{b^2 - 4ac}}{2a}$$

Or,

$$x = \frac{-b \pm \sqrt{b^2 - 4ac}}{2a}$$

This is known as the *quadratic formula*. You should memorize it.

When the expression $b^2 - 4ac$ is negative the equation has no solutions, except for so-called "imaginary" numbers, to be studied in a later course. It is helpful when using the quadratic formula to first find the standard form so that the coefficients a, b, and c can be determined easily.

Example 1. Solve $5x^2 - 8x = -3$.

First find the standard form and determine a, b, and c:

$5x^2 - 8x + 3 = 0$

$a = 5, b = -8, c = 3$

Then use the quadratic formula:

$$x = \frac{-b \pm \sqrt{b^2 - 4ac}}{2a}$$

$$x = \frac{-(-8) \pm \sqrt{(-8)^2 - 4 \cdot 5 \cdot 3}}{2 \cdot 5}$$

$$x = \frac{8 \pm \sqrt{64 - 60}}{10}$$

$$x = \frac{8 \pm \sqrt{4}}{10}$$

$$x = \frac{8 \pm 2}{10}$$

$$x = \frac{8 + 2}{10} \quad \text{or} \quad x = \frac{8 - 2}{10}$$

$$x = \tfrac{10}{10} = 1 \quad \text{or} \quad x = \tfrac{6}{10} = \tfrac{3}{5}$$

Check: for $\tfrac{3}{5}$ $\qquad 5x^2 - 8x = -3$

$$
\begin{array}{c|c}
5(\tfrac{3}{5})^2 - 8(\tfrac{3}{5}) & -3 \\
\dfrac{5 \cdot 9}{25} - \dfrac{24}{5} & \\
\dfrac{9}{5} - \dfrac{24}{5} & \\
-\dfrac{15}{5} & \\
-3 &
\end{array}
$$

The solutions are 1 and $\tfrac{3}{5}$.

Do exercise 14 at the right.

14. Solve $2x^2 = 4 - 7x$ using the quadratic formula.

15. Solve $5x^2 - 8x = 3$ using the quadratic formula.

Example 2. Solve $3x^2 = 7 - 2x$.

First find the standard form and determine a, b, and c:

$$3x^2 + 2x - 7 = 0$$

$$a = 3, b = 2, c = -7$$

Then use the quadratic formula:

$$x = \frac{-b \pm \sqrt{b^2 - 4ac}}{2a}$$

$$x = \frac{-2 \pm \sqrt{2^2 - 4 \cdot 3 \cdot (-7)}}{2 \cdot 3}$$

$$x = \frac{-2 \pm \sqrt{4 + 84}}{6}$$

$$x = \frac{-2 \pm \sqrt{88}}{6}$$

$$x = \frac{-2 \pm \sqrt{4 \cdot 22}}{6}$$

$$x = \frac{-2 \pm 2\sqrt{22}}{6} = \frac{2(-1 \pm \sqrt{22})}{2 \cdot 3} = \frac{-1 \pm \sqrt{22}}{3}$$

16. Use the square root table to approximate the solutions to Exercise 15. Round to the nearest hundredth.

The solutions are $\dfrac{-1 + \sqrt{22}}{3}$ and $\dfrac{-1 - \sqrt{22}}{3}$.

When such solutions arise in a practical problem we can find approximations using the square root table as illustrated below.

$$\frac{-1 + \sqrt{22}}{3} \approx \frac{-1 + 4.690}{3}. \qquad\qquad \frac{-1 - \sqrt{22}}{3} \approx \frac{-1 - 4.690}{3}$$

$$\approx \frac{3.69}{3} \qquad\qquad\qquad\qquad \approx \frac{-5.69}{3}$$

$$\approx 1.23 \text{ to the nearest} \qquad\qquad \approx -1.90 \text{ to the nearest}$$
$$\text{hundredth} \qquad\qquad\qquad\qquad\qquad \text{hundredth}$$

Do exercises 15 and 16 at the left.

Do exercise set 11.3, p. 401.

11.4 FRACTIONAL EQUATIONS

We can solve some fractional equations by first deriving a quadratic equation.

Example. Solve $\dfrac{3}{x-1} + \dfrac{5}{x+1} = 2$

L.C.M. $= (x-1)(x+1)$

$(x-1)(x+1)\dfrac{3}{x-1} + (x-1)(x+1)\dfrac{5}{x+1} = 2(x-1)(x+1)$

$3(x+1) + 5(x-1) = 2(x-1)(x+1)$

$3x + 3 + 5x - 5 = 2(x^2 - 1) = 2x^2 - 2$

$-2x^2 + 8x = 0$

$x(-2x + 8) = 0$ (factoring)

$x = 0$ or $-2x + 8 = 0$ (principle of zero products)

$x = 0$ or $-2x = -8$

$x = 0$ or $x = 4$

Check:

       for 0:   for 4:

$$\frac{3}{x-1} + \frac{5}{x+1} = 2 \qquad\qquad \frac{3}{x-1} + \frac{5}{x+1} = 2$$

$$\frac{3}{0-1} + \frac{5}{0+1} \;\bigg|\; 2 \qquad\qquad \frac{3}{4-1} + \frac{5}{4+1} \;\bigg|\; 2$$

$$\frac{3}{-1} + \frac{5}{1} \qquad\qquad\qquad\quad \frac{3}{3} + \frac{5}{5}$$

$$-3 + 5 \qquad\qquad\qquad\qquad\quad 1 + 1$$

$$2 \qquad\qquad\qquad\qquad\qquad\quad 2$$

Both numbers check. The solutions are 0 and 4.

Do exercise 17 at the right.

Do exercise set 11.4, p. 403.

You should be able to solve certain fractional equations by first deriving a quadratic equation.

17. Solve

$$\frac{20}{x+5} - \frac{1}{x-4} = 1.$$

You should be able to solve radical equations by first using the principle of squaring to derive a quadratic equation and then solving the quadratic equation.

18. $\sqrt{x + 2} + x = 4$

11.5 RADICAL EQUATIONS

We can solve some radical equations by first using the principle of squaring to find a quadratic equation and then solving the quadratic equation. We must be sure to check.

Example 1. Solve $x - 5 = \sqrt{x + 7}$.

$$(x - 5)^2 = x + 7 \quad \text{(principle of squaring)}$$

$$x^2 - 10x + 25 = x + 7$$

$$x^2 - 11x + 18 = 0$$

$$(x - 9)(x - 2) = 0$$

$$x - 9 = 0 \quad \text{or} \quad x - 2 = 0$$

$$x = 9 \quad \text{or} \quad x = 2$$

Check:

$x - 5 = \sqrt{x + 7}$		$x - 5 = \sqrt{x + 7}$	
$9 - 5$	$\sqrt{9 + 7}$	$2 - 5$	$\sqrt{2 + 7}$
4	4	-3	3

The number 9 checks, but 2 does not. Thus the solution is 9.

Example 2. Solve $\sqrt{27 - 3x} + 3 = x$

$\sqrt{27 - 3x} = x - 3$	(adding -3, to get radical alone on one side)
$27 - 3x = (x - 3)^2$	(principle of squaring)
$27 - 3x = x^2 - 6x + 9$	
$0 = x^2 - 3x - 18$	(it is all right for 0 to be on the left)
$0 = (x - 6)(x + 3)$	(factoring)
$x = 6 \quad \text{or} \quad x = -3$	(principle of zero products)

Check:

$\sqrt{27 - 3x} + 3 = x$		$\sqrt{27 - 3x} + 3 = x$	
$\sqrt{27 - 3 \cdot 6} + 3$	6	$\sqrt{27 - 3 \cdot -3} + 3$	-3
$\sqrt{9} + 3$		$\sqrt{27 + 9} + 3$	
$3 + 3$		$\sqrt{36} + 3$	
6		$6 + 3$	
		9	

There is only one solution, 6.

Do exercise 18 at the left.

Do exercise set 11.5, p. 405.

11.6 FORMULAS

Let us use the principles of equations to solve a given formula for a given letter. We try to get an equation with the letter alone on one side.

Example 1. Solve $v = \sqrt{\dfrac{2gE}{m}}$, for E.

$$v^2 = \frac{2gE}{m}$$

$$mv^2 = 2gE$$

$$\frac{mv^2}{2g} = E$$

Do exercise 19 at the right.

In most formulas the letters represent non-negative numbers, so we usually do not need to use absolute values when taking principal square roots in formulas.

Example 2. Solve $h = \dfrac{V^2}{2g}$, for V.

$$2gh = V^2$$

$$\sqrt{2gh} = V$$

Do exercise 20 at the right.

Example 3. Solve $S = gt + 16t^2$, for t.

$$16t^2 + gt - S = 0$$

$$a = 16,\ b = g,\ c = -S$$

$$t = \frac{-b \pm \sqrt{b^2 - 4ac}}{2a}$$

$$t = \frac{-g \pm \sqrt{g^2 - 4 \cdot 16 \cdot (-S)}}{2 \cdot 16}$$

$$t = \frac{-g \pm \sqrt{g^2 + 64S}}{32}$$

Do exercise 21 at the right.

Do exercise set 11.6, p. 407.

19. Solve $c = \sqrt{\dfrac{E}{m}}$, for E.

20. Solve $A = \pi r^2$, for r.

21. Solve $h = vt + 8t^2$, for t.

You should be able to solve applied problems involving quadratic equations.

22. A rectangular garden is 80 ft by 60 ft. Part of the garden is torn up to install a sidewalk of uniform width around the garden. The new area of the garden is 800 square ft. How wide is the sidewalk?

11.7 APPLIED PROBLEMS

Let us solve some applied problems involving quadratic equations.

Example 1. The frame of a picture is 14 inches by 16 inches around the outside and is of uniform width. Find the width of the frame if 48 square inches of picture shows.

First make a drawing:

Let x represent the width of the frame. Then:

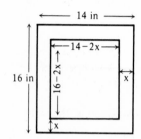

$$(16 - 2x)(14 - 2x) = 48$$

$$224 - 60x + 4x^2 = 48$$

$$4x^2 - 60x + 176 = 0$$

$$x^2 - 15x + 44 = 0$$

$$(x - 11)(x - 4) = 0$$

$$x - 11 = 0 \quad \text{or} \quad x - 4 = 0$$

$$x = 11 \quad \text{or} \quad x = 4$$

Checking in the original problem we see that 11 is not a solution because when $x = 11$, $16 - 2x = -6$, and the width of the picture cannot be negative. Thus the width of the frame is 4 inches.

Do exercise 22 at the left.

Example 2. The hypotenuse of a right triangle is 5 m long. One leg is 1 m longer than the other. Find the lengths of the legs of the triangle.

First make a drawing:

Let x represent the length of one leg. Then $x + 1$ represents the length of the other leg.

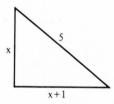

Then by the Pythagorean property we have:

$$x^2 + (x + 1)^2 = 5^2$$

$$x^2 + x^2 + 2x + 1 = 25$$

$$2x^2 + 2x - 24 = 0$$

$$x^2 + x - 12 = 0$$

$$(x + 4)(x - 3) = 0$$

$$x = -4 \text{ or } x = 3$$

Since the length of a leg cannot be negative, one leg is 3 m long and the other is 4 m long.

Do exercise 23 at the right.

Example 3. The current in a stream travels at a speed of 2 km/hr. A boat travels 24 km upstream and 24 km downstream in a total time of 5 hours. What is the speed of the boat in still water?

First make a drawing:

```
   24 km            Upstream
←───────────────────────────•
   r − 2 km/hr    t₁ time
```

```
   24 km            Downstream
•───────────────────────────→
   r + 2 km/hr    t₂ time
```

Let r represent the speed of the boat in still water. Then when traveling upstream the speed of the boat is $r - 2$. When traveling downstream the speed of the boat is $r + 2$. We see from the drawing that

$$t_1 = \frac{24}{r - 2} \text{ and } t_2 = \frac{24}{r + 2}$$

Since $t_1 + t_2 = 5$ it follows that

$$\frac{24}{r - 2} + \frac{24}{r + 2} = 5$$

Solving for r, we get $r = 10$ or $r = -\frac{2}{5}$. Since speed cannot be negative $-\frac{2}{5}$ cannot be a solution. But 10 checks, so the speed of the boat in still water is 10 km/hr.

Do exercise 24 at the right.

Do exercise 11.7, p. 409.

23. The hypotenuse of a right triangle is 13 cm long. The length of one leg is 7 cm less than the other. Find the lengths of the legs of the triangle.

24. The speed of a boat in still water is 12 km/hr. The boat travels 45 km upstream and 45 km downstream in a total time of 8 hours. What is the speed of the stream? (Hint: let r represent the speed of the stream. Then $12 - r$ is the speed upstream and $12 + r$ is the speed downstream.)

You should be able to:

a) Graph an equation of the type
$y = ax^2 + bx + c$.

b) Given an equation of the type
$y = ax^2 + bx + c$, tell whether its graph opens upward or downward depending on whether a is positive or negative.

c) Given an equation of the type
$0 = ax^2 + bx + c$, solve it by graphing the equation
$y = ax^2 + bx + c$

and finding the x-coordinates of the
x-intercepts.

11.8 GRAPHS OF QUADRATIC EQUATIONS, $y = ax^2 + bx + c$.

Graphs of quadratic equations,
$y = ax^2 + bx + c$ (where $a \neq 0$) are always cup-shaped, like the following. They all have a *line of symmetry* like the dotted line at the right. If you fold on this line the two halves will match. The arrows show that the curve goes on forever.

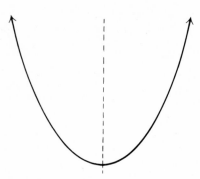

These curves are called *parabolas*. Some parabolas are thin and others are wide, but they all have the same general shape.

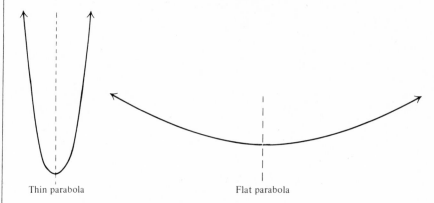

Thin parabola   Flat parabola

Let us graph some quadratic equations. We shall choose some numbers for x and compute the corresponding values of y.

$y = x^2 + 2x - 3$

x	1	0	-1	-2	-3	-4	2
y	0	-3	-4	-3	0	5	5

$y = -2x^2 + 3$

x	0	1	-1	2	-2
y	3	1	1	-5	-5

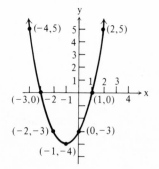

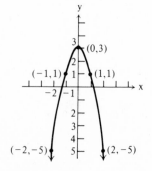

Note that the graph at the left opens upward and the coefficient of x^2 is 1, which is positive. The graph at the right opens downward and the coefficient of the x^2 term is -2, which is negative. We can tell whether a graph opens upward or downward by examining the coefficient of x^2.

Graphs of quadratic equations $y = ax^2 + bx + c$ are all parabolas. They are *smooth* cup-shaped symmetric curves, with no sharp points or kinks in them.

The graph of $y = ax^2 + bx + c$ opens upward if $a > 0$. It opens downward if $a < 0$.

In drawing parabolas, be sure to plot enough points to see the general shape of the graph.

If your graphs look like any of the following, they are incorrect.

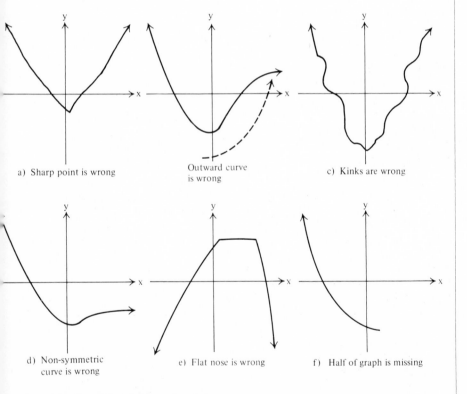

a) Sharp point is wrong

Outward curve is wrong

c) Kinks are wrong

d) Non-symmetric curve is wrong

e) Flat nose is wrong

f) Half of graph is missing

Do exercises 25 and 26 at the right.

Consider the graph on the left, at the bottom of page 394. The x-intercepts of the graph are $(-3, 0)$ and $(1, 0)$. Note also that the second coordinates are 0. Thus the x-coordinates are solutions of the equation $0 = x^2 + 2x - 3$. Thus we can graph an equation $y = ax^2 + bx + c$, and see where the graph crosses the x-axis to find solutions of the equation $0 = ax^2 + bx + c$. From the graph on the right at the bottom of page 394 we can find approximate x-intercepts; $(-1.3, 0)$ and $(1.3, 0)$. So the solutions of $0 = -2x^2 + 3$ are about -1.3 and 1.3.

25. a) Without graphing tell whether the graph of

$y = x^2 + 6x + 9$

opens upward or downward.

b) Now use graph paper and graph the equation.

26. a) Without graphing tell whether the graph of the equation

$y = -3x^2 + 6x$

opens upward or downward.

b) Now use graph paper and graph the equation.

27. Solve $x^2 - 3x - 4 = 0$, by graphing. Use graph paper. Be sure to check.

To see if they are solutions we substitute in the equation.

$$-2(1.3)^2 + 3 = -2 \times 1.69 + 3$$
$$= -3.38 + 3$$
$$= -.38$$
$$-2(-1.3)^2 + 3 = -2 \times 1.69 + 3$$
$$= -3.38 + 3$$
$$= -.38$$

We do not get 0, but we are close. So the numbers are approximate solutions of $0 = -2x^2 + 3$.

Example. Solve $x^2 + x - 6 = 0$, by graphing.

We graph the equation $y = x^2 + x - 6$, find its x-intercepts, and find their x-coordinates.

28. Solve $-2x^2 - 5x + 3 = 0$, by graphing. Use graph paper. Be sure to check.

x	0	1	-1	2	-2
y	-6	-4	-6	0	-4

The x-intercepts seem to be

$(-3, 0)$ and $(2, 0)$ so the solutions

of $0 = x^2 + x - 6$ seem to be -3 and 2.

These check.

Do exercises 27 and 28 at the left.

Do exercise set 11.8, p. 411.

EXERCISE SET 11.1 (pp. 381–383)

Write each equation in standard form and determine a, b, c.

1. $x^2 - 3x + 2 = 0$ **2.** $2x^2 - 3 = 0$ **3.** $5 = -2x^2 + 3x$

4. $7x^2 = 4x$ **5.** $2x = x^2 - 5$ **6.** $2x - 1 = 3x^2 - 7$

7. $x^2 - 3x + 2 = 7x - 5$ **8.** $2x^2 - 1 + 6x = 5x^2 - x + 8$

Solve.

9. $-3x^2 = 0$ **10.** $x^2 = 16$ **11.** $y^2 = 11$

12. $2x^2 - 10 = 0$ **13.** $2s^2 - 4s = 0$ **14.** $x^2 + 7x = 0$

15. $3x^2 + 5 = 26$ **16.** $a^2 + 2 = 3$ **17.** $r^2 + 5 = 1$

NAME

CLASS

ANSWERS

1. _____

2. _____

_____

3. _____

_____

4. _____

5. _____

6. _____

7. _____

8. _____

9. _____

10. _____

11. _____

12. _____

13. _____

14. _____

15. _____

16. _____

17. _____

ANSWERS

18. $3x^2 = 144$ **19.** $x^2 = 5x$ **20.** $5y^2 = 3$

18. _____

19. _____

21. $x^2 = \dfrac{1}{2}x$ **22.** $4m^2 - 3 = 9$ **23.** $x^2 = 27$

20. _____

21. _____

24. $x^2 - 16x + 48 = 0$ **25.** $3n^2 - n = 15 - n$

22. _____

23. _____

26. $2x^2 - 13x + 15 = 0$ **27.** $6a^2 + a - 2 = 0$

24. _____

25. _____

26. _____

28. $3b^2 - 10b - 8 = 0$ **29.** $9x^2 + 4 - 15x = 0$

27. _____

28. _____

29. _____

30. $(2x - 3)(x + 1) = 4(2x - 3)$ **31.** $(3x - 1)(2x + 1) = 3(2x + 1)$

30. _____

31. _____

EXERCISE SET 11.2 (pp. 384–385)

Solve by completing the square. Show your work.

1. $x^2 + 6x - 55 = 0$

2. $x^2 + 4x = 21$

3. $x^2 - 9 = -8x$

4. $x^2 - 8x = -7$

5. $x^2 + 3x = \frac{7}{4}$

6. $x^2 - \frac{5}{2}x = -1$

7. $3y^2 - 5y - 2 = 0$

8. $2x^2 - 9x - 5 = 0$

9. $x^2 + 7x + 1 = 0$

10. $3x^2 + 4x = 1$

1.

2.

3.

4.

5.

6.

7.

8.

9.

10.

ANSWERS

11. $6x^2 = 10 - 11x$

12. $4n^2 + 9 = 12n$

11. _____

12. _____

13. $x^2 - \dfrac{2}{3}x - 1 = 0$

14. $2a^2 + 3a = 17$

13. _____

14. _____

15. $2x^2 - 8x = 20$

16. $3x^2 - 4x - 3 = 0$

15. _____

16. _____

17. $5x^2 + 16x + 6 = 0$

18. $2a^2 - 5a - 12 = 0$

17. _____

18. _____

19. _____

19. $3b^2 + 5b - 1 = 0$

20. $3x^2 + 2x - 3 = 0$

20. _____

NAME

CLASS

ANSWERS

EXERCISE SET 11.3 (pp. 386–388)

Solve using the quadratic formula.

1. $3x^2 - 2x = 8$ **2.** $x^2 - 4x = 21$

3. $3x^2 - 5x - 4 = 0$ **4.** $4x^2 + 12x = 7$

5. $3x^2 = 7x - 4$ **6.** $2x^2 - 5x = 1$

7. $3x^2 - 4x - 2 = 0$ **8.** $3n^2 + 8n + 2 = 0$

9. $y^2 - 3y = 5$ **10.** $2x^2 + 3x = 1$

1. _____

2. _____

3. _____

4. _____

5. _____

6. _____

7. _____

8. _____

9. _____

10. _____

ANSWERS

11. $5x^2 = 3 + 11x$

12. $4x^2 - 4x - 1 = 0$

11.

_____

12.

_____

13. $x^2 - 2x - 2 = 0$

14. $5x + x(x - 7) = 0$

13.

_____

14.

15. $3 - x(x - 3) = 4$

16. $5x^2 - 7x = 1$

_____

15.

_____

16.

17. $(x + 4)(x + 3) = 15$

18. $(x + 5)(x - 1) = 27$

_____

17.

_____

18.

_____

Use the table on p. 423 to approximate the solutions to the nearest hundredth.

19. $3x^2 + 4x - 1 = 0$

20. $3x^2 - 10x = -2$

19.

_____

20.

_____

EXERCISE SET 11.4 (p. 389)

Solve.

1. $\dfrac{2}{y+1} = \dfrac{1}{y-2}$

2. $\dfrac{3}{y+1} = \dfrac{5}{y+3}$

3. $1 + \dfrac{12}{x^2-4} = \dfrac{3}{x-2}$

4. $\dfrac{5}{t-3} - \dfrac{30}{t^2-9} = 1$

5. $\dfrac{1}{x-1} + 3 = \dfrac{1}{x-1} - 1$

6. $\dfrac{8}{r+2} + \dfrac{8}{r-2} = 3$

1. _____

2. _____

3. _____

4. _____

5. _____

6. _____

ANSWERS

7. $\dfrac{n-5}{n-4} = \dfrac{3n}{3n+1}$

8. $\dfrac{x^2}{x+3} - \dfrac{5}{x+3} = 0$

7.

_____

8.

_____

9. $\dfrac{1}{y} + \dfrac{2}{y+1} = \dfrac{3}{y+2}$

10. $\dfrac{-x+4}{x-4} + \dfrac{x+3}{x-3} = 0$

9.

_____

10.

_____

11. $\dfrac{8}{x+1} + \dfrac{x}{1-x} = \dfrac{-2}{x^2-1}$

12. $\dfrac{y+2}{y} = \dfrac{1}{y+2}$

11.

_____

12.

_____

13. $\dfrac{6}{x-1} = \dfrac{12}{x^2-1} - 2$

14. $\dfrac{r}{r-1} + \dfrac{2}{r^2-1} = \dfrac{8}{r+1}$

13.

_____

14.

_____

NAME

CLASS

ANSWERS

EXERCISE SET 11.5 (p. 390)

Solve.

1. $2 - y = 2\sqrt{y + 1}$

2. $\sqrt{x - 5} = x - 7$

3. $2\sqrt{a - 1} = a - 1$

4. $\sqrt{27 - 3x} = x - 3$

5. $\sqrt{5x + 21} = x + 3$

6. $x + 4 = 4\sqrt{x + 1}$

1. _____

2. _____

3. _____

4. _____

5. _____

6. _____

7. $x - 1 = 6\sqrt{x - 9}$

8. $\sqrt{x^2 + 6} - x + 3 = 0$

9. $-3\sqrt{x + 6} = x + 2$

10. $\sqrt{(p + 6)(p + 1)} - 2 = p + 1$

11. $\sqrt{(4x + 5)(x + 4)} = 2x + 5$

12. $x - 9 = \sqrt{x - 3}$

13. $\sqrt{2x - 1} = x - 2$

14. $\sqrt{x + 7} = x - 5$

EXERCISE SET 11.6 (p. 391)

Solve for the indicated variable.

1. $c = \sqrt{a^2 + b^2}$; a

2. $v = \frac{1}{2}\sqrt{1 + \frac{T}{l}}$; l

3. $S = 4\pi r^2$; r

4. $F = \frac{Wv^2}{gr}$; v

5. $\frac{1}{d} = \sqrt{\frac{6F}{d} - 3}$; d

6. $\pi r^2 + 2\pi rh - A = 0$; r

1.

2.

3.

4.

5.

6.

7. $V = \pi r^2 h$; r

8. $x^2 = 9a^2$; a

7.

_____

8.

_____

9. $s = \frac{1}{2}gt^2$; t

10. $mx^2 = 1$; x

9.

_____

10.

_____

11. $\frac{1}{d} = \sqrt{\frac{6F}{d} - 3}$; F

12. $k = \frac{4ac - b^2}{4a}$; b

11.

_____

12.

_____

13. $F = h^2 + k^2 - r^2$; solve for the positive value of r.

14. $b = \frac{c}{e}\sqrt{1 - e^2}$; e. Do not rationalize the denominator.

13.

_____

14.

_____

NAME

CLASS

ANSWERS

EXERCISE SET 11.7 (pp. 392–393)

Solve.

1. If n represents the lesser of two consecutive integers, then $n + 1$ represents the greater integer. What is n, if the product of the integers is 132?

1.

2. The number of diagnals of a polygon is given by the formula, $d = \dfrac{n^2 - 3n}{2}$, where d is the number of diagonals for a polygon of n sides. If a polygon has 27 diagonals, how many sides does it have?

2.

3. A 10 km trip upstream took Jack $3\frac{1}{3}$ hrs longer than his return trip downstream. His rowing speed is 4 km/hr in still water. What was the speed of the current?

3.

4. The school auditorium has 5152 seats. There are 36 more seats per row than there are rows of seats. How many rows of seats are in the auditorium?

4.

ANSWERS

5. One number exceeds another by 8. The square of the larger number exceeds the square of the smaller by 208. Find the numbers.

5. _____

6. The hypotenuse of a right triangle is 50 centimeters long. One leg of the triangle is 10 centimeters longer than the other. Find the lengths of the legs.

6. _____

7. The area of a rectangle is 78 ft². The length is 1 ft greater than twice the width. Find the dimensions of the rectangle.

7. _____

8. The square of a certain number is 45 more than 4 times the number. Find the number.

8. _____

9. A rectangular lawn is 40 m by 60 m. The grass is cut in a uniform strip around the edge so that one-third of the grass remains uncut. How wide is the strip?

9. _____

NAME

CLASS

ANSWERS

EXERCISE SET 11.8 (pp. 394–396)

Determine whether the following graphs open upward or downward.

1. $y = 7 + 4x - x^2$ **2.** $y = 3x^2 + 9x + 31$

3. $y - 25 = x^2 + 8x$ **4.** $y + x^2 + 10x = -27$

5. $y + 4x = -(5 + x^2)$ **6.** $y = 2x^2 - 4x - 3$

Graph these equations. Use graph paper.

7. $y = x^2 + x - 6$ **8.** $y = 8 - x - x^2$

9. $y = 2x^2$ **10.** $y = 2x^2 - 7x + 4$

11. $y = x^2 + 7x - 5$ **12.** $y = -x^2 + 3x$

1.

2.

3.

4.

5.

6.

ANSWERS

13. $y = x^2 - x - 12$ **14.** $y = x^2 - 9$

19.

15. $y = 2x^2 - x + 3$ **16.** $y = 2x^2 + 3x - 5$

20.

17. $y = 6 - x^2 - x$ **18.** $y = \dfrac{x^2}{2}$

Approximate the solutions of the following equations by solving them graphically.

21.

19. $x^2 - x - 2 = 0$ **20.** $2x^2 - 3x = 5$

21. $3x^2 - 8x = 3$ **22.** $2x^2 - x = 2$

22.

NAME_____

CHAPTER 11 TEST

CLASS_____SCORE_____GRADE_____

Before taking the test *be sure* to allow a day or so for review. Use the objectives listed in the margins to guide your study. The test will evaluate your progress and aid your preparation for a possible classroom test. Allow about an hour for the test. Remove the test from the book. When you finish read the test analysis on the answer page at the end of the book.

ANSWERS

1. _____

Solve by any method.

2. _____

1. $x^2 + 2x = 48$

2. $x^2 + x - 5 = 0$

3. _____

4. _____

3. $4x^2 - 8x + 1 = 0$

4. $x^2 + 6x = 91$

5. _____

5. $x^2 - 2x = 10$

6. $2y^2 - 10y = -9$

6. _____

7. _____

7. $3y^2 + 5y = 2$

8. $0 = x^2 + 6x - 9$

8. _____

9. $\sqrt{-x + 22} = x - 2$

10. $\sqrt{2r^2 - 5r + 7} = r + 1$

9. _____

10. _____

11. $\dfrac{10}{4-r} - \dfrac{10}{4+r} = 3\dfrac{1}{3}$

11. _____

12. Solve for T: $v = \dfrac{1}{2}\sqrt{1 + \dfrac{T}{l}}$

12. _____

13. A diagonal of a rectangle is 100 centimeters long and the width is 30 centimeters. Find the length.

13. _____

14. One automobile travels 240 miles in 1 hour less time than a second automobile by going 12 miles an hour faster. Find the speed of the slower automobile.

14. _____

Use graph paper.

16.

15. Graph: $y = x^2 - 4x - 2$

16. Solve graphically: $x^2 - 2x = 15$

_____

NAME_____

FINAL EXAMINATION

CLASS_____SCORE_____GRADE_____

Before taking the final examination *be sure* to allow several days for review. Use the Objectives and Chapter Tests to guide your study. The final exam will evaluate your learning and aid your preparation for a possible classroom final exam. Remove the exam from the book. When you finish read the test analysis on the answer page at the end of the book.

Chapter 1

1. Write an expanded numeral, with exponents, for 67,305.

2. What does x^3 represent if x stands for 2?

3. Write a fractional numeral for 13.6.

4. Write a decimal numeral for 32.7%.

5. Name $\frac{17}{25}$ using %.

6. What is the reciprocal of $\frac{3}{x}$?

Chapter 2

7. Which symbol, $<$ or $>$, should be inserted to make a true statement?

$$-7 \qquad -3$$

8. Simplify $|-7|$.

9. Add.

$$-6 + 12 + (-4) + 7$$

10. Subtract.

$$2.8 - (-12.2)$$

ANSWERS

1. _____

2. _____

3. _____

4. _____

5. _____

6. _____

7. _____

8. _____

9. _____

10. _____

ANSWERS

11. _____

12. _____

13. _____

14. _____

15. _____

16. _____

17. _____

18. _____

19. _____

20. _____

11. Divide and simplify.

$$-\frac{3}{8} \div \frac{5}{2}$$

12. Multiply.

$$(-3)\left(\frac{2}{3}\right)(-2)(-1)\left(-\frac{1}{2}\right)$$

13. Rename, using a negative exponent.

$$\frac{1}{3^4}$$

14. Simplify.

$$x^{-6} \cdot x^2$$

15. Simplify.

$$\frac{y^3}{y^{-4}}$$

Chapter 3

16. Solve.

$$3x = -24$$

17. Solve.

$$3x + 7 = 2x - 5$$

18. Solve.

$$3(y - 1) - 2(y + 2) = 0$$

19. Solve for t.

$$A = \frac{4b}{t}$$

20. The sum of two consecutive integers is 49. What are the integers?

Chapter 4

21. Collect like terms and arrange in descending order.

$$2x - 3 + 5x^3 - 2x^3 + 7x^3 + x$$

22. Add.

$$4x^3 + 3x^2 \qquad - 5$$
$$3x^4 \qquad - 5x^2 + 4x - 12$$
$$-7x^3 + 12x^2 - 6x + 4$$

23. Subtract.

$$(6x^2 - 4x + 1) - (-2x^2 + 7)$$

24. Multiply.

$$-2x^2(4x^2 - 3x + 1)$$

25. Multiply.

$$(2x - 3)(3x^2 - 4x + 2)$$

Chapter 5

26. Factor completely.

$$8x^2 - 4x$$

27. Factor completely.

$$25x^2 - 4$$

28. Factor completely.

$$6x^2 - 5x - 6$$

29. Factor completely.

$$x^2 - 8x + 16$$

30. Factor completely.

$$x(8 - x) - x^2(8 - x)$$

21.

_____

22.

_____

23.

_____

24.

_____

25.

_____

26.

_____

27.

_____

28.

_____

29.

_____

30.

_____

31.

_____

32.

_____

33.

_____

34.

_____

35.

_____

36.

_____

37.

_____

38.

_____

39.

_____

40.

_____

31. What must be added to $x^2 + 20x$ to make it a perfect square trinomial?

32. Solve.

$$x^2 - 8x + 15 = 0$$

33. One less than a certain number times two more than that number is 28. Find the number.

34. The square of a certain positive number plus the number itself is 20. Find the number.

Chapter 6

35. In which quadrant is the graph of the point $(-2, 3)$?

36. What is the x-intercept of

$$2x + 3y = 6?$$

37. Complete a table for

$$y = x^2 - 2$$

x	-2	-1	0	1	2
y					

38. What is the y-intercept of

$$y = 3x - 2?$$

39. Which of these equations are not linear?

a) $2y + 3x = 2x + 1$

b) $2x + 3xy = 5$

c) $4x^2 = 2y^2 + 1$

d) $\dfrac{3}{x} = 2y + 5$

40. Solve graphically. Attach your graph paper to this sheet.

$$y - x = 1$$

$$y = -x + 3$$

Chapter 7

41. Which pairs of lines are parallel?

a) $x = 3$
   $y = 3$

b) $4x - 6y = 12$
   $10x - 6 = 15y$

c) $2x + 4y = 5$
   $2x + 4y = 10$

d) $y = 3$
   $y = -5$

42. What is the x-intercept of $3y - 2x = 6$?

41. _____

42. _____

43. Solve.

$x + y = 17$
$x - y = 7$

44. Solve.

$4x - 3y = 3$
$3x - 2y = 4$

43. _____

44. _____

45. How far from the fulcrum of a seesaw should a 60-kg weight be placed to balance a 100-kg weight which is 60 cm from the fulcrum?

46. The sum of two numbers is 24. Three times the first number minus the second number is 20. Find the numbers.

45. _____

46. _____

Chapter 8

47. _____

47. What is the degree of the first term of

$3x^2y^3 - 2xy^2 + 5xy - 4$?

48. Arrange in descending powers of x.

$x^2 + 3ax^3 + 5a^5x^4 + 2a$

48. _____

49. Combine like terms.

$5x^2yz^3 - 3xy^2 + 2x^2yz^3 + 5x^2y^2 - 2xy^2$

50. Combine like terms.

$4(7x + 3y) + 2(3y - 2x) - 3(x + y)$

49. _____

50. _____

51.

_____

52.

_____

53.

_____

54.

_____

55.

_____

56.

_____

57.

_____

58.

_____

59.

_____

60.

_____

51. Add.

$$(x^3y - 5x^2y^2 + xy^3 + 2)$$
$$+ (x^3y - 2x^2y^2 + 4)$$
$$+ (10 - xy + 2x^2y^2)$$

52. Subtract.

$$(15x^2y^3 + 10xy^2 + 5)$$
$$- (5xy^2 - x^2y^2 - 2)$$

53. Multiply.

$$(x^2 - 2y)(x^2 + 2y)$$

54. Multiply.

$$(3x + 4y^2)(2x - 3y^2)$$

55. Multiply.

$$3x^2(3x + 4y - 5)$$

56. Multiply.

$$(x - 2y + 3)(x - 2y + 3)$$

57. Factor.

$$25a^2b^2 - 1$$

58. Factor.

$$(x - y)^2 - 25$$

59. Factor.

$$9x^2 + 30xy + 25y^2$$

60. Factor.

$$15x^2 + 14xy - 8y^2$$

61. Factor.

$(a - b)^2 + 3(a - b) - 10$

62. Factor.

$2ac - 6ab - 3db + dc$

61. _____

62. _____

63. Factor.

$16x^4 - y^8$

64. Multiply.

$[(a - b) - 3][(a - b) + 1]$

63. _____

64. _____

65. Factor.

$3a^4 + 6a^2 - 72$

66. Solve for r.

$$\frac{A}{e} = \frac{r + R}{R}$$

65. _____

66. _____

Chapter 9

67. Multiply and simplify.

$$\frac{4}{2x - 6} \cdot \frac{x + 2}{3 - x}$$

68. Divide and simplify.

$$\frac{3a^4}{a^2 - 1} \div \frac{2a^3}{a^2 - 2a + 1}$$

67. _____

68. _____

69. Add.

$$\frac{3}{3x - 1} + \frac{4}{5x}$$

70. Subtract.

$$\frac{2}{x^2 - 16} - \frac{x - 3}{x^2 - 9x + 20}$$

69. _____

70. _____

71.

_____

72.

_____

73.

_____

74.

_____

75.

_____

76.

_____

77.

_____

78.

_____

80.

_____

71. Solve.

$$\frac{3y + 2}{2} + \frac{3y + 3}{3} = 7$$

72. Solve.

$$\frac{8}{x} - \frac{3}{x + 5} = \frac{5}{3}$$

Chapter 10

73. Simplify.

$$\sqrt{50}$$

74. Simplify.

$$\frac{\sqrt{72}}{\sqrt{45}}$$

75. Simplify.

$$4\sqrt{12} + 2\sqrt{12}$$

76. Solve.

$$3 - x = \sqrt{x^2 - 3}$$

Chapter 11

77. Solve.

$$x^2 - x - 6 = 0$$

78. Solve.

$$x^2 + 3x = 5$$

79. Graph.

$$y = x^2 + 2x + 1$$

Attach your graph to this paper.

80. The length of a rectangle is 7 more than the width. The length of a diagonal is 13. Find the length.

SQUARE ROOT APPROXIMATIONS

N	$\sqrt{N}$	N	$\sqrt{N}$
2	1.414	51	7.141
3	1.732	52	7.211
4	2	53	7.280
5	2.236	54	7.349
6	2.450	55	7.416
7	2.646	56	7.483
8	2.828	57	7.550
9	3	58	7.616
10	3.162	59	7.681
11	3.317	60	7.746
12	3.464	61	7.810
13	3.606	62	7.874
14	3.742	63	7.937
15	3.873	64	8
16	4	65	8.062
17	4.123	66	8.124
18	4.243	67	8.185
19	4.359	68	8.246
20	4.472	69	8.307
21	4.583	70	8.367
22	4.690	71	8.426
23	4.796	72	8.485
24	4.899	73	8.544
25	5	74	8.602
26	5.099	75	8.660
27	5.196	76	8.718
28	5.292	77	8.775
29	5.385	78	8.832
30	5.477	79	8.888
31	5.568	80	8.944
32	5.657	81	9
33	5.745	82	9.055
34	5.831	83	9.110
35	5.916	84	9.165
36	6	85	9.220
37	6.083	86	9.274
38	6.164	87	9.327
39	6.245	88	9.381
40	6.325	89	9.434
41	6.403	90	9.487
42	6.481	91	9.540
43	6.557	92	9.592
44	6.633	93	9.644
45	6.708	94	9.695
46	6.782	95	9.747
47	6.856	96	9.798
48	6.928	97	9.849
49	7	98	9.899
50	7.071	99	9.950
		100	10

APPENDIX THE METRIC SYSTEM

The *metric* system is used in most countries of the world. It differs from the American system of inches, feet, pounds, and so on. In fact, the United States is fast moving toward complete use of the metric system. An advantage of the metric system is that it is easier to convert from one unit to another. This is because it is based on the number 10.

LENGTH

The basic unit of length in the metric system is the *meter*. It is about 39.37 inches long, or just over a yard. The other units of length in the metric system are multiples of the length of a meter (10 times a meter, 100 times a meter, 1000 a meter, and so on) or fractions of the length of a meter ($\frac{1}{10}$ of a meter, $\frac{1}{100}$ of a meter, $\frac{1}{1000}$ of a meter, and so on.

1 *kilo*meter (km)	=	1000 meters (m)
1 *hecto*meter (hm)	=	100 meters (m)
1 *deka*meter (dam)	=	10 meters (m)
1 *deci*meter (dm)	=	$\frac{1}{10}$ meter (m)
1 *centi*meter (cm)	=	$\frac{1}{100}$ meter (m)
1 *milli*meter (mm)	=	$\frac{1}{1000}$ meter (m)

These names and abbreviations should be memorized. Think of *kilo*-, for 1000, *hecto*-, for 100, and so on. We will use these abbreviations again later. Here is a segment whose length is 1 cm. One inch is 2.54 cm.

_____   _____

1 cm   2.54 cm (1 in)

Go back to the inside front cover of this text and find other comparisons of Metric and American units of length.

You need not memorize these comparisons!

They are meant to give you an idea of metric lengths.

Do the familiarization exercises on the inside front cover.

CHANGING METRIC UNITS

To change units we can make substitutions, treating unit names just as we treat symbols for numbers.

Example 1. Complete. 4 km = ___ m

4 km = 4·1 km
    = 4·1000 m   (substituting 1000 m for 1 km)
    = 4000 m

Do exercises 1 through 3 at the right.

OBJECTIVES

You should be able to change from one metric unit of

a) Length to another.

b) Area to another.

c) Capacity to another.

d) Weight to another.

Complete.

1. 23 km = ___ m

2. 4 hm = ___ m

3. 78 dam = ___ m

Complete.

4. 23 dm = _____ m

5. 67 cm = _____ m

6. 478 m = _____ dm

7. 478 m = _____ cm

Complete.

8. 6755 m = _____ km

9. 6755 m = _____ hm

Example 2. Complete. 3 dm = _____ m

First note that 10 m = 1 dm, so we can divide by 10 and get

$1 \text{ m} = \dfrac{1}{10} \text{ dm.}$

Then

$3 \text{ dm} = 3 \cdot 1 \text{ dm}$

$\qquad = 3 \cdot \dfrac{1}{10} \text{ m} \qquad \left(\text{substituting } \dfrac{1}{10} \text{ m for } 1 \text{ dm} \right)$

$\qquad = \dfrac{3}{10} \text{ m}$

$\qquad = .3 \text{ m.}$

Example 3. Complete. 432 m = _____ dm

$432 \text{ m} = 432 \cdot 1 \text{ m}$

$\qquad = 432 \cdot 10 \text{ dm} \qquad \left(\text{since } 1 \text{ dm} = \dfrac{1}{10} \text{ m, } 10 \text{ dm} = 1 \text{ m, so we can} \right.$

$\left. \qquad\qquad\qquad\qquad \text{substitute } 10 \text{ dm for } 1 \text{ m} \right)$

$\qquad = 4320 \text{ dm.}$

Example 4. Complete. 93.4 m = _____ mm

$93.4 \text{ m} = 93.4 \cdot 1 \text{ m}$

$\qquad = 93.4 \cdot 1000 \text{ mm} \qquad \left(\text{since } 1 \text{ mm} = \dfrac{1}{1000} \text{ m, } 1000 \text{ mm} = 1 \text{ m,} \right.$

$\left. \qquad\qquad\qquad\qquad \text{so we substitute } 1000 \text{ mm for } 1 \text{ m} \right)$

$\qquad = 93{,}400 \text{ mm.}$

Do exercises 4 through 7 at the left.

Sometimes the property of 1 for multiplication helps in making conversions. We know, for example, that 1 km = 1000 m, so

$\dfrac{1 \text{ km}}{1000 \text{ m}} = 1 \qquad \text{and} \qquad \dfrac{1000 \text{ m}}{1 \text{ km}} = 1.$

Example 5. Complete. 2347 m = _____ km

$2347 \text{ m} = 2347 \text{ m} \cdot \dfrac{1 \text{ km}}{1000 \text{ m}} \qquad \left(\text{we multiply by 1, or } \dfrac{1 \text{ km}}{1000 \text{ m}} \right)$

$\qquad = \dfrac{2347 \text{ m}}{1000 \text{ m}} \cdot 1 \text{ km}$

$\qquad = \dfrac{2347}{1000} \cdot \dfrac{\text{m}}{\text{m}} \cdot 1 \text{ km}$

$\qquad = 2.347 \cdot 1 \text{ km} \qquad \left(\text{the } \dfrac{\text{m}}{\text{m}} \text{ acts like 1, so we omit it} \right)$

$\qquad = 2.347 \text{ km}$

Do exercises 8 and 9 at the left.

Sometimes we multiply by 1 more than once.

Example 6. Complete. 84.3 hm = ____ dm

Method 1.

$$84.3 \text{ hm} = 84.3 \text{ hm} \cdot \frac{100 \text{ m}}{1 \text{ hm}} \cdot \frac{10 \text{ dm}}{1 \text{ m}} \quad \left(\begin{array}{l} 1 \text{ hm} = 100 \text{ m, and } 1 \text{ m} = 10 \text{ dm,} \\ \text{since } \frac{1}{10} \text{ m} = 1 \text{ dm} \end{array} \right)$$

$$= \frac{84.3 \times 100 \times 10}{1} \frac{\text{hm}}{\text{hm}} \cdot \frac{\text{m}}{\text{m}} \cdot \text{dm}$$

$$= 84{,}300 \text{ dm.}$$

Do exercises 10 and 11 at the right.

MENTAL CONVERSION

Perhaps you have noticed that changing from one unit to another in the metric system amounts to only a movement of a decimal point, if decimal numerals are used. This is because the metric system and our system of naming numbers are based on 10. Consider this table.

1000	100	10	1	$\frac{1}{10}$	$\frac{1}{100}$	$\frac{1}{1000}$
km	hm	dam	m	dm	cm	mm

Each place in the table has a value one-tenth that to the left or ten times that to the right. Thus moving one place in the table corresponds to one decimal place. Let us consider Example 6 mentally.

Example 6. Complete. 84.3 hm = ____ dm

Method 2. Think: To go from "hm" to "dm" in the table is a move of 3 places to the right. Thus we move the decimal point 3 places to the right:

84.3 hm = 84,300 dm   (1 hm = 1000 dm)

Example 7. Complete. 786 cm = ____ km

Think: To go from "cm" to "km" in the table is a move of 5 places to the left. Thus we move the decimal point 5 places to the left:

786 cm = .00786 km   (1 cm = .00001 km)

Example 8. Complete. 1 m = ____ cm.

Think: To go from "m" to "cm" in the table is a move of 2 places to the right. Thus we move the decimal point 2 places to the right:

1 m = 100 cm

Make metric conversions mentally as much as possible!

The fact that this can be done so easily is an important advantage of the metric system.

Do exercises 12 through 16 at the right.

Complete.

10. 3.94 hm = ____ dm

11. 67 km = ____ cm

Complete. Try to use the table and mentally move a decimal point.

12. 6780 m = ____ km

13. 9.7 cm = ____ mm

14. 1 mm = ____ cm

15. .26 km = ____ dam

16. 845.1 mm = ____ dm

17. Find the area of a rectangle 4 dm by 7 dm.

AREA

The metric unit of area is a *square centimeter*, shown below is a *square inch* for comparison.

Square Centimeter

1 cm

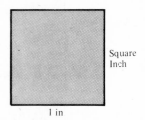

Square Inch

1 in

Example 1. Find the area of a rectangle 4 cm by 3 cm.

We multiply length and width, treating the dimension symbols just as if they were numerals or variables.

18. Find the area of a rectangle 2.4 km by 7.3 km.

$$A = (4 \text{ cm})(3 \text{ cm})$$
$$= 4 \cdot 3 \cdot \text{cm} \cdot \text{cm}$$
$$= 12 \text{ cm}^2$$

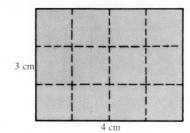

3 cm

4 cm

We read "12 cm²" as "12 square centimeters."

Example 2. Find the area of a square whose sides have length 20.3 m.

19. Find the area of a square whose sides have length 12 km.

$$A = (20.3 \text{ m})(20.3 \text{ m})$$
$$= 20.3 \times 20.3 \times \text{m} \times \text{m}$$
$$= 412.09 \text{ m}^2$$

Do exercises 17 through 20 at the left.

20. Find the area of a square whose sides have length 16.5 cm.

Let's do some conversions from one metric unit of area to another.

Example 1. Complete. 1 km² = _____ m²

$$1 \text{ km}^2 = 1 \cdot (1000 \text{ m})^2 \quad \text{(substituting 1000 m for km)}$$
$$= (1000 \text{ m}) \cdot (1000 \text{ m})$$
$$= 1000 \cdot 1000 \cdot \text{m} \cdot \text{m}$$
$$= 1,000,000 \text{ m}^2$$

Example 2. Complete. 1 m² = _____ cm²

1 m² = 1·(100 cm)²

= (100 cm)·(100 cm)

= 10,000 cm²

Do exercises 21 and 22 at the right.

Some other metric units of area are as follows.

1 *hect*are (ha)	= 100 are (a), or 10,000 m²	(100 m × 100 m)
1 are (a)	= 100 m²	(10 m × 10 m)
1 *cent*are (ca)	= .01 are, or 1 m²	(1 m × 1 m)

The word "are" is pronounced *air*. One hectare is a square hectometer and is about $2\frac{1}{2}$ acres. One centare is about 1 square yard.

Example 3. Complete. 30.4 ha = _____ m²

30.4 ha = 30.4·1 ha

= 30.4·10,000 m²

= 304,000 m²

Do exercises 23 and 24 at the right.

CAPACITY

To answer questions like "How much pop is in your bottle?" we need measures of *capacity*. Instead of ounces, quarts, and gallons, the metric system has a unit called a *liter*. A liter is just a bit more than a quart. More precisely it is 1.057 qt. The liter is defined as follows.

1 liter (ℓ) = 1000 cubic centimeters (1000 cm³, or 1000 cc).

The metric prefixes are also used with liters. The one used the most is

milli. The milliliter (mℓ) is, then, $\frac{1}{1000}$ liter.

Complete.

21. 1 hm² = _____ m²

22. 1 m² = _____ dm²

Complete.

23. 4.7 ha = _____ m²

24. 3800 ca = _____ a

25. Complete.

a) $1\ell =$ _____ $m\ell =$ _____ cc

b) $.001\ell =$ _____ $m\ell =$ _____ cc

Complete.

26. $.97\ell =$ _____ $m\ell$

27. 8990 $m\ell =$ _____ ℓ

Thus

$1\ell = 1000\ m\ell = 1000$ cc

$.001\ \ell = \qquad 1\ m\ell = 1$ cc

A milliliter is about $\frac{1}{5}$ of a teaspoon.

Example 1. Complete. $4.5\ \ell =$ _____ $m\ell$

$$4.5\ \ell = 4.5 \times (1\ \ell)$$
$$= 4.5 \times (1000\ m\ell)$$
$$= 4500\ m\ell$$

Example 2. Complete. $280\ m\ell =$ _____ ℓ

$$280\ m\ell = 280 \times (1\ m\ell)$$
$$= 280 \times \left(\frac{1}{1000}\ell\right)$$
$$= .28\ \ell$$

Do exercises 25 through 27 at the left.

WEIGHT

The *metric system of weights* has a basic unit called a *gram*. It is the weight of 1 cubic centimeter (1 $m\ell$) of water. Since a cubic centimeter is small, a gram is a small unit of weight.

1 gram = weight of 1 cubic centimeter
(1 cc, or 1 $m\ell$) of water.

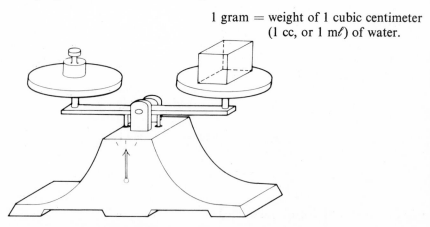

The table given below shows the metric units of weight. Note that the prefixes are the same as those studied before.

1 metric ton (MT or t)	=	1000 kilogram (kg)
1 kilogram (kg)	=	1000 grams (g)
1 hectogram (hg)	=	100 g
1 dekagram (dkg)	=	10 g
1 decigram (dg)	=	$\frac{1}{10}$ g
1 centigram (cg)	=	$\frac{1}{100}$ g
1 milligram (mg)	=	$\frac{1}{1000}$ g

One gram is about the weight of one raisin or 1 paperclip. One kilogram is about 2.2 pounds, so 1 metric ton is 2200 lb, which is just a little more than an American ton.

Example 1. Complete. 8 kg = ___ g

$8 \text{ kg} = 8 \cdot 1 \text{ kg}$

$\qquad = 8 \cdot 1000 \text{ g} \qquad$ (substituting 1000 g for 1 kg)

$\qquad = 8000 \text{ g}$

Example 2. Complete. 4235 g = ___ kg

$4235 \text{ g} = 4235 \text{ g} \cdot \dfrac{1 \text{ kg}}{1000 \text{ g}} \left(\text{multiplying by 1, or } \dfrac{1 \text{ kg}}{1000 \text{ g}} \right)$

$\qquad = \dfrac{4235 \text{ g}}{1000 \text{ g}} \cdot 1 \text{ kg}$

$\qquad = \dfrac{4235}{1000} \cdot \dfrac{\text{g}}{\text{g}} \cdot 1 \text{ kg}$

$\qquad = 4.235 \text{ kg}$

Do exercises 28 through 31 at the right.

Complete.

28. 6.2 km = ___ g

29. 304.8 g = ___ kg

30. 7 hg = ___ g

31. 0.9042 g = ___ hg

Complete.

32. 67 dg = ___ g

33. 678 g = ___ dg

34. 78 hg = ___ cg

Complete. Try to use the table and mentally reposition a decimal point.

35. .76 kg = ___ g

36. 87 kg = ___ g

37. 2344 g = ___ kg

38. 7.7 cg = ___ mg

39. 12.46 mg = ___ cg

40. 1.99 kg = ___ dag

Example 3. Complete. 25 dg = ___ g

$$25 \text{ dg} = 25 \cdot 1 \text{ dg}$$

$$= 25 \cdot \frac{1}{10} \text{ g} \qquad \left(\text{substituting } \frac{1}{10} \text{ g for } \boxed{1 \text{ dg}} \right)$$

$$= \frac{25}{10} \text{ g}$$

$$= 2.5 \text{ g}$$

Example 4. Complete. 6.89 dag = ___ cg

We may multiply by 1 as follows

$$6.89 \text{ dag} = 6.89 \text{ dag} \cdot \frac{10 \text{ g}}{1 \text{ dag}} \cdot \frac{100 \text{ cg}}{1 \text{ g}}$$

$$= \frac{6.89 \times 10 \times 100}{1} \cdot \frac{\text{dag}}{\text{dag}} \cdot \frac{\text{g}}{\text{g}} \cdot \text{cg}$$

$$= 6890 \text{ cg}.$$

Do exercises 32 through 34 at the left.

MENTAL CONVERSION

Perhaps you have noticed again that changing from one metric unit of weight to another amounts to only a repositioning of a decimal point. Consider the following table:

1000	100	10	1	$\frac{1}{10}$	$\frac{1}{100}$	$\frac{1}{1000}$
kg	hg	dag	g	dg	cg	mg

Now let us consider Example 4 again.

Example 4. Complete. 6.89 dag = ___ cg.

Method 2. Think: To go from "dag" to "cg" in the table is a move of 3 places to the right. Thus we move the decimal point 3 places to the right:

6.89 dag = 6890 cg   (1 dag = 1000 cg)

Example 5. Complete. 867.4 mg = ___ hg

Think: To go from "mg" to "hg" in the table is a move of 5 places to the left. Thus we move the decimal point 5 places to the left:

867.4 mg = .008674 hg (1 mg = .00001 hg)

Make metric conversions mentally as much as possible!

Do exercises 35 through 40 at the left.
Do the following exercise set.

NAME

CLASS

ANSWERS

EXERCISE SET

LENGTH

Complete. Do as much as possible mentally.

1. a) 1 km = ――― m

   b) 1 m = ――― km

2. a) 1 hm = ――― m

   b) 1 m = ――― hm

3. a) 1 dam = ――― m

   b) 1 m = ――― dam

4. a) 1 dm = ――― m

   b) 1 m = ――― dm

5. a) 1 cm = ――― m

   b) 1 m = ――― cm

6. a) 1 mm = ――― m

   b) 1 m = ――― mm

7. 6.7 km = ――― m

8. 9 hm = ――― m

9. 98 dam = ――― m

10. .233 dam = ――― m

11. 89.21 km = ――― m

12. 677 hm = ――― m

13. 5666 m = ――― km

14. 5.666 m = ――― hm

15. 5666 m = ――― dam

16. 435 dm = ――― m

17. 477 cm = ――― m

18. 3.45 mm = ――― m

19. 6.88 m = ――― cm

20. 6.88 m = ――― dm

21. 6.88 m = ――― dam

ANSWERS

1. a) _____
   b) _____
2. a) _____
   b) _____
3. a) _____
   b) _____
4. a) _____
   b) _____
5. a) _____
   b) _____
6. a) _____
   b) _____
7. _____
8. _____
9. _____
10. _____
11. _____
12. _____
13. _____
14. _____
15. _____
16. _____
17. _____
18. _____
19. _____
20. _____
21. _____

ANSWERS

22.

_____

23.

_____

24.

_____

25.

_____

26.

_____

27.

_____

28.

_____

29.

_____

30.

_____

31.

_____

32.

_____

33.

_____

34.

_____

35.

_____

36.

_____

37.

_____

Complete. Do as much as possible mentally.

22. 1 mm = _____ cm   **23.** 1 dm = _____ mm

24. 1 km = _____ dam   **25.** 1 hm = _____ cm

26. 1 cm = _____ dm   **27.** 1 m = _____ cm

28. 8.2 cm = _____ mm   **29.** 9.13 km = _____ m

30. 392 dam = _____ km   **31.** 7 km = _____ hm

32. .024 mm = _____ m   **33.** 80,000 mm = _____ dam

34. 2.3 dam = _____ dm   **35.** 6,000,000 m = _____ km

36. 4500 mm = _____ cm   **37.** .013 mm = _____ dm

NAME

CLASS

ANSWERS

AREA

Find the area of each rectangle.

38. 5 m by 10 m   **39.** 2.45 km by 100 km

40. 34.67 cm by 4.9 cm   **41.** 10.1 dm by 20.5 dm

Find the area of each square.

42. 22 m on a side   **43.** 56.9 km on a side

44. 45.5 m on a side   **45.** 17.2 mm on a side

Complete.

46. 1 ha = ___ m²   **47.** 1 ha = ___ a   **48.** 1 ca = ___ a

49. 1 ca = ___ m²   **50.** 1.8 ha = ___ m²   **51.** 20.1 ha = ___ m²

52. 4 ha = ___ m²   **53.** 6 ca = ___ m²   **54.** 8.99 a = ___ m²

55. 7.65 a = ___ m²   **56.** 20 km² = ___ m²   **57.** 65 km² = ___ m²

58. .014 m² = ___ cm²   **59.** .028 m² = ___ dm²

38.
39.
40.
41.
42.
43.
44.
45.
46.
47.
48.
49.
50.
51.
52.
53.
54.
55.
56.
57.
58.
59.

ANSWERS

CAPACITY

Complete.

60.
_____

60. $1\ \ell =$ ___ $m\ell =$ ___ $cm^3 =$ ___ cc

61.
_____

62.
_____

61. ___ $\ell = 1\ m\ell =$ ___ $cm^3 =$ ___ cc

63.
_____

62. $87\ \ell =$ ___ $m\ell$ **63.** $901\ \ell =$ ___ $m\ell$

64.
_____

65.
_____

69. $49\ m\ell =$ ___ ℓ **65.** $17\ m\ell =$ ___ ℓ

66.
_____

67.
_____

66. $.401\ m\ell =$ ___ ℓ **67.** $.013\ m\ell =$ ___ ℓ

68.
_____

68. $7.031\ \ell =$ ___ $m\ell$ **69.** $8.006\ \ell =$ ___ $m\ell$

69.
_____

70.
_____

70. $78.1\ \ell =$ ___ cc   **71.** $49.2\ \ell =$ ___ cc

71.
_____

NAME _____

CLASS _____

WEIGHT

Complete.

72. 1 MT = ___ kg   **73.** 1 kg = ___ g   **74.** 1 hg = ___ g

75. 1 dag = ___ g   **76.** 1 dg = ___ g   **77.** 1 cg = ___ g

78. 1 mg = ___ g   **79.** 1 g = ___ mg   **89.** 1 g = ___ cg

81. 1 g = ___ dg   **82.** 25 kg = ___ g   **83.** 234 hg = ___ g

84. 678 dag = ___ g   **85.** .809 kg = ___ hg   **87.** 67 hg = ___ kg

87. 45 cg = ___ g   **88.** .502 dg = ___ g   **89.** .0025 cg = ___ hg

90. 678 hg = ___ g   **91.** 5677 g = ___ kg   **92.** 6.9 dg = ___ g

93. 4.55 dg = ___ g   **94.** 678 cg = ___ g   **95.** 10 mg = ___ g

96. 7 g = ___ hg   **97.** 8 kg = ___ g   **98.** 1 cg = ___ dag

ANSWERS

72. _____
73. _____
74. _____
75. _____
76. _____
77. _____
78. _____
79. _____
80. _____
81. _____
82. _____
83. _____
84. _____
85. _____
86. _____
87. _____
88. _____
89. _____
90. _____
91. _____
92. _____
93. _____
94. _____
95. _____
96. _____
97. _____
98. _____

ANSWERS

PRETEST ANALYSIS

Below are the answers to the test, along with page numbers (in brackets) showing where to find the text material for each question. If you missed an item, you should study the chapter to which it refers. Should you get all the items for a chapter correct, you could probably skip that chapter after lightly reading it. If you get more than 30 correct answers, you should study the book *Intermediate Algebra: A Modern Approach*, rather than this book.

1. [6] 9 **2.** [12] 11 **3.** [15] $\frac{2}{x}$ **4.** [17] $<$ **5.** [61] $<$ **6.** [61] 12 **7.** [61] -4 **8.** [64] 32 **9.** [63] 3

10. [67] 25 **11.** [68] $-\frac{21}{25}$ **12.** [67] -3 **13.** [74] 4^{-3} **14.** [75] x^{-2} **15.** [75] y^7 **16.** [98] 3 **17.** [99] 2

18. [101] 31, 33 **19.** [126] $-2x^2 + 5x + 2$ **20.** [128] $5x^4 - 4x^2 + 3x - 2$ **21.** [134] $4x^3 + 9x^2 - 14x + 1$
22. [132] $11x^2 - 5x - 2$ **23.** [130] $3y^4 + 9y^2 - 3y + 2$ **24.** [135] $6x^3 + 15x$ **25.** [139] $x^3 - 7x + 6$
26. [142] $y^2 - 9$ **27.** [165] $6(x - 2)$ **28.** [172] $(3y + 1)(3y - 1)$ **29.** [168] $(x - 2)(x + 5)$ **30.** [168] $(2y + 3)(y - 2)$
31. [175] 2, 3 **32.** [176] 5, -5 **33.** [193] IV **34.** [175] 1, 6 **35.** [198] 4 **36.** [198] 6 **37.** [197] See graph below
38. [204] $x = 7, y = 6$ **39.** [207] $x = 4, y = -3$ **40.** [212] 18, 5

MARGIN EXERCISES CHAPTER 1

1. $\frac{4}{6}, \frac{20}{30}$, etc. **2.** $\frac{2}{4}, \frac{5}{10}, \frac{12}{24}$, etc. **3.** $\frac{3}{3}, \frac{5}{5}, \frac{1}{1}$, etc. **4.** $\frac{4}{1}, \frac{20}{5}, \frac{100}{25}$, etc.

5. $\frac{8}{10}, \frac{12}{15}, \frac{16}{20}$, etc. **6.** $\frac{16}{14}, \frac{24}{21}, \frac{32}{28}$, etc. **7.** $\frac{2}{3}$ **8.** $\frac{19}{9}$ **9.** $\frac{1}{2}$ **10.** 4

11. $\frac{54}{125}$ **12.** $\frac{16}{5}$ **13.** $\frac{22}{15}$ **14.** $\frac{29}{36}$ **15.** $\frac{2}{15}$ **16.** $\frac{7}{36}$

17. $50,000 + 2000 + 300 + 70 + 4$, $5 \times 10,000 + 2 \times 1000 + 3 \times 100 + 7 \times 10 + 4$,
$5 \times 10 \times 10 \times 10 \times 10 + 2 \times 10 \times 10 \times 10 + 3 \times 10 \times 10 + 7 \times 10 + 4$
18. $200,000 + 30,000 + 4000 + 800 + 9$, $2 \times 100,000 + 3 \times 10,000 + 4 \times 1000 +$
$8 \times 100 + 9$, $2 \times 10 \times 10 \times 10 \times 10 \times 10 + 3 \times 10 \times 10 \times 10 \times 10 + 4 \times 10 \times$
$10 \times 10 + 8 \times 10 \times 10 + 9$
19. $7 \times 10^5 + 5 \times 10^4 + 2 \times 10^3 + 3 \times 10^2 + 9 \times 10 + 8$
20. $2 \times 10^4 + 3 \times 10^2 + 4 \times 10 + 7$ **21.** $5 \times 5 \times 5 \times 5$ **22.** $x\,x\,x\,x\,x$
23. n^5 **24.** $y\,y\,y$ **25.** 8 **26.** 10,000 **27.** $3 \times 10^3 + 5 \times 10^2 + 6 \times 10^1$

$+ 2 \times 10^0$ **28.** 5; 1 **29.** $20 + 3 + \frac{6}{10} + \frac{7}{100} + \frac{8}{1000}$ **30.** $\frac{2}{10} + \frac{7}{100}$ **31.** $4 + \frac{6}{1000} + \frac{7}{10,000}$ **32.** $\frac{162}{100}$

33. $\frac{35431}{1000}$ **34.** .875 **35.** .8 **36.** $.818\ldots$ or $0.81\overline{81}$ **37.** $2.55\ldots$ or $2.5\overline{5}$ **38.** 27 **39.** 27 **40.** 22 **41.** 30

42. 26 **43.** 30 **44.** 27 **45.** 23 **46.** 25 **47.** Comm., mult. **48.** Assoc., add **49.** Comm., add
50. Comm. and assoc., mult. **51.** a) 28 b) 28 **52.** a) 77 b) 77 **53.** a) 54 b) 54 **54.** 23 **55.** 15 **56.** 16
57. 32 **58.** $4(x + y)$ **59.** $5(a + b)$ **60.** $7(p + q + r)$ **61.** $5(x + y)$, 35 **62.** $7(x + y)$, 49 **63.** $5(x + 2)$
64. $3(4 + x)$ **65.** $3(2x + 4 + 3y)$ **66.** $5(x + 2y + 1)$ **67.** $5y + 15$ **68.** $4x + 8y + 20$ **69.** $am + an + ap$
70. a) $8y$ b) $8x$ c) $1.03x$ **71.** $18p + 9q$ **72.** $11x + 8y$ **73.** $16y$ **74.** $7s + 13w$ **75.** $9x + 10y$ **76.** 7

77. 1 **78.** $\frac{2}{3}$ **79.** 1 **80.** $\frac{11}{4}$ **81.** $\frac{7}{15}$ **82.** $\frac{1}{5}$ **83.** 3 **84.** $\frac{21}{20}$ **85.** $\frac{10}{12}$ or $\frac{5}{6}$ **86.** $\frac{45}{28}$ **87.** $\frac{8}{21}$ **88.** $\frac{12}{35}$

89. $\frac{14}{45}$ **90.** (number line with point at $\frac{6}{5}$, marks 0 1 2 3) **91.** (number line with point at $\frac{17}{18}$, marks 0 1 2 3) **92.** $\frac{1}{2} > \frac{1}{4}$

93. $2.7 < 4.5$ **94.** $\frac{99}{100} < 1$ **95.** $<$ **96.** $=$ **97.** $>$ **98.** $\frac{7}{11}$ **99.** $>$ **100.** $<$ **101.** $>$ **102.** $>$

103. $3 + 2 = 5$, etc.   **104.** $5 + 4 = 1$, etc.   **105.** $x + 3 = 12$, etc.   **106.** $78 + 9$ and $80 + 7$ name the same number.

107. $5 + \frac{1}{2}$ and $9 - 2$ name the same number.   **108.** 3, 2, 15, etc.   **109.** 7   **110.** 6   **111.** 4   **112.** 1

113. $17 - 12 = 5, 17 = 5 + 12$ **114.** $10 - 4 = 5, 10 - 5 = 4$ **115.** $3 - 1 = x, 3 = 1 + x$

116. $10.5 - x = 2, 10.5 - 2 = x$ **117.** $18 = 6 \cdot 3, \frac{18}{3} = 6$ **118.** $\frac{35}{5} = 7, \frac{35}{7} = 5$ **119.** $5 = \frac{35}{7}, 7 = \frac{35}{5}$

120. $\frac{39}{13} = x, 39 = 13x$ **121.** $5 = \frac{27}{x}, x = \frac{27}{5}$ **122.** Yes   **123.** No   **124.** Yes   **125.** No   **126.** No   **127.** No

128. No   **129.** 8   **130.** 7   **131.** 4   **132.** 157   **133.** 2   **134.** 5   **135.** 8   **136.** $\frac{42}{15}$ **137.** $\frac{5}{2}$ **138.** $\frac{29}{24}$

139. 48   **140.** $37 + x = 73, x = 36$ **141.** $44 = \frac{2}{3}x, x = 66$ **142.** $224 - x = \frac{3}{4} \cdot 224$ **143.** $1.5H = .320$

144. .462   **145.** 1   **146.** $\frac{67}{100}$ **147.** $\frac{456}{1000}$ **148.** $\frac{\frac{1}{4}}{100}$ or $\frac{1}{400}$ **149.** 677%   **150.** 99.44%   **151.** 25%

152. 37.5%   **153.** $66\frac{2}{3}$%   **154.** $x = 23\% \times 48 = 11.04$ **155.** $25\% \times 40 = x, x = 10$ **156.** $15 = x\% \times 60, x = 25$

157. $x\% \times 50 = 16, x = 32$ **158.** $45 = 20\%x, x = 225$ **159.** $120\%x = 60, x = 50$
160. $G = 25\% \times 3,615,000 = 903,750$ **161.** $5 = x\% \times 35, x = 14.3$ **162.** $A = 21$ sq. m., $P = 20$ m
163. $A = 25$ sq. m, $P = 20$ m   **164.** 60 cm²   **165.** 38 m²   **166.** 35 cm²   **167.** 64°   **168.** 200 m³   **169.** 18 m
170. 56.52 cm   **171.** 254.34 ft²

EXERCISE SET 1.1, p. 35

1. $\frac{8}{6}$, etc.   **3.** $\frac{12}{22}$, etc.   **5.** $\frac{4}{22}$, etc.   **7.** $\frac{25}{5}, \frac{20}{4}$, etc.   **9.** $\frac{4}{3}$ **11.** $\frac{1}{2}$ **13.** $\frac{2}{1}$ **15.** $\frac{10}{3}$ **17.** $\frac{1}{8}$ **19.** $\frac{51}{8}$ **21.** 1

(Answers will vary)

23. $\frac{7}{6}$ **25.** $\frac{36}{35}$ **27.** 0   **29.** $\frac{5}{18}$ **31.** $\frac{31}{60}$

EXERCISE SET 1.2, p. 37

1. $5000 + 600 + 70 + 7, 5 \times 1000 + 6 \times 100 + 7 \times 10 + 7, 5 \times 10 \times 10 \times 10 + 6 \times 10 \times 10 + 7 \times 10 + 7$
3. $900,000 + 8,000 + 500 + 60 + 3, 9 \times 100,000 + 8 \times 1000 + 5 \times 100 + 6 \times 10 + 3, 9 \times 10 \times 10 \times 10 \times 10 \times 10 + 8 \times$
$10 \times 10 \times 10 + 5 \times 10 \times 10 + 6 \times 10 + 3$ **5.** $3 \times 10^3 + 4 \times 10^2 + 6 \times 10^1 + 6 \times 10^0$ **7.** $4 \times 4 \times 4 \times 4 \times 4$

9. 1   **11.** n^3 **13.** r^2 **15.** $yyyy$ **17.** 10   **19.** 1000   **21.** 27   **23.** 1   **25.** 8   **27.** $10 + 6 + \frac{3}{10} + \frac{3}{100}$

29. $\frac{467}{100}$ **31.** $\frac{31,415}{10,000}$ **33.** .25   **35.** .6   **37.** .444…

EXERCISE SET 1.3, p. 39

1. 22   **3.** 89   **5.** 1000   **7.** Assoc. law of mult.   **9.** Assoc.   **11.** 122   **13.** 15   **15.** 180   **17.** 12
19. $6(x + y)$ **21.** $9(t + e)$ **23.** $10(a + b + c)$ **25.** $4(x + y); 60$ **27.** $10(x + y); 150$ **29.** $3(a + b); 27$
31. $5(p + q); 45$ **33.** $10(p + q); 90$

EXERCISE SET 1.4, p. 41

1. $2(x + 2)$ **3.** $3(x + 2y)$ **5.** $6(x + 4)$ **7.** $5(x + 2 + 3y)$ **9.** $7(1 + 2b + 8w)$ **11.** $d(b + g + h)$
13. $3x + 3$ **15.** $5x + 25$ **17.** $4 + 4y$ **19.** $7x + 28 + 42y$ **21.** $72x + 40y + 640$ **23.** $at + au + av$ **25.** $5x$

27. $11a$ **29.** $8x + 9z$ **31.** $43a + 150c$ **33.** $46x + 8b$ **35.** $33u + 5t$ **37.** $24t + 7h$ **39.** $6t + 8y + 50$ **41.** $3\frac{1}{4}y$

EXERCISE SET 1.5, p. 43

1. 4   **3.** 2   **5.** $\frac{1}{8}$ **7.** $\frac{10}{7}$ **9.** $\frac{55}{56}$ **11.** $\frac{1}{2}$ **13.** $\frac{99}{140}$ **15.** $\frac{136}{18}$ or $\frac{68}{9}$ **17.** $\frac{3}{40}$

19.

21. $=$ **23.** $<$ **25.** $=$ **27.** $>$

29. $<$ **31.** $>$

EXERCISE SET 1.6, p. 45

1. $3 + 2 = 5$, etc.   **3.** $x + 2 = 7$, etc.   **5.** $7 - 2$ and $1 + 4$ name the same number.   **7.** 4   **9.** 8   **11.** 5   **13.** 10

15. 7   **17.** $15 = 7 + 8, 15 - 8 = 7$ **19.** $8 = x + 7, 8 - 7 = x$ **21.** $9 - 4 = x, 9 - x = 4$ **23.** $21 = 7 \cdot 3, \frac{21}{3} = 7$

25. $\frac{56}{7} = 8, \frac{56}{8} = 7$ **27.** $5 = 1 \cdot x, \frac{5}{x} = 1$ **29.** $x = 2 + 17, x - 17 = 2$ **31.** $\frac{45}{3/4} = x, \frac{45}{x} = 3/4$ **33.** Yes   **35.** No

37. No.

EXERCISE SET 1.7, p. 47

1. 8   **3.** 5   **5.** 2   **7.** 35   **9.** 33.5   **11.** 4   **13.** 16   **15.** 96   **17.** .24   **19.** $\frac{13}{40}$ **21.** $\frac{5}{9}$ **23.** 16.09

25. $\frac{8}{15}$ **27.** 3.2

EXERCISE SET 1.8, p. 49

1. $48 = \frac{2}{3} \cdot x; x = 72$ **3.** $15 - x = 7; x = 8$ **5.** $x = 4 + 5; x = 9$ **7.** $1945 - 60 = x$ **9.** $30,160 = 4S$

11. $125 = \frac{2}{3} \cdot x$ **13.** $35 = \frac{2}{5} \cdot P$ **15.** 1885   **17.** 7,540 sq. mi.   **19.** $187\frac{1}{2}$ lb   **21.** $87\frac{1}{2}$ words per min

EXERCISE SET 1.9, p. 51

1. .76   **3.** 1   **5.** 454%   **7.** 75%   **9.** $\frac{20}{100}$ **11.** $\frac{125}{1000}$ **13.** 25%   **15.** $33\frac{1}{3}$%   **17.** $x = 65\% \times 840 = 546$

19. $x\% \times 80 = 100; 125\%$ **21.** $76 = x\% \cdot 88; 86.36\%$ **23.** $85 = 20\% \cdot x; \$425$ **25.** $x = 5\% \times 428.86; \$21.44$
27. $(6\% - 5.4\%) \cdot 9600 = x; \$57.60$

EXERCISE SET 1.10, p. 53

1. $A = 1100$ cm², $P = 138$ cm   **3.** $A = 7500$ sq. m, $P = 350$ m   **5.** 200 in²   **7.** 1.86 m²   **9.** 11.25 sq. ft

11. 326.69 cm²   **13.** 4.8 m   **15.** $1\frac{1}{2}$ in   **17.** 64.06 cm   **19.** 384 m³   **21.** 64.48 cm³   **23.** 75°   **25.** 12.3 cm

27. 7.2

EXERCISE SET 1.11, p. 55

1. .38   **2.** .35   **3.** .721   **4.** .356   **5.** .654   **6.** .826   **7.** .0325   **8.** .0532   **9.** .0824   **10.** .0945
11. .0061   **12.** .0083   **13.** .0043   **14.** .0073   **15.** .00012   **16.** .00023   **17.** .00045   **18.** .00053
19. .000035   **20.** .000041   **21.** 1.25   **22.** 1.35   **23.** 2.4   **24.** 3.2   **25.** 62%   **26.** 73%   **27.** 85%

28. 91%   **29.** 62.3%   **30.** 74.1%   **31.** 81.2%   **32.** 73.2%   **33.** 720%   **34.** 830%   **35.** 350%   **36.** 260%
37. 200%   **38.** 300%   **39.** 400%   **40.** 500%   **41.** 7.2%   **42.** 8.5%   **43.** 1.3%   **44.** 4.5%   **45.** .13%
46. .57%   **47.** .73%   **48.** .68%   **49.** 30/100   **50.** 40/100   **51.** 70/100   **52.** 80/100   **53.** 136/1000
54. 178/1000   **55.** 734/1000   **56.** 825/1000   **57.** 32/1000   **58.** 48/1000   **59.** 84/1000   **60.** 76/1000
61. 120/100   **62.** 140/100   **63.** 250/100   **64.** 370/100   **65.** 35/10,000   **66.** 53/10,000   **67.** 48/10,000
68. 59/10,000   **69.** 42/100,000   **70.** 35/100,000   **71.** 83/100,000   **72.** 74/100,000   **73.** 17%   **74.** 35%
75. 119%   **76.** 173%   **77.** 70%   **78.** 30%   **79.** 80%   **80.** 90%   **81.** 35%   **82.** 55%   **83.** 28%
84. 48%   **85.** 50%   **86.** 150%   **87.** 25%   **88.** 75%   **89.** 60%   **90.** 80%   **91.** 34%   **92.** 62%

93. $33\frac{1}{3}$%   **94.** $66\frac{2}{3}$%   **95.** 37.5%   **96.** 62.5%   **97.** 95   **98.** 150.4   **99.** 31.62   **100.** 12.32   **101.** 25%

102. $33\frac{1}{3}$%   **103.** 175   **104.** 64

CHAPTER 1   TEST ANALYSIS

Below are the answers to the test. Each item you missed should be reviewed. (See page numbers given in brackets with each answer.) If you got 27 correct (your instructor may have another standard), you are prepared to continue after reviewing those items missed. If you missed more than 12 questions, *carefully* review the entire chapter, giving special emphasis to the items missed; then continue.

CHAPTER 1   TEST

1. [3] Answers will vary; $\frac{10}{12}$ etc.   **2.** [4] $\frac{2}{7}$ **3.** [5] $\frac{15}{32}$ **4.** [5] $\frac{19}{24}$ **5.** [5] $\frac{1}{6}$ **6.** [6] $3 \times 10^4 + 4 \times 10^3 + 8 \times 10^2 + 9$

7. [6] $9 \times 9 \times 9 \times 9$ **8.** [7] x^4 **9.** [7] yyy **10.** [7] 125   **11.** [7] $\frac{247}{10}$ **12.** [8] .125   **13.** [11] 130

14. [9] Associative law for addition.   **15.** [7] 126   **16.** [12] $5(3y + 1)$ **17.** [13] $8(3x + 2 + y)$ **18.** [13] $a(x + y + b)$

19. [13] $90x + 30y$ **20.** [13] $63m + 14x + 7$ **21.** [14] $26x + 10b$ **22.** [15] $\frac{7}{8}$ **23.** [16] $\frac{225}{144}$

24. [17]

$$\xrightarrow{\;\;\;|\;\;\;|\;\;\;|\;\;\bullet\;|\;\;\;}$$
$$\begin{array}{cccc} 0 & 1 & 2 & 3 \end{array} \quad \overset{\frac{8}{3}}{}$$

25. [17] $<$ **26.** [19] Answers may vary.   **27.** [23] 3   **28.** [23] 6

29. [26] $x\% \cdot 75 = 5$; $6\frac{2}{3}$%   **30.** [26] $\frac{5}{4}x = 75$; $x = 60$ **31.** [26] .458   **32.** [27] 56%   **33.** [30] 161 m²

34. [32] 30 cm²   **35.** [31] 69 sq. ft   **36.** [28] 80 percent   **37.** [32] 124.3 ft   **38.** [34] 84.91 m²   **39.** [33] 1000 cm³

MARGIN EXERCISES   CHAPTER 2

1. $>$ **2.** $>$ **3.** $<$ **4.** a) 5 b) 3 c) 7 d) 120   **5.** a) 2 b) 7 c) 3 d) 0   **6.** 1   **7.** -2 **8.** 0   **9.** -4
10. -7 **11.** 0   **12.** a) -5 b) 13 c) 12 d) 0   **13.** a) 2 b) 7 c) 0 d) -4 **14.** a) -3 b) 12 c) 5 d) 0
15. -9 **16.** -15 **17.** -15 **18.** -12 **19.** -17 **20.** -19 **21.** 2   **22.** -4 **23.** -2 **24.** -12
25. a), b), c), d)

$$\xrightarrow[\begin{array}{cccccccccc} -4 & -3 & -2 & -1 & 0 & 1 & 2 & 3 & 4 & 5 \end{array}]{\;\;\bullet\;\;\;\bullet\;\;|\;\;|\;\;|\;\;|\;\;|\;\;\bullet\;\;|\;\;\bullet\;\;|\;\;}$$

with labels -3.5, $-\frac{8}{3}$, $\frac{5}{2}$, 4.1

26. -6.2 **27.** $-\frac{2}{9}$ **28.** $-.6$ **29.** a) 3 b) 5 c) -6 **30.** a) 3; 3 b) 10; 10 c) -8; -8 d) -2; -2
31. a) -3 b) -9 c) 7   **32.** a) $(a - b) + b = a$ by definition of subtraction b) $(a + -b) + b = a + (-b + b) = a + 0 = a$
33. Comm., addition   **34.** Assoc., addition   **35.** Property of inverses   **36.** Property of zero
37. $2 \cdot 10 = 20$ $\quad -1 \cdot 10 = -10$ **38.** $4(-3 + 5) = 4(2) = 8$ **39.** -15 **40.** -28 **41.** -30
$\quad\;\; 1 \cdot 10 = 10$ $\quad -2 \cdot 10 = -20$ $\quad\;\; 4(-3) + 4(5) = -12 + 20 = 8$
$\quad\;\; 0 \cdot 10 = 0$ $\quad -3 \cdot 10 = -30$

42. $-\frac{10}{27}$ **43.** $-\frac{28}{15}$ **44.** -30.033 **45.** $2(-10) = -20$ $\quad -1(-10) = 10$ **46.** 12   **47.** 50   **48.** 64
$\qquad\qquad\qquad\qquad\qquad\qquad\qquad\qquad\qquad\;\;\; 1(-10) = -10$ $\quad -2(-10) = 20$
$\qquad\qquad\qquad\qquad\qquad\qquad\qquad\qquad\qquad\;\;\; 0(-10) = 0$ $\quad\;\;\; -3(-10) = 30$

49. $\dfrac{20}{63}$　**50.** $\dfrac{8}{15}$　**51.** 13.455　**52.** -2　**53.** -4　**54.** 2　**55.** $-\dfrac{4}{5}$　**56.** $-\dfrac{4}{3}$　**57.** $-\dfrac{1}{3}$　**58.** -5　**59.** $-\dfrac{20}{21}$

60. $-\dfrac{12}{5}$　**61.** $\dfrac{52}{21}$　**62.** a) 8　b) 8　**63.** a) -6　b) -6　**64.** a) -25　b) -25　**65.** a) -20　b) -20

66. $5x,\ -4y,\ 3$　**67.** $-4y,\ -2x,\ 3z$　**68.** $4(x-2)$　**69.** $3(x-2y+3)$　**70.** $b(x+y-z)$　**71.** $7(p+q-2t)$
72. $3x-15$　**73.** $5x-5y+20$　**74.** $-2x+6$　**75.** $bx-2by+bz$　**76.** a) $3x$　b) $6x$　c) $-5x$　**77.** $3x+3y$
78. $-4x+2y-3z$　**79.** $-2x-5y$　**80.** $-x-2$　**81.** $-4x+5y-2$　**82.** $2x-9$　**83.** $3y+2$　**84.** 4

85. -10　**86.** $5x-y-8$　**87.** $x+6y-8$　**88.** $\dfrac{1}{4^3}$　**89.** $\dfrac{1}{5^2}$　**90.** $\dfrac{1}{2^4}$　**91.** 3^{-2}　**92.** 5^{-4}　**93.** 7^{-3}

94. $\dfrac{1}{5^3}$　**95.** $\dfrac{1}{7^5}$　**96.** $\dfrac{1}{10^4}$　**97.** 3^8　**98.** 5^2　**99.** 6^{-7}　**100.** 5^{-5}　**101.** 7^4　**102.** $4^2 3^2$　**103.** 4^3　**104.** 5^6

105. 3^7　**106.** 7　**107.** 6^3　**108.** 8^{-7}　**109.** 3^{20}　**110.** x^{-12}　**111.** y^{15}　**112.** $-\dfrac{2}{5}$　**113.** $-\dfrac{20}{27}$

114. $48=\dfrac{3}{4}p;\ p=64\cancel{c}$

EXERCISE SET 2.1, p. 79

1. $5>0$　**3.** $-9<5$　**5.** $-6<6$　**7.** $-5>-8$　**9.** 17　**11.** 5　**13.** 10　**15.** 9　**17.** 6　**19.** 3
21. 43　**23.** 11　**25.** 33　**27.** 5　**29.** -12　**31.** 8　**33.** -41　**35.** 11　**37.** -18　**39.** -6　**41.** 31
43. -12　**45.** -15　**47.** 11　**49.** 8　**51.** 0　**53.** 0　**55.** 52　**57.** 251　**59.** 2,739

EXERCISE SET 2.2, p. 81

1. H　**3.** F　**5.** C　**7.** -1.8　**9.** $\dfrac{4}{5}$　**11.** 0　**13.** 1.7　**15.** $-\dfrac{3}{8}$　**17.** -8.1　**19.** 5.1　**21.** 8　**23.** 5

25. -14　**27.** $\dfrac{1}{12}$　**29.** $-\dfrac{17}{12}$　**31.** 13　**33.** -3.2　**35.** $\dfrac{1}{15}$　**37.** B　**39.** A

EXERCISE SET 2.3, p. 83

1. -28　**3.** -38.95　**5.** 16.2　**7.** $-\dfrac{2}{5}$　**9.** $-\dfrac{1}{2}$　**11.** -132　**13.** $-\dfrac{1}{12}$　**15.** 1　**17.** -945　**19.** 360

21. $\dfrac{1}{5}$　**23.** 4　**25.** $\dfrac{3}{2}$　**27.** $\dfrac{3}{5}$　**29.** -3　**31.** 4　**33.** -6　**35.** $-\dfrac{9}{8}$　**37.** $\dfrac{5}{3}$　**39.** $-\dfrac{9}{14}$

EXERCISE SET 2.4, p. 85

1. $7x-21$　**3.** $-3t+21$　**5.** $16x-72$　**7.** $-12x+44$　**9.** $-108+36x$　**11.** $5(x-3)$　**13.** $-8(x-3)$
15. $2(3x-5)$　**17.** $-4(3x-7)$　**19.** $a(x-7)$　**21.** $8x$　**23.** $5n$　**25.** $4x+2y$　**27.** $7x+y$　**29.** $-2x+5y$

31. $5x+12y-8$　**33.** $x+y$　**35.** $\dfrac{3}{5}x+\dfrac{3}{5}y$

EXERCISE SET 2.5, p. 87

1. $-2x-7$　**3.** $-4a+3b-7c$　**5.** $-6x-8y+5$　**7.** $3x,\ 5y,\ 6$　**9.** $-8x,\ -6y,\ -4$　**11.** $5x-3$　**13.** $7a-9$
15. $5x-6$　**17.** $-5x-2y$　**19.** 6　**21.** 65　**23.** 38　**25.** 19　**27.** 30　**29.** $2x+10$　**31.** $6x-35$
33. $12x-2$　**35.** $3x+30$　**37.** $-4x-68$

EXERCISE SET 2.6, p. 89

1. $\dfrac{1}{3^2}$ or $\dfrac{1}{9}$　**3.** $\dfrac{1}{2^3}$ or $\dfrac{1}{8}$　**5.** $\dfrac{1}{10^1}$ or $\dfrac{1}{10}$　**7.** 4^{-3}　**9.** 3^{-4}　**11.** x^{-3}　**13.** a^{-4}　**15.** $\dfrac{1}{7^3}$　**17.** $\dfrac{1}{2^4}$　**19.** $\dfrac{1}{a^3}$

21. $\frac{1}{y^4}$ **23.** 2^7 **25.** 5 **27.** 7^5 **29.** 10^4 **31.** 4^{-2} or $\frac{1}{16}$ **33.** x^3 **35.** 7^3 **37.** 10^4 **39.** 2^{-7} or $\frac{1}{2^7}$

41. x^{-3} **43.** n^7 **45.** 3^{12} **47.** x^{-6} **49.** a^9

EXERCISE SET 2.7, p. 91

1. 3 **3.** -5 **5.** 1.6 **7.** 10 **9.** 14 **11.** 8 **13.** 3 **15.** $\frac{1}{4}$ **17.** 4 **19.** $\frac{1}{2}$ **21.** 4 **23.** 12 **25.** 4.5

27. 3

CHAPTER 2   TEST ANALYSIS

Below are the answers to the test. Each item you missed should be reviewed. (See page numbers given in brackets with each answer.) If you got 30 correct (your instructor may have another standard), you are prepared to continue after reviewing those items missed. If you missed more than 11 questions, *carefully* review the entire chapter, giving special emphasis to the items missed; then continue.

CHAPTER 2   TEST

1. [61] $>$ **2.** [61] $<$ **3.** [61] $>$ **4.** [61] 7 **5.** [61] 9 **6.** [62] 6 **7.** [62] 0 **8.** [62] 8 **9.** [62] -4

10. [64] -11.3 **11.** [64] 1.7 **12.** [64] -3.7 **13.** [64] 19.6 **14.** [64] $-\frac{7}{9}$ **15.** [64] $\frac{19}{15}$ or $1\frac{4}{15}$ **16.** [67] 54

17. [67] -9.18 **18.** [68] -3 **19.** [68] $-\frac{2}{7}$ **20.** [68] $-\frac{3}{4}$ **21.** [68] $\frac{3}{2}$ **22.** [68] -210 **23.** [68] 1

24. [71] $15x - 35$ **25.** [71] $-8x + 10$ **26.** [71] $54 - 12x$ **27.** [71] $2(x - 7)$ **28.** [71] $3(2x - 3)$ **29.** [71] $-7(x - 3)$

30. [71] $7a - 3b$ **31.** [71] $-2x + 5y$ **32.** [72] $-3a + 9$ **33.** [72] $9y - 9$ **34.** [72] $2x + 12$ **35.** [72] $10x - 16$

36. [74] 4^{-2} **37.** [74] 5^{-3} **38.** [75] 6^5 **39.** [75] x^{-4} **40.** [76] 7^{-3} **41.** [76] x^5

MARGIN EXERCISES   CHAPTER 3

1. -5 **2.** 8 **3.** -2 **4.** 10.8 **5.** 5 **6.** 12 **7.** $-\frac{7}{4}$ **8.** -3 **9.** 4 **10.** 4 **11.** 3 **12.** 1 **13.** 2

14. 2 **15.** $\frac{17}{2}$ **16.** 10 **17.** 2 **18.** -2 **19.** 3, -4 **20.** 0, $\frac{17}{3}$ **21.** 3 ft, 5 ft **22.** 5 **23.** $60°, 90°, 30°$

24. 18, 20 **25.** 10 m, 20 m **26.** $I = \frac{E}{R}$ **27.** $t = \frac{d}{r}$

EXERCISE SET 3.1, p. 105

1. 4 **3.** -20 **5.** $-\frac{3}{2}$ **7.** -1 **9.** $\frac{41}{24}$ **11.** $\frac{1}{2}$ **13.** -5.1 **15.** 10.9 **17.** 16 **19.** -4 **21.** 16

23. $\frac{2}{3}$ **25.** $1\frac{5}{6}$ **27.** $6\frac{2}{3}$ **29.** $-7\frac{6}{7}$

EXERCISE SET 3.2, p. 107

1. 6 **3.** 8 **5.** -6 **7.** 63 **9.** $\frac{3}{5}$ **11.** -20 **13.** -50 **15.** $\frac{10}{3}$ **17.** 10 **19.** -2 **21.** -8 **23.** 13.38

EXERCISE SET 3.3, p. 109

1. 5 **3.** 10 **5.** -8 **7.** 6 **9.** 5 **11.** -20 **13.** 6 **15.** 30 **17.** 7 **19.** 7 **21.** 3 **23.** 5 **25.** 2

27. 4 **29.** 10 **31.** 5 **33.** 7 **35.** $\frac{1}{2}$ **37.** 12 **39.** 1

EXERCISE SET 3.4, p. 113

1. 3   **3.** 2   **5.** 6   **7.** 8   **9.** 4   **11.** 1   **13.** 17   **15.** −8   **17.** $-\dfrac{5}{3}$ **19.** −3   **21.** $-\dfrac{5}{2}$ **23.** 5

EXERCISE SET 3.5, p. 117

1. −3, −2   **3.** 3, −2   **5.** 0, −10   **7.** $\dfrac{5}{2}$, 4   **9.** $\dfrac{1}{2}$, −2   **11.** 2, $\dfrac{1}{2}$ **13.** 3, −3   **15.** $\dfrac{1}{9}$, $\dfrac{1}{10}$ **17.** 10, 3

19. 0, 5.3   **21.** 2, −2   **23.** $\dfrac{1}{3}$, 20   **25.** 0, 1000   **27.** 5000, 50

EXERCISE SET 3.6, p. 119

1. 19   **3.** 20 ft, 40 ft, 120 ft   **5.** 20°, 80°, 80°   **7.** 35, 36, 37   **9.** $22\dfrac{1}{2}^{\circ}$

EXERCISE SET 3.7, p. 121

1. $I = \dfrac{E}{R}$ **3.** $t = \dfrac{d}{r}$ **5.** $w = \dfrac{p-21}{2}$ **7.** $t = \dfrac{I}{pr}$ **9.** $x = \dfrac{ac}{b}$ **11.** $x = a - b$ **13.** $b = \dfrac{2A}{h}$ **15.** $t = \dfrac{3k}{v}$

17. $h = \dfrac{2A}{b}$ **19.** $E = \dfrac{6}{L}$ **21.** $b = \dfrac{3V}{h}$ **23.** $S = \dfrac{tv}{2}$

CHAPTER 3   TEST ANALYSIS

Below are the answers to the test. Each item you missed should be reviewed. (See page numbers in brackets with each answer.) If you got 17 correct (your instructor may have another standard), you are prepared to continue after reviewing those items missed. If you missed more than 6 questions, carefully review the entire chapter, giving special emphasis to the items missed; then continue.

CHAPTER 3   TEST

1. [95] 26   **2.** [95] 8   **3.** [95] −3   **4.** [96] 7   **5.** [96] −6   **6.** [96] 4   **7.** [97] −12   **8.** [99] 7   **9.** [99] 2

10. [99] $-\dfrac{1}{5}$ **11.** [99] 5   **12.** [99] −5   **13.** [99] $-\dfrac{5}{6}$ **14.** [99] 7   **15.** [100] −5, 3   **16.** [100] 0, $\dfrac{8}{3}$

17. [104] $r = \dfrac{A}{2\pi h}$ **18.** [104] $A = \dfrac{bh}{2}$ **19.** [104] $t = \dfrac{3h}{A}$ **20.** [101] 21   **21.** [101] 33, 35   **22.** [102] 30°, 60°, 90°

23. [103] 7 cm, 11 cm

MARGIN EXERCISES   CHAPTER 4

1. Answers will vary.   **2.** −19   **3.** −104   **4.** −3   **5.** −6   **6.** $-9x^3 + {}^-4x^5$ **7.** $-2x^3 + 3x^7 + {}^-7x$

8. $3x^2, 6x, \dfrac{1}{2}$ **9.** $-4y^5, 7y^2, -3y, -2$ **10.** $7x^4, -5x^2, 3x, -5$ **11.** $4x^3$ and $-1x^3$ **12.** $4x^2$ and $-5x^2$

13. $4x^4$ and $-7x^4$, $-9x^3$ and $10x^3$ **14.** $8x^2$ **15.** $2x^3 + 2$ **16.** $7x^4 + 4x^3$ **17.** $-\dfrac{1}{4}x^5 + 2x^2$ **18.** $-4x^3$ **19.** 0

20. $5x^3$ **21.** $-3x^5 + 25$ **22.** $6x$ **23.** $4x^3 + 4$ **24.** $2x^2 + 7x^4$ **25.** $-\dfrac{1}{4}x^3 + 4x^2 + 7$ **26.** $x^3 + 3x^2$

27. $6x^7 + 3x^5 - 2x^4 + 4x^3 + 5x^2 + x$ **28.** $7x^5 - 5x^4 + 2x^3 + 4x^2 - 3$ **29.** $14x^7 - 10x^5 + 7x^2 - 14$

30. $-2x^2 - 3x + 2$ **31.** $10x^4 - 8x - \dfrac{1}{2}$ **32.** 4, 2, 1, 0   **33.** 5, 6, 1, −1, 4   **34.** a) x b) x^3, x^2, x and constant

c) x^2, x d) x^3 **35.** a) $2x^3 + 4x^2 + 0x - 2$ c) $x^3 + 0x^2 + 0x + 1$ d) $x^4 + 0x^3 - x^2 + 3x + \dfrac{1}{4}$ **36.** c **37.** a

38. b **39.** 2   **40.** -3 **41.** -1 **42.** $x^2 + 7x + 3$ **43.** $-4x^5 + 7x^4 + 3x^3 + 2x^2 + 4$ **44.** $24x^4 + 5x^3 + x^2 + 1$

45. $2x^3 + 3\dfrac{1}{3}$ **46.** $2x^2 - 3x - 1$ **47.** $8x^3 - 2x^2 - 8x + \dfrac{5}{2}$ **48.** $-2x^3 + 5x^2 - 2$ **49.** $-2x^3 + 5x^2 - 2$

50. $-3x^2 + x - \dfrac{7}{4}$ **51.** $-3x^2 + x - \dfrac{7}{4}$ **52.** 0   **53.** 0   **54.** 0   **55.** $-12x^4 + 3x^2 - 4x$ **56.** $4x^4 - 3x^2 + 4x$

57. $13x^6 - 2x^4 + 3x^2 - x + \dfrac{5}{13}$ **58.** $7x^3 - 2x^2 + x - 3$ **59.** $-4x^3 + 6x - 3$ **60.** $-5x^4 - 3x^2 - 7x + 5$

61. $-14x^{10} + \dfrac{1}{2}x^5 - 5x^3 + x^2 - 3x$ **62.** $3x^2 + 3$ **63.** $2x^3 + 2x + 8$ **64.** $x^2 - 6x - 2$ **65.** $-8x^4 - 5x^3 + 8x^2 - 1$

66. $-2x^2 - 2x + 2$ **67.** $-2x^6 + 5x^4 - 2x^2 + 3$ **68.** $x^3 - x^2 - \dfrac{4}{3}x + .1$ **69.** $-\dfrac{6}{5}x + 1.3$

70. $-8x^4 + 4x^3 + 12x^2 + 5x - 8$ **71.** $-x^3 + x^2 + 3x + 3$ **72.** $2x^3 + 5x^2 - 2x - 5$ **73.** $-x^5 - 2x^3 + 3x^2 - 2x + 2$
74. $-15x$ **75.** $-x^2$ **76.** x^2 **77.** $-x^5$ **78.** $12x^7$ **79.** $-8x^{11}$ **80.** $7x^5$ **81.** 0   **82.** $8x^2 + 16x$
83. $-15x^3 + 6x^2$ **84.** $x^2 + 13x + 40$ **85.** $x^2 + x - 20$ **86.** $5x^2 - 17x - 12$ **87.** $6x^2 - 19x + 15$
88. $x^4 + 3x^3 + x^2 + 15x - 20$ **89.** $6x^5 - 20x^3 + 15x^2 + 14x - 35$ **90.** $3x^3 + 13x^2 - 6x + 20$
91. $20x^5 - 16x^4 + 40x^3 - 40x^2 - 16$ **92.** $8x^3 - 12x^2 + 16x$ **93.** $10x^6 + 8x^5 - 10x^4$ **94.** $x^4 + 2x^3 + 4x + 8$
95. $x^4 + x^3 + 2x + 2$ **96.** $x^2 + 7x + 12$ **97.** $x^2 - 2x - 15$ **98.** $2x^2 + 9x + 4$ **99.** $2x^3 - 4x^2 - 3x + 6$
100. $12x^5 + 10x^3 + 6x^2 + 5$ **101.** $x^6 - 49$ **102.** $-2x^7 + x^5 + x^3$ **103.** $x^2 + x - 6$ **104.** $3x^2 - 2x - 8$
105. $x^2 + 6x - 16$ **106.** $8x^2 - 14x + 5$ **107.** $-4x^2 - 11x - 6$ **108.** $x^2 - 25$ **109.** $4x^2 - 9$ **110.** $x^2 - 4$
111. $x^2 - 49$ **112.** $9x^2 - 25$ **113.** $4x^6 - 1$ **114.** $x^2 + 6x + 9$ **115.** $4x^2 - 16x + 16$ **116.** $x^2 - 10x + 25$
117. $16x^2 + 16x + 4$ **118.** $9x^2 - 24x + 16$ **119.** $x^2 + 4x + 4$ **120.** $x^2 - 8x + 16$ **121.** $4x^2 + 20x + 25$
122. $16x^4 - 24x^3 + 9x^2$ **123.** $x^2 + 18x + 81$ **124.** $9x^4 - 30x^2 + 25$ **125.** $x^2 + 11x + 30$ **126.** $x^2 - 16$

127. $-8x^5 + 20x^4 + 40x^2$ **128.** $81x^4 + 18x^2 + 1$ **129.** $4x^2 + 6x - 40$ **130.** $4x^2 - 2x + \dfrac{1}{4}$

EXERCISE SET 4.1, p. 145

1. -18 **3.** 148   **5.** 19   **7.** 2   **9.** 4   **11.** -6 **13.** $7x + {}^-1$ **15.** $-7x + \dfrac{-2}{3}$ **17.** $2 + {}^-3x + x^2$

19. $5x^2, -6x, -3$ **21.** $-2, -3x, 5x^4$ **23.** x and $-5x$, -3 and -2 **25.** x^5 and $4x^5$, $-6x^3$ and $7x^3$ **27.** $11x^3 + 4$

29. $-6x^5 + 4x^4 - 4$ **31.** $4x^4 + 5$ **33.** $-4x^3 - x^2 + 5$ **35.** $\dfrac{3}{4}x^5 - 2x - 5$ **37.** $\dfrac{3}{10}$

EXERCISE SET 4.2, p. 147

1. $x^5 + 6x^3 + 2x^2 + x + 1$ **3.** $15x^9 + 7x^8 + 5x^3 - x^2 + x$ **5.** $x^6 + x^4$ **7.** $12x^4 - 2x + \dfrac{1}{4}$ **9.** 3, 2, 1, 0

11. 6, 7, -8, -2 **13.** a) x b) x^4, x^3, x^2, constant c) x^2, x, constant d) x **15.** c **17.** a, f **19.** -5 **21.** 3
23. $-x + 5$ **25.** $x^2 - 5x - 1$ **27.** $3x^5 + 13x^2 + 6x - 3$ **29.** $-4x^4 + 6x^3 + 6x^2 + 2x + 4$ **31.** $12x^2 + 6$

33. $5x^4 - 2x^3 - 7x^2 - 5x$ **35.** $9x^8 + 8x^7 - 3x^4 + 2x^2 - 2x + 5$ **37.** $-\dfrac{1}{2}x^4 + \dfrac{2}{3}x^3 + x^2$

39. $\dfrac{2}{15}x^9 - \dfrac{2}{5}x^5 + \dfrac{1}{4}x^4 + \dfrac{1}{4}x^2 + \dfrac{15}{2}$

EXERCISE SET 4.3, p. 151

1. $-4x^4 + 7x^2 - 2$ **3.** $5x^5 + 4x^3 + 2x^2 - 11$ **5.** $x^2 + 6$ **7.** $6x^3 + x^2 + x$ **9.** 0   **11.** 0

13. $-4x^3 + 6x^2 + 8x - 1$ **15.** $-4x^2 + 3x - 2$ **17.** $4x^3 + 6x^2 - \dfrac{3}{4}x + 8$ **19.** $7x^3 - 9x^2 - 2x + 10$

21. $-x^2 - 2x + 4$ **23.** $7x^3 + 3x^2 + 2x - 1$ **25.** $\dfrac{3}{5}x^3 - .11$ **27.** $.1x^4 - .9$

EXERCISE SET 4.4, p. 153

1. $-3x^4 + 3x^2 + 4x$ **3.** $3x^5 - 3x^4 - 3x^3 + x^2 + 3x$ **5.** $5x^3 - 9x^2 + 4x - 7$ **7.** $\frac{1}{4}x^4 - \frac{1}{4}x^3 + \frac{3}{2}x^2 + 6\frac{3}{4}x + \frac{1}{4}$

9. $-x^4 + 3x^3 + 2x + 1$ **11.** $x^4 + 4x^2 + 12x - 1$ **13.** $x^5 - 6x^4 + 4x^3 - x^2 + 1$ **15.** $7x^4 - 2x^3 + 7x^2 + 4x + 9$

17. $3x^5 + x^4 + 10x^3 + x^2 + 3x - 6$ **19.** $1.05x^4 + .36x^3 + 14.22x^2 + x + .97$ **21.** $13x^4 - 3x^3 - x^2 + 15x - 4$

23. $4x^4 + 3x^3 - 7x^2 - 3x + 5$ **25.** $-x^5 - 6x^4 + 6x^3 - 2x^2 - \frac{1}{2}x - \frac{1}{4}$ **27.** $5x^4 + 12x^3 - 9x^2 - 8x - 8$

29. $-4x^5 + 9x^4 + 6x^2 + 16x + 6$

EXERCISE SET 4.5, p. 157

1. $-12x$ **3.** $42x^2$ **5.** $30x$ **7.** $-2x^3$ **9.** x^6 **11.** $6x^6$ **13.** 0 **15.** $-.02x^{10}$ **17.** $8x^2 - 12x$

19. $-6x^4 - 6x^3$ **21.** $4x^6 - 24x^5$ **23.** $-x^2 - 4x + 12$ **25.** $2x^2 - 15x + 25$ **27.** $9x^2 - 25$ **29.** $2x^2 + \frac{5}{2}x - \frac{3}{4}$

31. $x^3 + 7x^2 + 7x + 1$ **33.** $-10x^3 - 19x^2 - x + 3$ **35.** $3x^4 - 6x^3 - 7x^2 + 18x - 6$

EXERCISE SET 4.6, p. 159

1. $4x^2 + 4x$ **3.** $-3x^2 + 3x$ **5.** $x^5 + x^2$ **7.** $6x^3 - 18x^2 + 3x$ **9.** $x^3 + x^2 + 3x + 3$ **11.** $x^4 + x^3 + 2x + 2$

13. $x^2 - x - 6$ **15.** $9x^2 + 15x + 6$ **17.** $5x^2 + 4x - 12$ **19.** $9x^2 - 1$ **21.** $4x^2 - 6x + 2$ **23.** $x^2 - \frac{1}{16}$

25. $x^2 - .01$ **27.** $2x^3 + 2x^2 + 6x + 6$ **29.** $-2x^2 - 11x + 6$ **31.** $x^2 + 14x + 49$ **33.** $1 - x - 6x^2$

35. $x^5 + 3x^3 - x^2 - 3$ **37.** $x^3 - x^2 - 2x + 2$ **39.** $3x^6 - 2x^4 - 6x^2 + 4$ **41.** $6x^7 + 18x^5 + 4x^2 + 12$

43. $8x^6 + 65x^3 + 8$ **45.** $4x^3 - 12x^2 + 3x - 9$ **47.** $4x^6 + 4x^5 + x^4 + x^3$

EXERCISE SET 4.7, p. 161

1. $x^2 - 16$ **3.** $4x^2 - 1$ **5.** $25x^2 - 4$ **7.** $4x^4 - 9$ **9.** $9x^8 - 16$ **11.** $x^{12} - x^4$ **13.** $x^8 - 9x^2$ **15.** $x^{24} - 9$

17. $4x^{16} - 9$ **19.** $x^2 + 4x + 4$ **21.** $9x^4 + 6x^2 + 1$ **23.** $x^2 - x + \frac{1}{4}$ **25.** $9 + 6x + x^2$ **27.** $x^4 + 2x^2 + 1$

29. $4 - 12x^4 + 9x^8$ **31.** $25 + 60x^2 + 36x^4$ **33.** $9 - 12x^3 + 4x^6$ **35.** $4x^3 + 24x^2 - 12x$ **37.** $4x^4 - 2x^2 + \frac{1}{4}$

39. $\frac{9}{16}x^2 + \frac{9}{4}x + 2$ **41.** $12x^5 + 6x^3 - 2x^2 - 1$ **43.** $36x^8 + 48x^4 + 16$ **45.** $12x^3 + 8x^2 + 15x + 10$

47. $64 - 96x^4 + 36x^8$

CHAPTER 4   TEST ANALYSIS

Below are the answers to the test. Each item you missed should be reviewed. (See page numbers in brackets with each answer.) If you got 22 correct (your instructor may have another standard), you are prepared to continue after reviewing those items missed. If you missed more than 8 questions, carefully review the entire chapter, giving special emphasis to the items missed; then continue.

CHAPTER 4   TEST

1. [126] a) 1, b) -7 **2.** [126] a) 5, b) 1   **3.** [126] a) 0, b) 4   **4.** [126] a) 1, b) 1   **5.** [127, 128] $-5x^2 + 2x + 1$

6. [127, 128] $\frac{1}{5}x^3 - 11\frac{3}{5}x$ **7.** [134] $3x^5 + x^3 - 5x^2 + x - 16$ **8.** [134] $2x^5 - 6x^4 + 2x^3 - 2x^2 + 2$

9. [134] $-5x^4 + 5x^3 - 8x^2 + 6x - 12$ **10.** [134] $x^5 - 3x^3 - 2x^2 + 8$ **11.** [135] $3x^3 + x^2 - 3x + 12$

12. [135] $\frac{11}{3}x^2 - 5x - 46$ **13.** [135] $x^2 - x - \frac{1}{4}$ **14.** [135] $x^2 + \frac{3}{2}x - 4$ **15.** [135] $.04x^2 - .3x - .005$

16. [135] $.2x^3 - .12x^2 - .9x + .7$ **17.** [130] $-x^3 + x^2 + x + 4$ **18.** [130] $-x^5 + 2x^4 + 11x^3 + x^2 + 7x - 6$
19. [133] $-2x^2 + 4x + 6$ **20.** [133] $2x^4 - 3x^3 - 4x^2 - x$ **21.** [136] $-18x^4 + 30x^3 + 12x^2$
22. [136] $12x^3 - 23x^2 + 13x - 2$ **23.** [140] $2x^2 - 11x - 21$ **24.** [143] $49x^2 - 14x + 1$
25. [140] $6x^5 + 9x^3 - 10x^2 - 15$ **26.** [142] $4x^6 - 49$ **27.** [136] $6x^3 + 5x^2 - 8x - 3$ **28.** [143] $9x^2 + 12x + 4$
29. [140] $15x^2 - 11x - 14$ **30.** [136] $15x^7 - 40x^6 + 50x^5 + 10x^4$

MARGIN EXERCISES CHAPTER 5

1. a) $12x^2$ b) $(3x)(4x)$ **2.** a) $(16x^3)$ b) $(2x)(8x^2)$ **3.** $(2x^2)(4x^2)$, $(8x)(x^3)$, $(8x^3)(x)$, etc.
4. $(3x)(5x)$, $(15x)(x)$, $(3x^2)(5)$, etc. **5.** $(2x^3)(3x^2)$, $(6x^4)(x)$, $(3x)(2x^4)$, etc. **6.** a) $3x + 6$ b) $3(x + 2)$
7. a) $2x^3 + 10x^2 + 8x$ b) $2x(x^2 + 5x + 4)$ **8.** $x(x + 3)$ **9.** $3x(x + 2)$ **10.** $x^2(3x^4 - 5x + 2)$ **11.** $4(5x^3 + 3x^2 - 4)$

12. $3x^2(3x^2 - 5x + 1)$ **13.** $11(2x^3 + x^2 - 3x - 4)$ **14.** $10x(2x^3 + 3x^2 - 2x + 5)$ **15.** $\frac{1}{4}(3x^3 + 5x^2 + 7x + 1)$

16. $7x^3(5x^4 - 7x^3 + 2x^2 - 9)$ **17.** $(x + 2)(x + 5)$ **18.** $(x - 4)(x + 3)$ **19.** $(x + 5)(x - 4)$ **20.** $(5x + 2)(x - 2)$
21. $(2x - 3)(x + 4)$ **22.** $(4x - 3)(4x + 5)$ **23.** $(x + 3)(x + 4)$ **24.** $(x + 5)(x + 7)$ **25.** $(x - 2)(x + 1)$
26. $(3x + 2)(2x + 1)$ **27.** $3(2x + 3)(x + 1)$ **28.** $(2x - 3)(x + 2)$ **29.** $(2x + 1)(3x + 7)$ **30.** $2(2x + 3)(x - 1)$
31. $(3x - 1)(2x - 1)$ **32.** a, b, d, f **33.** $(x + 1)(x + 1)$ **34.** $(x - 1)(x - 1)$ **35.** $(x + 2)(x + 2)$
36. $(5x - 7)(5x - 7)$ **37.** $(4x - 7)(4x - 7)$ **38.** $(x + 3)(x - 3)$ **39.** $(x + 8)(x - 8)$ **40.** $5(1 + 2x^3)(1 - 2x^3)$
41. $(9x^2 + 1)(3x + 1)(3x - 1)$ **42.** $x^4(7 + 5x^3)(7 - 5x^3)$ **43.** $(x + 5)(x - 3)$ **44.** $(x + 8)(x - 6)$
45. $(x + 7)(x - 3)$ **46.** $(x + 7)(x - 9)$ **47.** $(x + 7)(x + 1)$ **48.** $(x + 18)(x - 2)$ **49.** 36 **50.** 9 **51.** 49
52. 121 **53.** $(x + 3)(x - 1)$ **54.** $(x - 4)(x - 2)$ **55.** $(x + 2)(x + 4)$ **56.** $(x + 3)(x + 7)$ **57.** $(x + 1)(x - 9)$

58. $-2, -3$ **59.** $0, 4$ **60.** $7, -4$ **61.** $\frac{4}{5}, -\frac{4}{5}$ **62.** $(x + 1)(x - 1) = 24, x = 5$ or -5

63. $(x - 7)(x - 8) = 0, x = 7$ or 8 **64.** $x(x - 1) = 0, x = 0$ or 1 **65.** $x^2 - x = 20, x = 5$ or -4
66. $2x^2 + 1 = 73, x = 6$ or -6 **67.** $x(x - 2) = 15, l = 5$ cm, $w = 3$ cm

EXERCISE SET 5.1, p. 179

1. a) $21x$ b) $7 \cdot 3x$ **3.** a) $16x^4$ b) $2 \cdot 8x^4$, $2x^2 \cdot 8x^2$, etc. **5.** $3x^2 \cdot 3x^2$, $3x \cdot 3x^3$, $9x^2 \cdot x^2$, etc.
7. $3x^2 \cdot 5x^3$, $3x^4 \cdot 5x$, $5x^4 \cdot 3x$, etc. **9.** $2x \cdot 12x^3$, $12x^2 \cdot 2x^2$, $24x^3 \cdot x$, etc. **11.** a) $6x^2 + 12x$ b) $6x(x + 2)$
13. a) $8x^4 - 12x^3 + 24x^2$ b) $4x^2(2x^2 - 3x + 6)$ **15.** a) $2x^4 - 12x^3 + 6x^2$ b) $2x^2(x^2 - 6x + 3)$ **17.** $x^2(x + 7)$

19. $17x(x^4 + 2x^2 + 3)$ **21.** $\frac{1}{7}(3x^2 - 6x^3 + 4)$ **23.** $(x + 5)(x + 2)$ **25.** $(a - 2)(a + 5)$ **27.** $3(x + 1)(3x - 2)$

29. $(6 - 5y)(4 - 3y)$ **31.** $(2x - 7)(4x - 3)$ **33.** $(2x^2 - 5)(x^2 - 3)$

EXERCISE SET 5.2, p. 181

1. $(x + 3)(x + 5)$ **3.** $(x - 5)(x + 3)$ **5.** $(x + 3)(x + 4)$ **7.** $(x - 3)(x + 5)$ **9.** $(y + 8)(y + 1)$
11. $(x^2 + 2)(x^2 + 3)$ **13.** $(x - 4)(x + 7)$ **15.** $(y + .2)(y - .4)$ **17.** $(x + 3)(x + 5)$ or $(3 + x)(5 + x)$
19. $(24x - 1)(x + 2)$ **21.** $2(-x + 5)(x + 2)$ or $2(5 - x)(2 + x)$ **23.** $(3x + 1)(x + 1)$ **25.** $(x + 3)(x - 8)$
27. $(2x + 1)(x - 1)$ **29.** $(3x - 2)(3x + 8)$ **31.** $(3x + 1)(x - 2)$ **33.** $(3x + 4)(4x + 5)$ **35.** $(7x - 1)(2x + 3)$
37. $(3x^2 + 2)(3x^2 + 4)$ **39.** $(3x - 7)(3x - 7)$ **41.** $(a - 1)(a - 12)$ **43.** $(x^2 + 2)(x^2 + 3)$

EXERCISE SET 5.3, p. 183

1. $(x + 3)(x + 3)$ **3.** $2(x - 1)(x - 1)$ **5.** $x(x - 9)(x - 9)$ **7.** $(y - 6)(y - 6)$ **9.** $(8 - y)(8 - y)$
11. $3(2x + 3)(2x + 3)$ **13.** $(x^3 - 8)(x^3 - 8)$ **15.** $(7 - 3x)(7 - 3x)$ **17.** $5(y^2 + 1)(y^2 + 1)$ **19.** $(x - 2)(x + 2)$
21. $2(2x - 7)(2x + 7)$ **23.** $(2x + 5)(2x - 5)$ **25.** $x(4 - 9x)(4 + 9x)$ **27.** $x^2(4 - 5x)(4 + 5x)$ **29.** $(7x^2 + 9)(7x^2 - 9)$

31. $a^2(a^5 + 2)(a^5 - 2)$ **33.** $(11a^4 + 10)(11a^4 - 10)$ **35.** $\left(\frac{1}{4} + y\right)\left(\frac{1}{4} - y\right)$ **37.** $\left(5 + \frac{1}{7}x\right)\left(5 - \frac{1}{7}x\right)$

39. $\left(6x + \frac{1}{5}y\right)\left(6x - \frac{1}{5}y\right)$ $(a^3 - 1)(a^3 - 1)$ or $(1 - a^3)(1 - a^3)$ **43.** $\left(\frac{1}{5}y + \frac{1}{4}\right)\left(\frac{1}{5}y - \frac{1}{4}\right)$

45. $\left(\frac{1}{9}x^3 - \frac{4}{3}\right)\left(\frac{1}{9}x^3 - \frac{4}{3}\right)$

EXERCISE SET 5.4, p. 185

1. $(x + 13)(x - 1)$ **3.** $(x + 11)(x + 3)$ **5.** $(x + 10)(x + 8)$ **7.** $(x^2 + 12)(x^2 + 8)$ **9.** $2(x^3 + 9)(x^3 - 1)$
11. $(a^2 - 2)(a + 2)(a - 2)$ **13.** $(x + 4)(x + 2)$ **15.** $(x - 6)(x - 2)$ **17.** $2(x + 8)(x - 16)$ **19.** $(a - 25)(a + 1)$
21. $10(b - 12)(b + 7)$ **23.** $(x - 18)(x + 4)$

EXERCISE SET 5.5, p. 187

1. $-1, -5$ **3.** $0, 5$ **5.** -3 **7.** $\frac{5}{3}, -1$ **9.** $4, -\frac{5}{3}$ **11.** $\frac{2}{3}, -\frac{1}{4}$ **13.** $0, \frac{3}{5}$ **15.** $4x^2 - x = 3; x = 1, -\frac{3}{4}$
17. $x^2 + 8 = 6x; x = 2, 4$ **19.** $x(x + 4) = 96; 12$ cm.   **21.** $x^2 = 28 + 12x; x = 14, -2$ **23.** $x(x + 2) = 168;$
    12, 14 or $-12, -14$ **25.** $x^2 + (x + 2)^2 = 74; 5, 7$

CHAPTER 5   TEST ANALYSIS

Below are the answers to the test. Each item you missed should be reviewed. (See page numbers in brackets with each answer.)
If you got 18 correct (your instructor may have another standard), you are prepared to continue after reviewing those items
missed. If you missed more than 7 questions, *carefully* review the entire chapter, giving special emphasis to the items missed;
then continue.

CHAPTER 5   TEST

1. [166] $3(x + 1)$ **2.** [166] $10(x - 2)$ **3.** [172] $(3x - 2)(3x + 2)$ **4.** [172] $2(x - 5)(x + 5)$ **5.** [168] $(x - 5)(x - 3)$
6. [168] $x(x - 3)(x - 4)$ **7.** [168] $3(x + 8)(x - 2)$ **8.** [168] $x(x - 3)(x - 5)$ **9.** [168] $(10x - 1)(x - 10)$
10. [171] $(x - 3)^2$ **11.** [171] $3(2x + 5)^2$ **12.** [171] $(3x - 5)^2$ **13.** [172] $(4x^2 + 1)(2x - 1)(2x + 1)$
14. [167] $-4(x + 7)$ **15.** [172] $(x - 4)(x + 6)$ **16.** [168] $(x - 7)(x + 1)$ **17.** [174] 100   **18.** [174] 225

19. [175] $5, -7$ **20.** [175] $\frac{3}{2}, -4$ **21.** [175] $-2, 8$ **22.** [175] $-2, 10$ **23.** [176] $3, -2$

24. [176] 16, 18; $-16, -18$ **25.** [176] 1, 12

MARGIN EXERCISES   CHAPTER 6

1. They are both negative.   **2.** The first coordinate is positive; the second negative.   **3.**

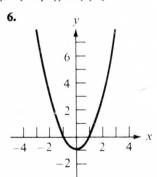

4. B $(-3, 5)$, C $(-4, -3)$, D $(2, -4)$, E $(1, 5)$, F $(-2, 0)$, G $(0, 3)$

5.   **6.**   **7.** a, b

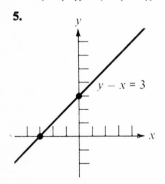

8. 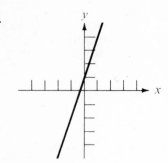   **9.** 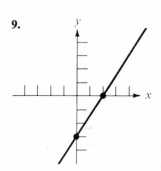   **10.**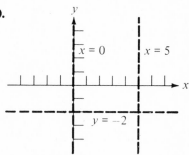

11. a) Yes b) No   **12.** $x + y = 115; x - y = 21$ **13.** $R = \frac{1}{2}T; T + 5 = 56$ **14.** $1 = w + 17, 21 + 2w = 76$

15. Yes   **16.** Yes   **17.** $(2, -3)$ **18.** No solution. Lines are parallel.   **19.** $(3, 2)$ **20.** $(4, 1)$ **21.** $(3, 2)$
22. $(1, -1)$ **23.** $(1, 4)$ **24.** $(3, 2)$ **25.** $(1, 1)$ **26.** $(1, 1)$ **27.** $(1, -1)$ **28.** $(2, -4)$ **29.** 4, 22   **30.** $d = rt$

31. $t = \frac{d}{r}$ **32.** 3 hr   **33.** 324 mi   **34.** 275 km/hr   **35.** $7Q, 13D$ **36.** 125 adults, 41 children   **37.** $22\frac{1}{2}$ kg, $7\frac{1}{2}$ kg
38. 30 lb, 20 lb   **39.** Yes   **40.** 4 ft, 6 ft

EXERCISE SET 6.1, p. 221

1. a) II b) I c) IV d) II e) IV f) III   **3.** a) second b) first
5. A: $(3, 3)$, B: $(0, -4)$, C: $(-5, 0)$, D: $(-1, -1)$, E: $(2, 0)$ **7.** $y = 2x - 3$

9.

x	0	$\frac{3}{2}$	-1	1	-2	2
y	-3	0	-5	-1	-7	1

11.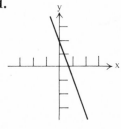

13.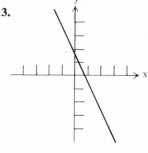

15.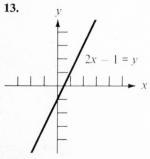

EXERCISE SET 6.2, p. 223

1. Not linear; contains a second degree term

3. Linear

5. Not linear; contains a product of variables (2nd degree)

7. $\left(\frac{1}{2}, 0\right); (0, -1)$

9. $\left(\frac{3}{5}, 0\right); \left(0, -\frac{3}{7}\right)$

11. No x-intercept; $(0, -4)$

13.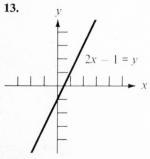

$2x - 1 = y$

15.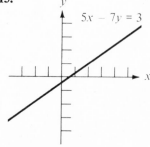

$5x - 7y = 3$

17.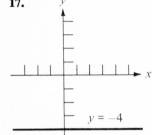

19. x – intercept $(0, 0)$; y – intercept: every point of the line   **21.** Yes

23. No (the graphs are the same line).   **25.** Of the form $y = -3x + c$ b) Of the form $x = a$

EXERCISE SET 6.3, p. 225

1. $l + s = 27, l = s + 3; l = 15, s = 12$ **3.** $l - s = 16, 3l = 7s; s = 12, l = 28$ **5.** $2l + 2w = 400, l = w + 40$; width, 80 m   **7.** $a + s = 180, a = 3s + 8; 43°, 137°$ **9.** $f = 2d, f + 4 = 3(d - 6)$; A1, 44 yrs; daughter, 22 yrs
11. $r = 2f + 2, r + f = 26$; reserves, 18; freshmen, 8   **13.** $n + d = 150, n = 2d$, 100 nickels, 50 dimes
15. $m = j + 3, m + j = 216 - 16$; Mary 101 - 1/2 lb; Jane 98 - 1/2 lb

EXERCISE SET 6.4, p. 229

1. Yes   **3.** Yes   **5.** No   **7.** $(-12, 11)$ **9.** $(4, 3)$ **11.** $(-6, -2)$

EXERCISE SET 6.5, p. 231

1. $(1, 3)$ **3.** $(1, 2)$ **5.** $(4, 1)$ **7.** $(2, 5)$ **9.** $(-1, -1)$ **11.** $\left(\frac{17}{3}, \frac{2}{3}\right)$ **13.** $\left(\frac{5}{2}, -\frac{7}{4}\right)$ **15.** $\left(\frac{17}{3}, \frac{16}{3}\right)$

EXERCISE SET 6.6, p. 233

1. $(9, 1)$ **3.** $\left(3, -\frac{1}{2}\right)$ **5.** $(4, 1)$ **7.** $(10, 2)$ **9.** $(2, 5)$ **11.** $(1, -1)$ **13.** $(4, -1)$ **15.** $(1, 3)$

17. Many solutions   **19.** $\left(5, \frac{1}{2}\right)$

EXERCISE SET 6.6A, p. 235

1. $(3, 1)$ **3.** $(3, -2)$ **5.** $\left(\frac{3}{2}, -1\right)$ **7.** $(50, 18)$ **9.** $(2, 1)$ **11.** $(3, 1)$ **13.** $(-2, -3)$ **15.** $(-9, 12)$
17. $(5, 5)$ **19.** $(5, 2)$

EXERCISE SET 6.7, p. 237

1. $f = 3s, f + 15 = 2(s + 15)$; first 45, second 15   **3.** $q + d = 27, d + 7 = q$; 17 quarters, 10 dimes
5. $4r_1 + 4r_2 = 72, r_1 = 2r_2$; Dicky 6 km/hr   **7.** $60t_1 = 40t_2, t_2 = t_1 + 2$; 240 mi; Micky 12 km/hr
9. $7t_1 = 3t_2, t_1 + t_2 = 10$; 21 miles

EXERCISE SET 6.8, p. 239

1. 70 dimes, 33 quarters   **3.** 22 quarters, 29 half dollars   **5.** 32 nickels, 13 dimes   **7.** 55.6 liters, 44.4 liters
9. 20 lb, 10 lb

EXERCISE SET 6.9, p. 241

1. $\frac{2}{3}$ ft from the object   **3.** Kay, 40 kg; her sister, 30 kg   **5.** Bob 7 ft, Bud 6 ft   **7.** 45 kg

CHAPTER 6   TEST ANALYSIS

Below are the answers to the test. Each item you missed should be reviewed. (See page numbers in brackets with each answer.) If you got 19 correct (your instructor may have another standard), you are prepared to continue after reviewing those items missed. If you missed more than 7 questions, *carefully* review the entire chapter, giving special emphasis to the items missed; then continue.

CHAPTER 6 TEST

1. [199] $-3x + 5$

2. [199] $y = \frac{4}{3} x + \frac{8}{3}$

3. [195]

x	-2	-1	0	1	2
y	1	-2	-3	-2	1

4. [195]

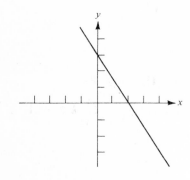

5. [198] x-intercept $(3,0)$

y-intercept $(0, \frac{6}{5})$

6. [197] (c)

7. [200] $x + y = 22$, $x = y + 8$
Answers may vary.

8. [200] $d = 4q$, $10d + 25q = 195$

9. [194] Yes

10. [194] No

11. [194] No

12. [204] $(18, -4)$

13. [207] $(-5, -4)$

14. [204] $(8, -2)$

15. [204] $(-\frac{5}{2}, 2)$

16. [207] $(-\frac{13}{10}, -\frac{22}{5})$

17. [207] $(6, 7)$

18. [211] Many solutions

19. [207] $(0, 1)$

20. [207] $(\frac{92}{19}, \frac{21}{19})$

21. [204] $(-1, 3)$

22. [207] $(-2, 9)$

23. [207] $(\frac{6}{7}, -\frac{23}{7})$

24. [213] 24 min or $\frac{2}{5}$ hr

25. [218] 5.4 m from the 60 kg wt

26. [212] 28, 12

MARGIN EXERCISES   CHAPTER 7

1. $x < 2$ **2.** $x > \dfrac{9}{8}$ **3.** $x > -1$ **4.** $y < -3$ **5.** $x < -\dfrac{1}{2}$ **6.** $y < \dfrac{2}{15}$ **7.** $x > \dfrac{11}{5}$ **8.** $x < \dfrac{1}{2}$ **9.** $x \leqslant 1$

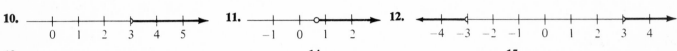

10.   **11.**   **12.**

13.   **14.**   **15.**

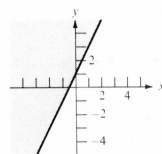 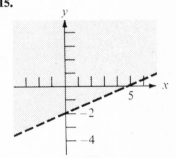

EXERCISE SET 7.1, p. 255

1. Number, inequality   **3.** F **5.** T **7.** T **9.** $x < 4$ **11.** $x \geqslant 3$ **13.** $x > -2$ **15.** $x > 5.483$

17. $x < \dfrac{17}{15}$ **19.** All x **21.** $t \geqslant -\dfrac{14}{3}$ **23.** $x \geqslant 20$ **25.** $z < -12$ **27.** All x **29.** $x \leqslant \dfrac{1}{70}$ **31.** $k \geqslant -11$

33. No solutions

EXERCISE SET 7.2, p. 257

1. $<$ **3.** T **5.** F **7.** $b \geqslant -2$ **9.** $m < -4$ **11.** $s \geqslant 0$ **13.** $c < -2$ **15.** $x > 4$ **17.** $k < -1$

19. $x < 2$ **21.** $x \geqslant 4$ **23.** $z < 5$ **25.** $a \geqslant \dfrac{10}{7}$ **27.** No solutions

EXERCISE SET 7.3, p. 259

1.   **3.**   **5.**   **7.**

9.   **11.**   **13.**

15.   **17.**   **19.**

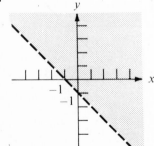 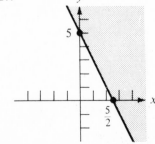 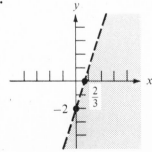

21.

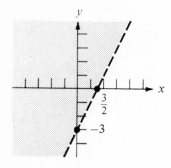

23.

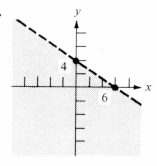

25.

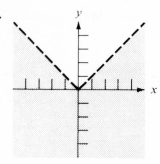

27.

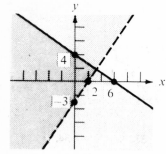

CHAPTER 7   TEST ANALYSIS

Below are the answers to the test. Each item you missed should be reviewed. (See page numbers in brackets with each answer.) If you got 9 correct (your instructor may have another standard), you have mastered the chapter sufficiently well. If you missed more than 4 questions, *carefully* review the entire chapter, giving special emphasis to the items missed; then continue.

1. [247] $<$ **2.** [247] $\leqslant$ **3.** [248] $x < 3$ **4.** [249] $x > 1$ **5.** [249] $n \leqslant -2$ **6.** [251]

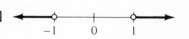

7. [251]   **8.** [252]   **9.** [252]

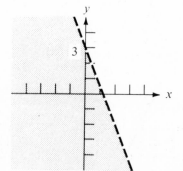

10. [253]   **11.** [253]   **12.** [254]

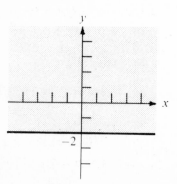

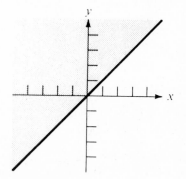

13. [254]

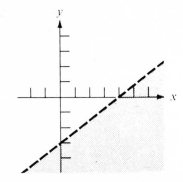

MARGIN EXERCISES   CHAPTER 8

1. $-3, 3, -2, 1, 2$ **2.** $3, 7, 1, 1, 0$ **3.** $2x^2y + 3xy$ **4.** $5pq + 4$ **5.** $5xy^4 - 3xy^3 - 7xy^2 + 3xy$
6. $-2 + 5xy^2z + 2x^2yz + 5x^3yz^2$ **7.** $14x^3y + 7x^2y - 3xy - 2y$ **8.** $-5p^2q^4 + 2p^2q^2 + 6pq^2 + 3p^2q + 3q + 5$
9. $-8s^4t + 6s^3t^2 + 2s^2t^3 - s^2t^2$ **10.** $-9p^4q + 10p^3q^2 - 4p^2q^3 - 9q^4$ **11.** $x^5y^5 + 2x^4y^2 + 3x^3y^3 + 6x^2$
12. $p^5q - 4p^3q^3 + 3pq^3 + 6q^4$ **13.** $3x^3y + 6x^2y^3 + 2x^3 + 4x^2y^2$ **14.** $2x^2 - 11xy + 15y^2$ **15.** $16x^2 + 40xy + 25y^2$
16. $9x^4 - 12x^3y^2 + 4x^2y^4$ **17.** $4x^2y^4 - 9x^2$ **18.** $9y^2 + 24y + 16 - 9x^2$ **19.** $16y^2 - 9x^2y^4$ **20.** $x^2y(x^2y + 2x + 3)$
21. $2p^4q^2(5p^2 - 2pq + q^2)$ **22.** $(a - b)(2x + 5 + y^2)$ **23.** $(a + b)(x^2 + y)$ **24.** $(3a + 4x^2)(3a - 4x^2)$
25. $(5xy^2 + 2a)(5xy^2 - 2a)$ **26.** $5(1 + xy^3)(1 - xy^3)$ **27.** $y^2(4 + 9x^2)(2 + 3x)(2 - 3x)$ **28.** $(x^2 + y^2)^2$

29. $-1(2x - 3y)^2$ **30.** $(xy + 4)(xy + 1)$ **31.** $2(x^2y^3 + 5)(x^2y^3 - 2)$ **32.** $y = \dfrac{b + 3}{a}$ **33.** $x = \dfrac{6}{b}$ **34.** $y = 1$

35. $x = \dfrac{2 - ab}{a + b}$ **36.** $x = -2a - 3b - 8$ **37.** $Q_2 = \dfrac{fr^2}{kQ_1}$ **38.** $R^2 = \dfrac{1}{3}\left(h^2 + \dfrac{3V}{\pi h}\right)$

EXERCISE SET 8.1, p. 275

1. $1, -2, 3, -5$ **3.** $17, -3, -7$ **5.** $4, 2, 2, 0$ **7.** $5, 5, 0$ **9.** $-a - 2b$ **11.** $3x^2y - 2xy^2 + x^2$
13. $8u^2v - 5uv^2$ **15.** $10a(2u + v)$ **17.** $-y^3 - xy^2 + 5x^2y$ **19.** $2t^3 - 2r^2t + p^4rt$ **21.** $3y^2 + 2xy + x^2$
23. $-11uv^2 + 7u^2v + 5uv$

EXERCISE SET 8.2, p. 277

1. $x^2 - 4xy + 3y^2$ **3.** $3r + 7$ **5.** $-x^2 - 8xy - y^2$ **7.** $2ab$ **9.** $-2a + 10b - 5c + 8d$ **11.** $6z^2 + 7zu - 3u^2$
13. $a^4b^2 - 7a^2b + 10$ **15.** $a^4 + a^3 - a^2y - ay + a + y - 1$ **17.** $a^6 - b^2c^2$ **19.** $y^6x + y^4x + y^4 + 2y^2 + 1$
21. $r^3 + r^2y + rs^2 + r^2s + rsy + s^3$ **23.** $5x^2y^2 + xy - 4$ **25.** $5y^2z^4 + 4yz^2 - 3$ **27.** $2r^2 + 3rs - rt - 2s^2 + 3st - t^2$

EXERCISE SET 8.3, p. 279

1. $12x^2y^2 + 2xy - 2$ **3.** $m^6n^2 + 2m^3n - 48$ **5.** $c^6 - 64d^2$ **7.** $9a^2 + 12ab + 4b^2$ **9.** $x^2 + xy^3 - 2y^6$

11. $4a^6 - 2a^3b^3 + \dfrac{1}{4}b^6$ **13.** $a^2b^2 - c^2d^4$ **15.** $x^2 + y^2 + 2xy + 6x + 6y + 9$ **17.** $20 - x - y - x^2 - 2xy - y^2$

19. $a^4 + 2a^2b + b^2 + 4b + 4a^2 + 4$ **21.** $1 + r - s - 2r^2 + 4rs - 2s^2$ **23.** $3a^3 - 12a^2b + 12ab^2$
25. $h^2 - k^2 + 8k - 16$ **27.** $x^4 + x^2y^2 + y^4$ **29.** $9x^2 - 16y^2 - 40yz - 25z^2$ **31.** $-2ab + 2b^2$ **33.** $4ab$
35. $4y - 4$ **37.** $a^2 - 2ab + b^2 - b + a$ **39.** $a^3 + 6a^2b + 11ab^2 + 6b^3 - 12a - 12b$

EXERCISE SET 8.4, p. 281

1. $12n^2(1 + 2n)$ **3.** $9xy(xy - 4)$ **5.** $(x + 1)(x - 1 - y)$ **7.** $xy(x - y)$ **9.** $\pi r(r + h)$ **11.** $(x + 1)(x + y)$
13. $(a^2b^2 + 4)(ab + 2)(ab - 2)$ **15.** $-3xy^3(4x^2 - 6xy - 9y^2)$ **17.** $9s^2(s + 1)(s - 1)$ **19.** $3(y + 4)(y - 4)$
21. $(a + y)(a - 3)$ **23.** $(3x^2y + b)(3x^2y - b)$ **25.** $(3y + p)(2y - 1)$ **27.** $(4x + 1)(1 + 2x^2)$
29. $(a + m - 1)(a - m + 1)$ **31.** $(a - 1)(y + 1)(y - 1)$ **33.** $(x - 3)(a + 1)(a - 1)$ **35.** $(4a + 3b)(2a - 3b)$
37. $(s - t)(s + t - 4)$ **39.** $(c^2 - x)(xy - c)$

EXERCISE SET 8.5, p. 283

1. Yes **3.** No **5.** Yes **7.** Yes **9.** Yes **11.** $4b^2$ **13.** $(3q+1)^2$ **15.** $(y^2+5z)^2$ **17.** $p(2p+q)^2$
19. $(r-s-1)^2$ **21.** $(t-2+s)(t-2-s)$ **23.** $-(k-18l)^2$ **25.** $(1+n+x)(1-n-x)$ **27.** $(3b+a)(b-6a)$
29. $(a+b)(a-b)(c+2)(c-2)$ **31.** $(m+20n)(m-18n)$ **33.** $2a(a-b)$ **35.** $(s^2+t^2)(k+3)(k-3)$

EXERCISE SET 8.6, p. 285

1. $x=\dfrac{2a-12}{3}$ **3.** $y=\dfrac{3b-1}{2}$ **5.** $y=\dfrac{9}{2c}$ **7.** $x=a-b$ **9.** $z=\dfrac{1}{6}(3c-5b+12)$ **11.** $x=b+2$

13. $x=\dfrac{3a-5}{1-2a}$ **15.** $x=\dfrac{7b}{4}$ **17.** $z=2$ **19.** $x=b+4$ **21.** $x=a+b$ **23.** $y=\dfrac{1}{a}$ **25.** $x=3a$

27. $x=d-1$ **29.** $x=\dfrac{c}{rt}$ **31.** $x=\dfrac{c}{2a-b}$ **33.** $x=\dfrac{5m}{2}$

EXERCISE SET 8.7, p. 287

1. $r=\dfrac{S}{2\pi k}$ **3.** $b=\dfrac{2A}{k}$ **5.** $n=\dfrac{S+360}{180}$ **7.** $b=\dfrac{3V-kB-4kM}{k}$ **9.** $r=\dfrac{S-a}{S-l}$ **11.** $h=\dfrac{2A}{b_1+b_2}$

13. $a=\dfrac{v^2pL}{r}$ **15.** $b_2=\dfrac{2A-hb_1}{h}$ **17.** $E=\dfrac{180A}{\pi r^2}$ **19.** $M=\dfrac{-V-hB-hc}{4h}$ **21.** $L=\dfrac{ay}{v^2p}$ **23.** $p=\dfrac{ar}{v^2L}$

25. $n=\dfrac{a}{c(1+b)}$ **27.** $F=\dfrac{9C+160}{5}$ **29.** $g=\dfrac{mf+t}{m}$

CHAPTER 8   TEST ANALYSIS

Below are the answers to the test. Each item you missed should be reviewed. (See page numbers in brackets with each answer.) If you got 17 correct (your instructor may have another standard), you are prepared to continue after reviewing those items missed. If you missed more than 8 questions, carefully review the entire chapter, giving special emphasis to the items missed; then continue.

CHAPTER 8 TEST

1. [263] a) 5, 1, -3 b) 8, 7, -3 **2.** [263] a) 16, 4, 5, 0 b) 2, 7, 4, 1 **3.** [264] $a^3+3a^2b+3ab^2+b^3$
4. [264] $25j+36m$ **5.** [264] $4x^2yz^3+x^2y^2$ **6.** [265] $2a^3b-9a^2b^2-ab^3-ab+23$
7. [266] $2y^4z^3-y^3z^2-4y^2z^2-4yz+3$ **8.** [265] $-4x-3y-5z$ **9.** [265] $18s^2t^3+6st^2+s^2t^2+7$
10. [268] c^4-d^2 **11.** [268] $a^2+4b^2+c^2+4ab-2ac-4bc$ **12.** [290] $(4xy+1)(4xy-1)$ **13.** [271] $(3p+7q)^2$
14. [271] $(7x-3y)(5x-y)$ **15.** [270] $(x+y+2)(x+y-2)$ **16.** [270] $(c+3+4x)(c+3-4x)$

17. [271] $(m-n+5)(m-n+3)$ **18.** [269] $(2u+v)(w-3x)$ **19.** [272] $b=-\dfrac{c}{2}$ **20.** [273] $x=\dfrac{r^2+2s^2-r-s}{r}$

21. [274] $p=\dfrac{A}{1+rt}$ **22.** [272] $x=-3a$ **23.** [270] $a(x^2+6)(x+2)(x-2)$ **24.** [270] $(4c^2+d^4)(2c+d^2)(2c-d^2)$

25. [267] $x^2-2xy+y^2-4x+4y-12$

MARGIN EXERCISES   CHAPTER 9

1. $\dfrac{(x+3)\cdot(x+2)}{5(x+4)}$ **2.** $\dfrac{-12}{4x^2-1}$ **3.** a) $\dfrac{2x^2+x}{3x^2-2x}$ b) $\dfrac{x^2+3x+2}{x^2-4}$ **4.** $\dfrac{8-x}{y-x}$ **5.** a) 5 b) $13x$ **6.** a) $\dfrac{3x+1}{x-1}$

b) $\dfrac{2x+1}{3x+2}$ c) $\dfrac{x+1}{2x+1}$ **7.** a) $x+2$ b) $\dfrac{y+2}{4}$ **8.** $\dfrac{a-2}{a-3}$ **9.** $\dfrac{15}{28}$ **10.** $\dfrac{20}{27}$ **11.** $\dfrac{8}{91}$ **12.** $\dfrac{25}{27}$ **13.** $\dfrac{2x^2-3}{x^3+5x}$

14. $\dfrac{x}{3x+1}$ **15.** $\dfrac{y+1}{y-1}$ **16.** $\dfrac{7}{9}$ **17.** $\dfrac{3+x}{x-2}$ **18.** $\dfrac{6x+4}{x-1}$ **19.** 0 **20.** $\dfrac{x-1}{x-3}$ **21.** $\dfrac{5}{2x-1}$ **22.** $\dfrac{4}{11}$

23. $\dfrac{3-x}{x+2}$ **24.** $\dfrac{2x+6}{x+1}$ **25.** $\dfrac{x^2+2x+1}{2x+1}$ **26.** $\dfrac{3x-1}{3}$ **27.** $\dfrac{4x-3}{x-2}$ **28.** 36 **29.** 120 **30.** $\dfrac{43}{72}$ **31.** $\dfrac{251}{588}$

32. $60x^3y^2$ **33.** $(y + 1)(y + 1)(y + 4)$ or $(y + 1)^2(y + 4)$ **34.** $3x(x - 1)(x + 1)(x + 1)$ or $3x(x + 1)^2(x - 1)$ or

$3x(1 - x)(x + 1)^2$ **35.** $\dfrac{4x^2 - x + 3}{x(x + 1)^2(x - 1)}$ **36.** $\dfrac{8x + 88}{(x + 1)(x + 8)(x + 16)}$ **37.** $\dfrac{-x - 7}{15x}$ **38.** $\dfrac{y^2 + y + 4}{(y - 1)(y + 1)(y - 6)}$

39. $\dfrac{2(3x^2 - x - 1)}{3x(x + 1)}$ **40.** $\dfrac{20}{21}$ **41.** $\dfrac{2(6 + x)}{5}$ **42.** $\dfrac{7x^2}{3(2 - x^2)}$ **43.** $b - x$ **44.** $x^2 + 3x + 2$ **45.** $2x^2 + x - \dfrac{2}{3}$

46. $4x^2 - \dfrac{3}{2}x + \dfrac{1}{2}$ **47.** $2x^2 - 3x + 5$ **48.** $x - 2$ **49.** $x + 4$ **50.** $x + 4R(-2)$; $x + 4 + \dfrac{-2}{x + 3}$ **51.** $x^2 + x + 1$

52. 3 **53.** $-\dfrac{4}{27}$ **54.** 1 **55.** 2 **56.** $-\dfrac{1}{8}$ **57.** -3 **58.** 40 km/hr, 50 km/hr   **59.** $p = \dfrac{n}{2 - m}$ **60.** $s = \dfrac{rt}{r + t}$

61. 675 **62.** $w = 20$ cm, $l = 28$ cm   **63.** 1600 **64.** 5 amp

EXERCISE SET 9.1, p. 317

1. $\dfrac{2x^2 + 5x + 3}{4x - 20}$ **3.** $\dfrac{30}{15x^2 - 2x - 24}$ **5.** $\dfrac{3x - 1}{2x + y}$ **7.** $\dfrac{3x + y}{2x - 3}$ **9.** $\dfrac{3(x - 2)}{2(x + 2)}$ **11.** $\dfrac{b}{x - y}$ **13.** $\dfrac{x + 2}{2(x - 4)}$

15. $\dfrac{x + 5}{x - 5}$ **17.** $\dfrac{6}{x - 3}$ **19.** $\dfrac{a^2 + 1}{a + 1}$ **21.** $\dfrac{3 - b}{-b - 4}$ **23.** $\dfrac{12a}{a - 2}$ **25.** $\dfrac{1}{a}$

EXERCISE SET 9.2, p. 319

1. $\dfrac{4}{7}$ **3.** $\dfrac{3}{5}$ **5.** $\dfrac{1}{9}$ **7.** $\dfrac{24}{35}$ **9.** $\dfrac{1}{18}$ **11.** $\dfrac{x}{y}$ **13.** $\dfrac{(a + 2)(a + 3)}{(a - 3)(a - 1)}$ **15.** $\dfrac{(x - 1)^2}{x}$ **17.** $\dfrac{1}{2}$ **19.** $\dfrac{3}{2}$ **21.** $\dfrac{(x + y)^2}{x^2 + y}$

23. $\dfrac{x + 3}{x - 5}$ **25.** $\dfrac{1}{(c - 5)^2}$ **27.** $\dfrac{y - 3}{2y - 1}$ **29.** $\dfrac{2a + 1}{2a + 3}$

EXERCISE SET 9.3, p. 321

1. a) -3 b) 4 c) $-x - 1$ or $-(x + 1)$ d) $1 - x$ e) $2x - 1$ f) $x - 3$ g) $x + 5$ h) $6 - 2x$ **3.** $\dfrac{8}{17}$ **5.** $\dfrac{10}{2x + 1}$

7. $\dfrac{4x - 3}{3x + 4}$ **9.** $\dfrac{x}{3x - 2}$ **11.** $\dfrac{x + 3}{4x - 3}$ **13.** $\dfrac{a + b}{x + y}$ **15.** $\dfrac{1}{y - z}$ **17.** 0 **19.** $a + b$ **21.** $\dfrac{10x + 2}{2x - 3}$ **23.** $\dfrac{12x + 7}{4 - 3x}$

25. $\dfrac{17}{(2x - 3)(x - 1)}$

EXERCISE SET 9.4, p. 323

1. $\dfrac{4}{11}$ **3.** $\dfrac{5}{3}$ **5.** $\dfrac{x + 4}{x - 2}$ **7.** $\dfrac{1}{a + b}$ **9.** $x - 4$ **11.** $\dfrac{4}{x - 1}$ **13.** $\dfrac{1}{x + 1}$ **15.** $\dfrac{3a}{a^2 - a - 6}$ **17.** $\dfrac{1}{x - 1}$

19. $\dfrac{-9}{2x - 3}$ or $\dfrac{9}{3 - 2x}$ **21.** $\dfrac{18x + 5}{x - 1}$ **23.** 0 **25.** $\dfrac{20}{2y - 1}$

EXERCISE SET 9.5, p. 325

1. 108 **3.** 72 **5.** 126 **7.** 360 **9.** 420 **11.** $300; \dfrac{59}{300}$ **13.** $120; \dfrac{71}{120}$ **15.** $180; \dfrac{23}{180}$ **17.** $8a^2b^2$ **19.** c^3d^2

21. $72(x^2y^2)$ **23.** $(a + 1)(a - 1)^2$ **25.** $(3k + 2)(3k - 2)$ **27.** $18x^3(x - 2)^2(x + 1)$

EXERCISE SET 9.6, p. 327

1. $\dfrac{2x + 5}{x^2}$ **3.** $\dfrac{x^2 + 4xy + y^2}{x^2y^2}$ **5.** $\dfrac{4x}{(x - 1)(x + 1)}$ **7.** $\dfrac{x^2 + 6x}{(x + 4)(x - 4)}$ **9.** $\dfrac{3x - 1}{(x - 1)^2}$ **11.** $\dfrac{x^2 + 5x + 1}{(x + 1)^2(x + 4)}$

13. $\dfrac{2x^2 - 4x + 34}{(x - 5)(x + 3)}$ **15.** $\dfrac{3a + 2}{(a + 1)(a - 1)}$ **17.** $\dfrac{2x + 6y}{(x + y)(x - y)}$ **19.** $\dfrac{3x^2 + 19x - 20}{(x + 3)(x - 2)^2}$

EXERCISE SET 9.7, p. 329

1. $\dfrac{-(x+4)}{6}$ **3.** $\dfrac{7z-12}{12z}$ **5.** $\dfrac{y-19}{4y}$ **7.** $\dfrac{2x-40}{(x+5)(x-5)}$ **9.** $\dfrac{3-5t}{2t(t-1)}$ **11.** $\dfrac{2s-st-s^2}{(t+s)(t-s)}$ **13.** $\dfrac{2}{y(y-1)}$

15. $\dfrac{z-3}{2z-1}$ **17.** $\dfrac{1-3x}{(2x-3)(x+1)}$ **19.** $\dfrac{1}{2c-1}$

EXERCISE SET 9.8, p. 331

1. $\dfrac{25}{4}$ **3.** $\dfrac{1}{3}$ **5.** $\dfrac{1+3x}{1-5x}$ **7.** -1 **9.** $\dfrac{5}{3y^2}$ **11.** c **13.** $-\dfrac{5}{ab}$ **15.** $\dfrac{x+y}{x}$ **17.** $\dfrac{x-2}{x-3}$

EXERCISE SET 9.9, p. 333

1. $1-2u-u^4$ **3.** $5t^2+8t-2$ **5.** $-4x^4+4x^2+1$ **7.** $1-2x^2y+3x^4y^5$ **9.** $3+n$ **11.** $x+2$ **13.** x^3-6
15. $x^4-x^3+x^2-x+1$ **17.** $2x^2-7x+4$

EXERCISE SET 9.10, p. 335

1. 3   **3.** 10   **5.** 5   **7.** 3   **9.** $\dfrac{17}{2}$ **11.** $\dfrac{5}{3}$ **13.** $\dfrac{2}{9}$ **15.** $\dfrac{1}{2}$ **17.** No solution

EXERCISE SET 9.11, p. 337

1. $\dfrac{2}{7}$ **3.** $\dfrac{8}{3}, \dfrac{20}{3}$ **5.** 28 km/hr   **7.** $p=\dfrac{fq}{q-f}$ **9.** $I=\dfrac{En}{Rn+r}$ **11.** $Q=\dfrac{Vr_1r_2}{r_2-r_1}$ **13.** $h_2=p\left(\dfrac{h_1}{q}-1\right)$

EXERCISE SET 9.12, p. 339

1. 80 miles   **3.** 42   **5.** 36 gal   **7.** 216   **9.** 22 lb   **11.** a) 1.92 tons   b) 14.4 kg.

CHAPTER 9   TEST ANALYSIS

Below are the answers to the test. Each item you missed should be reviewed. (See page numbers in brackets with each answer.) If you got 14 correct (your instructor may have another standard), you are prepared to continue after reviewing those items missed. If you missed more than 5 questions, *carefully* review the entire chapter, giving special emphasis to the items missed; then continue.

CHAPTER 9   TEST

1. [293] $\dfrac{2y}{x-2y}$ **2.** [293] $\dfrac{7x+3}{x-3}$ **3.** [294] $\dfrac{x^2-5x+6}{x^2+5x+4}$ **4.** [294] $\dfrac{a-6}{5}$ **5.** [296] $\dfrac{2x(x-1)}{x+1}$

6. [296] $\dfrac{(2x-1)(x-1)}{3x(2x+3)}$ **7.** [301] -1 **8.** [301] $\dfrac{40z-5}{4z(4z-1)}$ **9.** [302] $\dfrac{26+x-x^2}{(x+5)(x-5)(x+1)}$ **10.** [303] 2

11. [305] $-6x^2-4x+2$ **12.** [306] $3k^2-7k+4+\dfrac{1}{2k+3}$ **13.** [308] 5   **14.** [309] 3, -5 **15.** [312] $n=\dfrac{rI}{E-RI}$

16. [312] $R=\dfrac{rE-re}{e}$ or $\dfrac{rE}{e}-r$ **17.** [310] $\dfrac{33}{28}$ **18.** [311] 20 mph; 30 mph   **19.** [311] 8

MARGIN EXERCISES   CHAPTER 10

1. 4   **2.** 7   **3.** 10   **4.** $45+x$ **5.** $\dfrac{x}{x+2}$ **6.** 6, -6 **7.** 9, -9 **8.** 10, -10 **9.** 10   **10.** -10 **11.** -6

12. 6   **13.** 0   **14.** a, c.   **15.** a) $x<3$ b) None   **16.** $|xy|$ **17.** $|xy|$ **18.** $|x-1|$ **19.** $|x+y|$ **20.** $5|x|$
21. Rational   **22.** Rational   **23.** Rational   **24.** Rational   **25.** Irrational   **26.** Irrational   **27.** Rational

28. Irrational   **29.** $\dfrac{7}{128}$ **20.** 2.646   **31.** 8.485   **32.** a) 8 b) 8   **33.** $\sqrt{21}$ **34.** $\sqrt{x(x+1)}$ **35.** $\sqrt{x^2-1}$

36. $4\sqrt{2}$ **37.** $2\sqrt{14}$ **38.** $4|y|$ **39.** $5|x|$ **40.** $\sqrt{x+1}\sqrt{x-1}$ **41.** 16.59   **42.** $\sqrt{6}\sqrt{17}\approx 10.10$ **43.** $4\sqrt{2}$
44. $4\sqrt{a}$ **45.** $\sqrt{3}|x-1|$ **46.** $|x^4|$ **47.** $|(x+2)^7|$ **48.** $4|x^3y^2|$ **49.** $5|xy^2|\sqrt{2xy}$ **50.** $12\sqrt{2}$ **51.** $5\sqrt{5}$

52. $-12\sqrt{10}$ **53.** $5\sqrt{6}$ **54.** $\sqrt{x+1}$ **55.** $\dfrac{4}{3}$ **56.** $\dfrac{1}{5}$ **57.** $\dfrac{1}{3}$ **58.** $-\dfrac{4}{5}$ **59.** $\dfrac{3}{4}$ **60.** $\dfrac{1}{5}\sqrt{15}$ **61.** $\dfrac{1}{4}\sqrt{10}$

62. .632   **63.** $3\sqrt{3}$ **64.** $\dfrac{3}{2}\sqrt{2}$ **65.** $\dfrac{2}{\sqrt{15}}$ or $\dfrac{2}{15}\sqrt{15}$ **66.** $\dfrac{1}{3}$ **67.** $\sqrt{\dfrac{1}{3}}$ or $\dfrac{1}{3}\sqrt{\ }$ **68.** $\dfrac{\sqrt{35}}{7}$ **69.** $\dfrac{\sqrt{xy}}{|y|}$

70. $c=\sqrt{65}\approx 8.062$ **71.** $b=\sqrt{44}\approx 6.633$ **72.** $5\sqrt{13}\approx 18.03$ ft   **73.** $\dfrac{64}{3}$ **74.** 66

EXERCISE SET 10.1, p. 359

1. $3x$ **3.** $\dfrac{3}{x+2}$ **5.** No.   **7.** All x negative or 0   **9.** 9, -9 **11.** 18, -18 **13.** 27, -27 **15.** -8
17. Meaningless   **19.** 12   **21.** -17 **23.** $x<-7$ **25.** None   **27.** None   **29.** None   **31.** 2   **33.** 12
35. $2|a|$ **37.** 5   **39.** $4|d|$ **41.** $|x-7|$ **43.** $|2x-5|$

EXERCISE SET 10.2, p. 361

1. Rational   **3.** Rational   **5.** Irrational   **7.** Rational   **9.** Rational   **11.** Rational   **13.** Irrational   **15.** Rational
17. Rational; integers   **19.** All of them   **21.** $\dfrac{2}{3}$ **23.** $\dfrac{41}{60}$ **25.** $\dfrac{31}{48}$ **27.** 6.557   **29.** 9.644   **31.** 7.937
33. 9.327

EXERCISE SET 10.3, p. 363

1. T **3.** T **5.** F **7.** 24   **9.** 24   **11.** 33   **13.** 3   **15.** $\sqrt{42}$ **17.** $\sqrt{x^2+x}$ **19.** $\sqrt{x^2-3x}$
21. $\sqrt{10x-5}$ **23.** $\sqrt{2x^2+9x+9}$ **25.** $\sqrt{x^2-4}$ **27.** $\sqrt{12x}$ or $2\sqrt{3x}$ **29.** $\sqrt{x^2-y^2}$

EXERCISE SET 10.4, p. 365

1. $2\sqrt{3}$ **3.** $2\sqrt{2}$ **5.** $4\sqrt{x}$ **7.** $\sqrt{2}\sqrt{3}$ **9.** $\sqrt{2}\sqrt{5}$ **11.** $\sqrt{2}\sqrt{7}\sqrt{x}$ **13.** $\sqrt{3}\sqrt{x-1}$ **15.** $|x|\sqrt{x-2}$
17. $2|x-1|$ **19.** $2|x|$ **21.** $\sqrt{3}|2x-3|$ **23.** 63   **25.** 10.25   **27.** 22.36   **29.** $x\leqslant 0$ **31.** $x\leqslant -3$

EXERCISE SET 10.5, p. 367

1. $2\sqrt{6}$ **3.** $11\sqrt{m}$ **5.** $5\sqrt{5y}$ **7.** $7x^2\sqrt{2}$ **9.** $6|m|\sqrt{m}$ **11.** $2a^2\sqrt{2a}$ **13.** $9|a^3|\sqrt{3}$ **15.** $|2x+1|\sqrt{2}$
17. $|x-1|\sqrt{y}$ **19.** $(y-2)^4$ **21.** $2(x+5)^2$ **23.** $4\sqrt{3}$ **25.** $|a|\sqrt{bc}$ **27.** $6|xy^3|\sqrt{3xy}$ **29.** $6|x^2y^2|\sqrt{x}$

EXERCISE SET 10.6, p. 369

1. $\sqrt{5}$; 3   **3.** $\sqrt{17}$; 5   **5.** $x=0$ or $y=0$, or both   **7.** $11\sqrt{3}$ **9.** $-2\sqrt{x}$ **11.** $8\sqrt{3}$ **13.** $19\sqrt{2}$
15. $8\sqrt{2}-2\sqrt{3}$ **17.** $(2+9|x|)\sqrt{x}$ **19.** $3\sqrt{2(x+1)}$ **21.** $(|x+3|+|y|)\sqrt{y}$ **23.** $(6|a+b|-1)\sqrt{2(a+b)}$

EXERCISE SET 10.7, p. 371

1. $\dfrac{3}{7}$ **3.** $\dfrac{1}{12}$ **5.** $\dfrac{6}{|x|}$ **7.** $\dfrac{3|a|}{25}$ **9.** $\dfrac{\sqrt{10}}{5}$ **11.** $\dfrac{\sqrt{6}}{4}$ **13.** $\dfrac{1}{3}$ **15.** $\dfrac{\sqrt{7}}{7}$ **17.** $\dfrac{2\sqrt{6}}{3}$ **19.** $\dfrac{\sqrt{2x}}{|x|}$ **21.** 0.632

23. 1.633   **25.** $9\sqrt{5}$ **27.** $3\sqrt{2}$ **29.** $\dfrac{26}{3}\sqrt{3}$ **31.** $\dfrac{\sqrt{6}}{6}$

EXERCISE SET 10.8, p. 373

1. 3   **3.** 2   **5.** 6   **7.** $\sqrt{2}$ **9.** 2   **11.** $3|y|$ **13.** $\dfrac{\sqrt{10}}{5}$ **15.** $\dfrac{3}{4}\sqrt{2}$ **17.** $\dfrac{\sqrt{21}}{9}$ **19.** $\dfrac{\sqrt{35}}{5}$ **21.** $\dfrac{7\sqrt{2}}{2}$

23. $\dfrac{3\sqrt{10a}}{4|a|}$ **25.** $\dfrac{|x|\sqrt{21ax}}{6|a|}$ **27.** $\dfrac{|x|\sqrt{y}}{|y|}$ **29.** $\dfrac{3|n|\sqrt{10}}{8}$

EXERCISE SET 10.9, p. 375

1. p **3.** c **5.** $\sqrt{34}$ **7.** 12   **9.** $\sqrt{31}$ **11.** $b = 24$ **13.** $c = 14$ **15.** $b = 6$ **17.** $a = 24$
19. $\sqrt{260}$ ft ≈ 16.12 ft

EXERCISE SET 10.10, p. 377

1. 25   **3.** 73   **5.** $\dfrac{69}{4}$ **7.** $\dfrac{1}{4}$ **9.** 3   **11.** $\dfrac{49}{9}$ **13.** No solution   **15.** $\dfrac{1}{49}$ **17.** $\dfrac{63}{5}$ **19.** No solution

21. No solution   **23.** 16

CHAPTER 10   TEST ANALYSIS

Below are the answers to the test. Each item you missed should be reviewed. (See page numbers in brackets with each answer.) If you got 21 correct (your instructor may have another standard), you are prepared to continue after reviewing those items missed. If you missed more than 9 questions, *carefully* review the entire chapter, giving special emphasis to the items missed; then continue.

CHAPTER 10   TEST

1. [343] $\dfrac{x}{2+x}$ **2.** [345] Irrational   **3.** [345] Rational   **4.** [345] Irrational   **5.** [345] Rational   **6.** [352] $\dfrac{5}{8}$

7. [352] $3\sqrt{3}$ **8.** [352] $\dfrac{\sqrt{2}}{2}$ **9.** [352] $\dfrac{\sqrt{6}}{4}$ **10.** [352] $\dfrac{\sqrt{3}}{|x|}$ **11.** [352] $\dfrac{\sqrt{ab}}{|b|}$ **12.** [352] $\dfrac{4|m|\sqrt{am}}{|a^3|}$

13. [352] $\dfrac{|x|\sqrt{10xy}}{2|y|}$ **14.** [350] $2\sqrt{3}$ **15.** [350] -6 **16.** [350] Meaningless   **17.** [350] $|x-3|$ **18.** [355] $\dfrac{\sqrt{15}}{5}$

19. [355] $\dfrac{\sqrt{5}}{5}$ **20.** [350] $\sqrt{x^2 - 16}$ **21.** [355] $\dfrac{|x|\sqrt{30}}{6}$ **22.** [354] $4\sqrt{3}$ **23.** [354] $12\sqrt{3}$ **24.** [354] $-8\sqrt{5}$

25. [354] $12\sqrt{3}$ **26.** [358] 4   **27.** [358] $-\dfrac{3}{2}$ **28.** [358] 0   **29.** [356] $3\sqrt{5}$ **30.** [356] 6 cm by 8 cm

MARGIN EXERCISES   CHAPTER 11

1. $x^2 - 7x = 0$; $a = 1, b = -7, c = 0$ **2.** $-x^2 - 9x + 3 = 0$; $a = -1, b = -9, c = 3$ **3.** $4x^2 + 2x + 4 = 0$;
$a = 4, b = 2, c = 4$ **4.** 0   **5.** $\sqrt{5}, -\sqrt{5}$ **6.** $\dfrac{3}{2}, -\dfrac{3}{2}$ **7.** $0, -\dfrac{5}{3}$ **8.** $\dfrac{2}{3}, -1$ **9.** 4, 1   **10.** $-4, -2$

11. $3 \pm \sqrt{11}$ **12.** $2 \pm \sqrt{7}$ **13.** $\dfrac{-1 \pm \sqrt{7}}{2}$ **14.** $\dfrac{1}{2}, -4$ **14.** $\dfrac{4 \pm \sqrt{31}}{5}$ **16.** $1.91, -.31$ **17.** 13, 5   **18.** 2

19. $E = mc^2$ **20.** $r = \sqrt{\dfrac{A}{\pi}}$ **21.** $t = \dfrac{-v \pm \sqrt{v^2 + 32h}}{16}$ **22.** 20 ft   **23.** 5 cm, 12 cm   **24.** 3 km/hr   **25.** Upward

26. a) Downward   b)

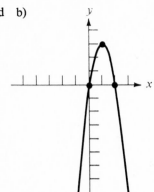

27. $-1, 4$ **28.** $-3, .5$

EXERCISE SET 11.1, p. 397

1. $x^2 - 3x + 2 = 0; a = 1, b = -3, c = 2$ **3.** $2x^2 - 3x + 5 = 0; a = 2, b = -3, c = 5$
5. $x^2 - 2x - 5 = 0; a = 1, b = -2, c = -5$ **7.** $x^2 - 10x + 7 = 0; a = 1, b = -10, c = 7$ **9.** 0 **11.** $\sqrt{11}, -\sqrt{11}$
13. $0, 2$ **15.** $\sqrt{7}, -\sqrt{7}$ **17.** No solution   **19.** $0, 5$ **21.** $0, \dfrac{1}{2}$ **23.** $3\sqrt{3}, -3\sqrt{3}$ **25.** $\sqrt{5}, -\sqrt{5}$

27. $\dfrac{1}{2}, -\dfrac{2}{3}$ **29.** $\dfrac{1}{3}, \dfrac{4}{3}$ **31.** $\dfrac{4}{3}, -\dfrac{1}{2}$

EXERCISE SET 11.2, p. 399

1. $5, -11$ **3.** $-9, 1$ **5.** $\dfrac{1}{2}, -\dfrac{7}{2}$ **7.** $2, -\dfrac{1}{3}$ **9.** $\dfrac{-7 \pm 3\sqrt{5}}{2}$ **11.** $\dfrac{2}{3}, -\dfrac{5}{2}$ **13.** $\dfrac{1 \pm \sqrt{10}}{3}$ **15.** $2 \pm \sqrt{14}$

17. $\dfrac{-8 \pm \sqrt{34}}{5}$ **19.** $\dfrac{-5 \pm \sqrt{37}}{6}$

EXERCISE SET 11.3, p. 401

1. $2, -\dfrac{4}{3}$ **3.** $\dfrac{5 \pm \sqrt{73}}{6}$ **5.** $1, \dfrac{4}{3}$ **7.** $\dfrac{2 \pm \sqrt{10}}{3}$ **9.** $\dfrac{3 \pm \sqrt{29}}{2}$ **11.** $\dfrac{11 \pm \sqrt{181}}{10}$ **13.** $1 \pm \sqrt{3}$ **15.** $\dfrac{3 \pm \sqrt{5}}{2}$

17. $\dfrac{-7 \pm \sqrt{61}}{2}$ **19.** $.22, -1.55$

EXERCISE SET 11.4, p. 403

1. 5 **3.** 1 **5.** No solution   **7.** $-\dfrac{5}{2}$ **9.** $-\dfrac{1}{2}$ **11.** 6 **13.** -4

EXERCISE SET 11.5, p. 405

1. 0 **3.** $5, 1$ **5.** 3 **7.** $25, 13$ **9.** -5 **11.** 5 **13.** 5

EXERCISE SET 11.6, p. 407

1. $a = \sqrt{c^2 - b^2}$ **3.** $r = \dfrac{\sqrt{S\pi}}{2\pi}$ **5.** $d = \dfrac{3F \pm \sqrt{9F^2 - 3}}{3}$ **7.** $r = \dfrac{\sqrt{V\pi h}}{\pi h}$ **9.** $t = \dfrac{\sqrt{2sg}}{g}$ **11.** $F = \dfrac{3d^2 + 1}{6d}$
13. $|r| = \sqrt{h^2 + k^2 - F}$

EXERCISE SET 11.7, p. 409

1. 11, −12   **3.** 2 km/hr   **5.** 9, 17   **7.** 6 ft by 13 ft   **9.** 10 m

EXERCISE SET 11.8, p. 411

1. Downward   **3.** Upward   **5.** Downward

7.

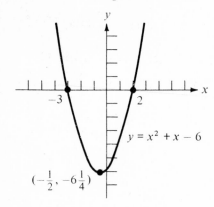

$y = x^2 + x - 6$

$(-\frac{1}{2}, -6\frac{1}{4})$

9.

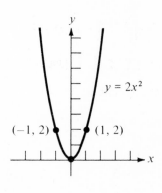

$y = 2x^2$

$(-1, 2)$ $(1, 2)$

11.

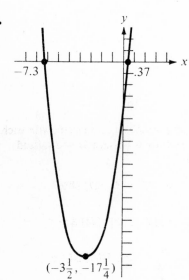

$(-3\frac{1}{2}, -17\frac{1}{4})$

13.

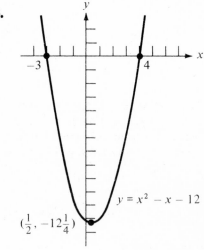

$y = x^2 - x - 12$

$(\frac{1}{2}, -12\frac{1}{4})$

15.

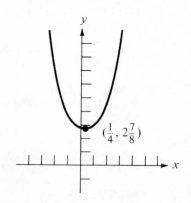

$(\frac{1}{4}, 2\frac{7}{8})$

17.

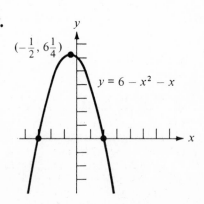

$(-\frac{1}{2}, 6\frac{1}{4})$

$y = 6 - x^2 - x$

19. 2, −1   **21.** $-\frac{1}{3}$, 3

CHAPTER 11   TEST ANALYSIS

Below are the answers to the test. Each item you missed should be reviewed. (See page numbers in brackets with each answer.) If you got 10 correct (your instructor may have another standard), you are prepared to continue after reviewing those items missed. If you missed more than 5 questions, *carefully* review the entire chapter, giving special emphasis to the items missed. Then you will be ready to review the entire book in preparation for taking the final examination.

CHAPTER 11   TEST

1. [383] $-8, 6$ **2.** [386] $\dfrac{-1 \pm \sqrt{21}}{2}$ **3.** [386] $\dfrac{2 \pm \sqrt{3}}{2}$ **4.** [383] $7, -13$ **5.** [386] $1 \pm \sqrt{11}$ **6.** [386] $\dfrac{5 \pm \sqrt{43}}{2}$

7. [383] $\dfrac{1}{3}, -2$ **8.** [386] $-3 \pm 3\sqrt{2}$ **9.** [390] 6   **10.** [390] 6, 1   **11.** [389] $2, -8$ **12.** [391] $T = 4v^2l - l$

13. [392] $10\sqrt{91}$ cm   **14.** [393] 48 mph   **15.** [394]   **16.** [396] $5, -3$

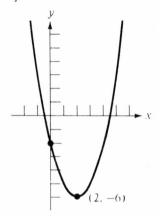

(2, −6)

FINAL EXAMINATION ANALYSIS

Below are the answers to the final examination. Each item missed should be reviewed. (See page numbers in brackets with each answer.) Fifty-six questions correct (your instructor may have another standard) would be considered passing. If you missed more than 24 questions, review the entire book, giving special emphasis to the items missed.

1. [6] $6 \times 10^4 + 7 \times 10^3 + 3 \times 10^2 + 5 \times 10^0$ **2.** [12] 8   **3.** [7] $\dfrac{136}{10}$ or $\dfrac{68}{5}$ **4.** [26] 0.327 **5.** [27] 68%

6. [15] $\dfrac{x}{3}$ **7.** [61] $<$ **8.** [61] 7   **9.** [63] 9   **10.** [67] 15.0 **11.** [68] $-\dfrac{3}{20}$ **12.** [67] 2   **13.** [74] 3^{-4}

14. [75] x^{-4} **15.** [75] y^7 **16.** [98] -8 **17.** [98] -12 **18.** [99] 7   **19.** [104] $t = \dfrac{4b}{A}$ **20.** [101] 24, 25

21. [126, 128] $10x^3 + 3x - 3$ **22.** [134] $3x^4 - 3x^3 + 10x^2 - 2x - 13$ **23.** [132] $8x^2 - 4x - 6$
24. [135] $-8x^4 + 6x^3 - 2x^2$ **25.** [139] $6x^3 - 17x^2 + 16x - 6$ **26.** [165] $4x(2x - 1)$ **27.** [172] $(5x + 2)(5x - 2)$
28. [168] $(3x - 4)(2x + 1)$ **29.** [171] $(x - 4)(x - 4)$ **30.** [167] $x(1 - x)(8 - x)$ **31.** [173] 100   **32.** [175] 3, 5

33. [176] $5, -6$ **34.** [176] 4   **35.** [193] II   **36.** [198] 3   **37.** [195]

x	−2	−1	0	1	2
y	2	−1	−2	−1	2

38. [198] -2

39. [197] b, c, d **40.** [202] (1, 2)   **41.** [200] b, c, d **42.** [198] -3 **43.** [204] (12, 5)   **44.** [207] (6, 7)
45. [212] 100 cm   **46.** [212] 11, 13   **47.** [263] 5   **48.** [264] $5a^5x^4 + 3ax^3 + x^2 + 2a$ **49.** [264] $7x^2yz^3 + 5x^2y^2 - 5xy^2$
50. [264] $21x + 15y$ **51.** [265] $2x^3y - 5x^2y^2 - xy + xy^3 + 16$ **52.** [266] $15x^2y^3 + x^2y^2 + 5xy^2 + 7$
53. [268] $x^4 - 4y^2$ **54.** [267] $6x^2 - xy^2 - 12y^4$ **55.** [266] $9x^3 + 12x^2y - 15x^2$
56. [268] $x^2 + 6x - 4xy - 12y + 4y^2 + 9$ **57.** [270] $(5ab + 1)(5ab - 1)$ **58.** [270] $(x - y + 5)(x - y - 5)$
59. [271] $(3x + 5y)^2$ **60.** [271] $(5x - 2y)(3x + 4y)$ **61.** [271] $(a - b + 5)(a - b - 2)$ **62.** [269] $(2a + d)(c - 3b)$
63. [270] $(2x + y^2)(2x - y^2)(4x^2 + y^4)$ **64.** [266] $a^2 - 2ab + b^2 - 2a + 2b - 3$ **65.** [271] $3(a^2 + 6)(a + 2)(a - 2)$

56. [274] $r = \dfrac{AR}{e} - R$ **67.** [292] $\dfrac{2(x + 2)}{(x - 3)(3 - x)}$ or $\dfrac{-2(x + 2)}{(x - 3)^2}$ **68.** [295] $\dfrac{3a(a - 1)}{2(a + 1)}$ **69.** [297] $\dfrac{27x - 4}{5x(3x - 1)}$

70. [298] $\dfrac{-x^2 + x + 2}{(x + 4)(x - 4)(x - 5)}$ **71.** [309] 2   **72.** [309] 4, −6   **73.** [349] $5\sqrt{2}$ **74.** [355] $\dfrac{2\sqrt{10}}{5}$ **75.** [351] $12\sqrt{3}$

76. [358] 2   **77.** [383] 3, −2   **78.** [386] $\dfrac{-3 \pm \sqrt{29}}{2}$ **79.** [394]   **80.** [392] 12

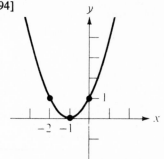

APPENDIX, MARGIN EXERCISES

1. 23,000   **2.** 400   **3.** 780   **4.** 2.3   **5.** .67   **6.** 4780
7. 47,800   **8.** 6.755   **9.** 67.55   **10.** 3940   **11.** 6,700,000
12. 6.78   **13.** 97   **14.** 0.1   **15.** 26   **16.** 8.451   **17.** 28 dm²
18. 17.52 km²   **19.** 144 km²   **20.** 272.25 cm²   **21.** 10,000
22. 100   **23.** 47,000   **24.** 38   **25.** a) 1000, 1000 b) 1, 1
26. 970   **27.** 8.99   **28.** 6200   **29.** .3048   **30.** 700   **31.** .009042
32. 6.7   **33.** 6780   **34.** 780,000   **35.** 760   **36.** 87,000
37. 2.344   **38.** 77   **39.** 1.246   **40.** 199

APPENDIX, EXERCISE SET, pp. 433–437

LENGTH

1. a) 1000 b) .001   **3.** a) 10 b) 0.1   **5.** a) 0.01 b) 100
7. 6700   **9.** 980   **11.** 89,210   **13.** 5.666   **15.** 566.6   **17.** 4.77
19. 688   **21.** .688   **23.** 100   **25.** 10,000   **27.** 100   **29.** 9130
31. 70   **33.** 8   **35.** 6000   **37.** .00013

AREA

39. 245 km²   **41.** 207.05 dm²   **43.** 3237.61 km²   **45.** 295.84 mm²
47. 100   **49.** 1   **51.** 201,000   **53.** 6   **55.** 765
57. 65,000,000   **59.** 2.8

CAPACITY

61. .001, 1, 1   **63.** 901,000   **65.** .017   **67.** .000013   **69.** 8006
71. 49,200

WEIGHT

73. 1000   **75.** 10   **77.** 0.01   **79.** 1000   **81.** 10   **83.** 23,400   **85.** 8.09
87. .45   **89.** .00000025   **91.** 5.677   **93.** .455   **95.** .01
97. 8000

INDEX